T0418820

VOLUME EIGHTY EIGHT

# ADVANCES IN FOOD AND NUTRITION RESEARCH

## Food Applications of Nanotechnology

VOLUME EIGHTY EIGHT

# ADVANCES IN FOOD AND NUTRITION RESEARCH

## Food Applications of Nanotechnology

Edited by

**LOONG-TAK LIM**
*Department of Food Science,*
*University of Guelph*

**MICHAEL ROGERS**
*Department of Food Science,*
*University of Guelph*

Academic Press is an imprint of Elsevier
50 Hampshire Street, 5th Floor, Cambridge, MA 02139, United States
525 B Street, Suite 1650, San Diego, CA 92101, United States
The Boulevard, Langford Lane, Kidlington, Oxford OX5 1GB, United Kingdom
125 London Wall, London, EC2Y 5AS, United Kingdom

First edition 2019



ISBN: 978-0-12-816073-2
ISSN: 1043-4526

For information on all Academic Press publications
visit our website at https://www.elsevier.com/books-and-journals

*Publisher:* Zoe Kruze
*Acquisition Editor:* Sam Mahfoudh
*Editorial Project Manager:* Leticia M. Lima
*Production Project Manager:* Denny Mansingh
*Cover Designer:* Greg Harris

Typeset by SPi Global, India

# Contents

# Contributors

**Rafael Auras**
School of Packaging, Michigan State University, East Lansing, MI, United States

**Dhimiter Bello**
Department of Biomedical and Nutritional Sciences, University of Massachusetts, Lowell, MA, United States

**B.M. Bohrer**
Department of Food Science, University of Guelph, Guelph, ON, Canada

**Sarah Caballero**
Nutrition and Food Science Department, California State Polytechnic University, Pomona, CA, United States

**Hongda Chen**
U.S. Department of Agriculture, National Institute of Food and Agriculture, Washington DC, United States

**Young-Hee Cho**
Department of Food Science, Purdue University, West Lafayette, IN, United States

**Ioannis S. Chronakis**
Nano-BioScience Research Group, DTU-Food, Technical University of Denmark, Lyngby, Denmark

**M.G. Corradini**
Arrell Food Institute, University of Guelph, Guelph, ON, Canada

**Gabriel Davidov-Pardo**
Nutrition and Food Science Department, California State Polytechnic University, Pomona, CA, United States

**Alvaro Renato Guerra Dias**
Department of Agroindustrial Science and Technology, Federal University of Pelotas, Pelotas, RS, Brazil

**L.M. Duizer**
Department of Food Science, University of Guelph, Guelph, ON, Canada

**J.M. Farber**
Department of Food Science, University of Guelph, Guelph, ON, Canada

**Qingrong Huang**
Department of Food Science, Rutgers University, New Brunswick, NJ, United States

**Owen Griffith Jones**
Department of Food Science, Purdue University, West Lafayette, IN, United States

**Dena Jones**
Nutrition and Food Science Department, California State Polytechnic University, Pomona, CA, United States

**I.J. Joye**
Department of Food Science, University of Guelph, Guelph, ON, Canada

**Dianini Hüttner Kringel**
Department of Agroindustrial Science and Technology, Federal University of Pelotas, Pelotas, RS, Brazil

**G. LaPointe**
Department of Food Science, University of Guelph, Guelph, ON, Canada

**Loong-Tak Lim**
Department of Food Science, University of Guelph, Guelph, ON, Canada

**Xiaobo Liu**
Department of Biomedical and Nutritional Sciences, University of Massachusetts, Lowell, MA, United States

**Ana C. Mendes**
Nano-BioScience Research Group, DTU-Food, Technical University of Denmark, Lyngby, Denmark

**M.A. Rogers**
Department of Food Science, University of Guelph, Guelph, ON, Canada

**Maria Rubino**
School of Packaging, Michigan State University, East Lansing, MI, United States

**Ikjot Singh Sohal**
Purdue University, Center for Cancer Research, West Lafayette, IN, United States

**P.A. Spagnuolo**
Department of Food Science, University of Guelph, Guelph, ON, Canada

**Yining Xia**
Institute of Quality Standard and Testing Technology for Agro-Products, Chinese Academy of Agricultural Sciences, Beijing, China

**Elessandra da Rosa Zavareze**
Department of Agroindustrial Science and Technology, Federal University of Pelotas, Pelotas, RS, Brazil

**Boce Zhang**
Department of Biomedical and Nutritional Sciences, University of Massachusetts, Lowell, MA, United States

**Jieyu Zhu**
Department of Food Science, Rutgers University, New Brunswick, NJ, United States

# Preface

This special volume is dedicated to reviewing the state-of-the-art-in existing and emerging topics pertaining to food nanotechnologies. Chapter 1 provides a broad-spectrum overview on food-grade components, production techniques, characterization, methodologies, existing/potential applications, and bioavailability topics related to food nanomaterials. Chapter 2 focuses specifically on protein assembly phenomena in nature and during processing, focusing on the effects of functional properties in foods when nanomaterials are deployed as fillers, hydrogels, and emulsion/foam stabilizers. Assembly of proteins is also discussed in the presence of polysaccharides. Various derivatives of polysaccharides have been investigated by researchers for the production of nanocrystals, nanocomplexes, nanoemulsion, nanocapsules, and nanofibers. Chapter 3 covers the recent advances in these areas. Targeting a specific application, Chapter 4 examines novel nanodelivery systems used to increase bioavailability, stability, and sensory attributes of functional bioactive. Nanoencapsulation methods for (−)-epigallocatechin gallate, resveratrol, flavonoids, curcumin, flavors, antimicrobials, and colorants are elaborated on. Chapter 5 delves into basic electrohydrodynamic principles for the production of electrospun fibers and electrosprayed particles. Selected proteins and polysaccharides deployed as carrier vehicles for various bioactives are discussed. Recent developments on self-assembled low-molecular weight building blocks for electrospinning are reviewed. Chapter 6 is dedicated to the discussions on bioaccessibility and bioavailability of several classes of bioactive (polyphenols, carotenoids, dietary fibers, plant sterols, bioactive lipids, bioactive peptides, and micronutrients) encapsulated in nanometric delivery systems (liposomes, micelles, micro/nanoemulsions, biopolymeric particles, polyelectrolyte complex, hydrogels, molecular inclusion). Jones et al. also summarized the existing *in vitro* and *in vivo* models for evaluating the bioaccessibility and bioavailability of bioactives. Being an integral component of preservation and distribution systems, nanotechnology has infiltrated into the packaging arena to enhance material properties of packaging materials. Among the nanotechnologies applied in food packaging, nanoclay-reinforced packaging nanocomposites have been at the forefront of commercialized nanomaterials. Chapter 7 reviews various types of nanoclay-based composites for packaging applications and addressing migration of nanofiller in simulated food systems and within the environment. And finally, in Chapter 8, Liu et al. address the safety aspects of nanoparticles, focusing on soft engineered nanoparticles derived from proteins, polysaccharides, and lipids. The authors discussed the physiological effects of these soft engineered nanoparticles on human health when incorporated in food systems and summarized the holistic methodologies currently available to assess their pathophysiological effects.

The evolution of expectation for food nanotechnolgy can be depicted with Gartner's Hype Cycle, which proceeds through five stages: (1) *technology trigger*; (2) *peak of inflated expectations*; (3) *trough of disillusionment*; (4) *slope of enlightenment*; and (5) *plateau of productivity* (Fenn, Raskimo, & Burton, 2017). Although the Gartner's Hype Cycle has been criticized for its accuracy on portraying the actual evolution of technologies, it has an immanent influence on many companies' strategic and investment decision (Steinert & Leifer, 2010). The hype of nanotechnology has been steadily increasing over the past two decades ago. In the context of food nanotechnology, it has progressed beyond stage 2, the peak of initial inflated expectations, of the Gartner's Hype Cycle. Whether it is on the *slope of enlightenment* or has entered the *plateau of productivity* will depend on the specific nanotechnology for food. For example, the application of nanocomposite for barrier food packaging and nanoencapsulation for drug delivery are proven and productive nanotechnologies. Few technologies, such as engineered nanostructures in fats and oils and nanocomposites used in food packages, are progressing along the *slope of enlightenment* as companies overcome technical manufacturing hurdles and begin to see the benefits. On the other hand, the potential of encapsulation of bioactive for tailored release profiles is yet to be utilized. These applications probably are emerging from the *trough of disillusionment*, as their strengths and limitations are becoming evident. However, not all food-related nanotechnologies will make it through the cycle to reach the *plateau of productivity*. Most imperative is that acute and chronic health risks constantly remain ahead on the Gartner's Hype Cycle compared to technological development. Hopefully, we have learned from allowing *trans* fats into our food supply before understanding their negative, chronic health implications.

In socioeconomic contexts, aspects that are most important to the agri-food industry are production of safer food, valorization of waste streams, enhancement of product quality/functionality, extension of shelf life, and increased profit margin. The ability of a nanotechnology to meet some, if not most of these requirements, is crucial in overcoming the impediments to adoption. Due to the immense complexity of human body and food systems, interactions between nanomaterials and these biological systems are still poorly understood and a considerable gap exists in establishing mechanistic understanding of nanotoxicology and methods to harmonize nanorisk assessment (Singh et al., 2019). Consumer perception of nanofoods is greatly influenced by uncertainties associated with safety, environmental impact, and media framing (Handford et al., 2015). Consumers are sensitive and rightly cautious concerning the introduction of these new technologies, especially related to scientific manipulation of food, as demonstrated by genetically modified (GM) foods (Dudo, Choi, & Scheufele, 2011). Therefore, besides addressing technical challenges through research and development, continual strategic engagement of the agri-food industry, scientists, governmental bodies, consumers, and media to increase awareness and improve accurate reporting of nanotechnologies are imperative to ensure commercial success (Boholm & Larsson, 2019; Frewer et al., 2014). To this end, we hope that this volume contributes to an increasing in our understanding of nanotechnology.

The compilation of this volume would not have possible without the support from the authors. We are indebted to their participation, patience, and hard work during the pursuit of this special issue.

LOONG-TAK LIM and MICHAEL ROGERS
University of Guelph

## References

Boholm, Å., & Larsson, S. (2019). What is the problem? A literature review on challenges facing the communication of nanotechnology to the public. *Journal of Nanoparticle Research*, *21*, 86.

Dudo, A., Choi, D.-H., & Scheufele, D. A. (2011). Food nanotechnology in the news. Coverage patterns and thematic emphases during the last decade. *Appetite*, 78–89.

Fenn, J., Raskimo, M., & Burton, B. (2017). *Understanding Gartner's Hype Cycles*. Gartner, Inc. G00251964.

Frewer, L. J., Gupta, N., George, S., Fischer, A. R. H., Giles, E. L., & Coles, D. (2014). Consumer attitudes towards nanotechnologies applied to food production. *Trends in Food Science and Technology*, *40*, 211–225.

Handford, C. E., Dean, M., Spence, M., Henchion, M., Elliott, C. T., & Campbell, K. (2015). Awareness and attitudes towards the emerging use of nanotechnology in the agri-food sector. *Food Control*, *57*, 24–34.

Singh, A. V., Laux, P., Luch, A., Sudrik, C., Wiehr, S., Wild, A.-M., et al. (2019). Review of emerging concepts in nanotoxicology: Opportunities and challenges for safer nanomaterial design. *Toxicology Mechanisms and Methods*, *29*, 378–387.

Steinert, M., & Leifer, L. (2010). Scrutinizing Gartner's hype cycle approach. In *PICMET 2010 technology management for global economic growth, 18–22 July* (pp. 1–13), Phuket, Thailand: IEEE.

CHAPTER ONE

# A comprehensive perspective of food nanomaterials

**I.J. Joye[a], M.G. Corradini[b], L.M. Duizer[a], B.M. Bohrer[a], G. LaPointe[a], J.M. Farber[a], P.A. Spagnuolo[a], M.A. Rogers[a,*]**
[a]Department of Food Science, University of Guelph, Guelph, ON, Canada
[b]Arrell Food Institute, University of Guelph, Guelph, ON, Canada
*Corresponding author: e-mail address: mroger09@uoguelph.ca

## Contents



## Abstract

Nanotechnology is a rapidly developing toolbox that provides solutions to numerous challenges in the food industry and meet public demands for healthier and safer food products. The diversity of nanostructures and their vast, tunable functionality drives their inclusion in food products and packaging materials to improve their nutritional quality through bioactive fortification and probiotics encapsulation, enhance their safety due to their antimicrobial and sensing capabilities and confer novel sensorial properties.

In this food nanotechnology state-of-the-art communication, matrix materials with particular focus on food-grade components, existing and novel production techniques, and current and potential applications in the fields of food quality, safety

*Advances in Food and Nutrition Research*, Volume 88
ISSN 1043-4526
https://doi.org/10.1016/bs.afnr.2019.01.001

and preservation, nutrient bioaccessibility and digestibility will be detailed. Additionally, a thorough analysis of potential strategies to assess the safety of these novel nanostructures is presented.

## Abbreviations

| | |
|---|---|
| **0D** | zero-dimensional |
| **1D** | one-dimensional |
| **2D** | two-dimensional |
| **3D** | three-dimensional |
| **ADME** | absorption, distribution, metabolism and excretion |
| **AgNPs** | silver nanoparticles |
| **AUC** | areas under the curve |
| **AuNPs** | gold nanoparticles |
| **bioMEMs** | biomedical microelectromechanical systems |
| **CAP** | cellulose acetate phthalate |
| **CFU** | colony forming units |
| **DNA** | deoxyribose nuclei acid |
| **ESBL** | extended-spectrum β-lactamase |
| **GIT** | gastrointestinal tract |
| **HASMC** | Human Aortic Smooth Muscle Cells |
| **LAS** | liquid antisolvent precipitation |
| **LBL** | layer-by-layer |
| **LOD** | limits of detection |
| **MRSA** | methicillin-resistant *Staphylococcus aureus* |
| **MRSE** | methicillin-resistant *Staphylococcus epidermidis* |
| **PCR** | polymerase chain reaction |
| **PK** | pharmacokinetic |
| **SERS** | Surface-Enhanced Raman Scattering |
| **USDA** | United States Department of Agriculture |
| **UV light** | ultra-violet light |
| **UV-Vis Spectroscopy** | ultra-violet and visible spectroscopy |
| **VRE** | vancomycin-resistant enterococci |

## 1. Introduction

Nano-market research in food and beverages is continuously expanding with a compound annual growth rate of 12.7% to reach $15.0 billion by 2020 (Trujillo, Avalos, Granda, Guerra, & Pais-Chanfrau, 2016). Currently, there are 633 nanomaterials available with 55 associated with the agricultural and food sectors (Belluco, Gallocchio, Losasso, & Ricci, 2018). In addition, estimates foresee that 50% of food products will be associated with nanotechnology by 2020 (Trujillo et al., 2016). The food sector faces challenges related to changing consumer attitudes, preferences, demographics, climate change and its associated economic and societal impacts, as well as the increasing share of e-commerce. The industry must continually adapt its strategies to cope with these challenges, not only through formulation and processing, but also through communication and marketing. This review will focus on nanotechnology and its potential applications in food formulation and processing.

Nanotechnology focuses on independent, self-standing structures with at least one dimension below 100 nm. 0D and 1D nanostructures are referred to as nanoparticles and nanofibers, respectively, while 2D structures are essentially named nanofilms. Liquid crystals also fall under 0D, 1D, and 2D structures; however, they will not be covered in this review, as their applications in the food industry are limited. In nature, nanostructures are omnipresent in our daily diet. Well-known natural nanostructures are protein aggregates (such as casein micelles, actin-myosin complexes, and oil bodies), nanoscale oil droplets, dietary fiber clusters, *etc.* (Rogers, 2016). Herein, emphasis will be placed on nanostructures synthesized *de novo* or which originate from naturally-occurring structures that served to template further functional modification. Nanostructures are designed to have specific functionalities during food processing, storage, and/or digestion. Their diversity and versatility reflect the range of matrix materials in which they are comprised, ranging from biopolymers such as proteins and polysaccharides to lipids. The most-researched applications for food nanostructures are: (i) production of nanosized ingredients or additives to change ingredient functionality, usually associated with quality improvement; (ii) bioactive fortification; (iii) probiotics encapsulation into nanocarriers for controlled release; (iv) formation of food structures or textures; and/or (v) smart packaging systems that detect and counteract microbiological, biochemical and/or chemical spoilage.

Nanotechnology holds promise to revolutionize the food industry. However, enormous care must be taken to include consumers in the developments by open and correct communication strategies. In this food nanotechnology state-of-the-art communication, matrix materials, production techniques, and applications will be detailed.

## 2. Matrix material

Different nanostructures exist using a variety of building materials; among these, lipid-based nanoemulsions are the most well-studied. Nanoemulsions are produced *via* conventional high-energy or novel low-energy methods. Although low-energy methods require less energy to produce the desired nanostructures, they need a higher emulsifier load which hampers their acceptability. Besides lipid-based nanoparticles, biopolymer-based systems are more versatile in terms of molecules that can be encapsulated and food products they can be added to.

## 2.1 Lipid-based systems

Foods, containing colloid fat crystal networks, have a hierarchal order of structures that exist at various length scales (Acevedo & Marangoni, 2015; Acevedo, Peyronel, & Marangoni, 2011; Mazzanti, Li, Marangoni, & Idziak, 2011). Nanoscale elements in traditional fat crystal networks are influenced by fatty acid composition and fatty acid position on the glycerol backbone, cooling rate, presence/absence of shear, and minor additives (Loisel, Keller, Bourgaux, & Ollivon, 1998; MacMillan et al., 2002; Sato & Ueno, 2005; Wille & Lutton, 1966). Specifically, the packing arrangement of the triglyceride molecules, within the lipid domains, alters the final physical properties of the fat, influencing the hardness, melting point and sensory properties (Sato & Ueno, 2005). For example, in chocolate, ideally lipid molecules arrange themselves into a triclinic arrangement, or the B form V polymorph, while bloomed chocolate packs into the B form VI polymorph resulting in a higher melting temperature and waxy mouthfeel. Furthermore, the heat and mass transfer conditions during crystallization impact both the length and width of the crystallite, which are both in the nanoscale size range; however, at this point it is unknown how this impacts the macroscale and sensory properties of the fat (Acevedo, Block, & Marangoni, 2012; Acevedo & Marangoni, 2010, 2015; Acevedo et al., 2011). These nano length-scale crystallites assemble into higher order crystals that are visible under low magnification and consequently are not typically discussed as nanostructures.

Most lipid-derived nanotechnologies focus on lipid-in-water dispersions (*i.e.*, oil-in-water emulsions or solid-lipid nanoparticles) as they have been shown to improve stability, solubility, control delivery, bioavailability, and effectiveness of food bioactives while reducing their side-effects (Mendes et al., 2016). Nanoparticles containing lipids have potential to improve not only the quality of the foods but also human health (de Souza et al., 2017). These delivery systems facilitate solubilization of hydrophobic bioactive molecules into the core of these droplets and protect them from biological and chemical degradation, increasing the bioactives' shelf-life (de Souza et al., 2017). The generation of lipid nanoparticles may be either facilitated *via* the use of top-down, energy-intensive, approaches such as spray drying, homogenization and extrusion, or bottom-up, low-energy processes such as coacervation and self-assembly/crystallization.

High energy, top-down methods are commonly used to produce nanoemulsions (McClements & Rao, 2011). High-energy mechanical approaches generate intensive disruptive forces that overcome the Laplace pressure, $\Delta P$, of the discontinuous phase. When the disruptive pressure is greater than $\Delta P$, which is dependent on the interfacial surface free energy, $\gamma$, and radius, $r$, of the particles in the dispersed phase (*i.e.*, $\Delta P = \gamma/2r$), particle size reduction occurs. As the radius of particles decreases, the Laplace pressure increases making them more difficult to disrupt and ultimately breaking the particles into smaller objects. A minimum particle size is achieved when the Laplace pressure and disruptive pressure are in equilibrium. The final particle size of the emulsion is dependent on the use of emulsifiers or any surface-active molecules. Emulsifiers reduce the surface free energy of the droplet, thereby decreasing the Laplace pressure, which in turn allows attaining smaller particles using the same disruptive pressures. Of the high-energy processes, the most commonly employed method is high pressure valve homogenization. Other top-down homogenization methods include microfluidization and ultrasonic homogenization (Schubert & Engel, 2004). The particle size is affected not only by the pressure of the homogenizer but also by the number of cycles through the homogenizer. The viscosity difference between the continuous and discontinuous phases, and the presence or absence of surfactants also plays an important role in affecting the particle size (McClements & Rao, 2011).

Low-energy approaches for the generation of emulsions still require a driving force, which is typically achieved by changing their environment (Yin, Chu, Kobayashi, & Nakajima, 2009). Self-emulsifying systems include spontaneous emulsification and phase inversion. The dimensionality of these systems is dictated by the chemical composition of the continuous and discontinuous phases, as well as their relative ratios, and the presence or absence of any surface-active material. Phase-inversion entails an oil-in-water emulsion converting to a water-in-oil emulsion, or *vice versa* (McClements & Rao, 2011). Phase-inversion may be triggered by modifying temperature or chemical composition. Phase-inversion, *via* temperature, relies on changes in solubility or molecular geometry of non-ionic surfactants (Israelachvili, 1991). As temperature changes, the driving force for phase-inversion is associated with the alterations of the physicochemical properties of the surfactant (Gutiérrez et al., 2008). Appropriate surfactant selection is of utmost importance, because the trigger for self-emulsification is dependent on the surfactant becoming more soluble in the opposite phase. Phase-inversions, driven by changes in chemical composition, are similar to temperature induced inversions whereby the packing geometry of the surfactant changes in response to a changing environment (Mayer, Weiss, & McClements, 2013; Ostertag, Weiss, & McClements, 2012; Yu, Li, Xu, Hao, & Sun, 2012). Changes in composition most often include alterations in pH or ionic strength. As an example, salt may either be added to trigger the transition, or the continuous phase may be diluted altering ionic strength below a critical level.

Irrespective of the technique employed, the major advantages of creating particles smaller than 100 nm are associated with the drastically increased surface area of the particles which enhances the ability to adsorb and carry active compounds (Wilczewska, Niemirowicz, & Markiewicz, 2012). Nanoscale particles can be used to increase the bioaccessibility of essential fatty acids (ω-3, 6 and 9), phenolic compounds (quercetin, curcumin, and catechin), and vitamins (A, D, E, and K). The functional properties of these nanomaterials may then be tailored based on their interfacial layer, which is often comprised of polysaccharides, proteins, phospholipids, or synthetic/natural emulsifiers, or on their core materials (*i.e.*, liquid oil, solid fat or a blend of solid and oil).

## 2.2 Biopolymer-based systems

Biopolymer-based nanostructures can be composed of proteins, polysaccharides, and protein-polysaccharide-composites. Protein and polysaccharide-based nanostructures, in contrast to most lipid-based systems, are formed *via* bottom-up approaches (Joye & McClements, 2016). Bottom-up approaches are inspired by mechanisms that drive the formation of naturally-occurring nanostructures (Wang, Sun, Li, Wu, & Wei, 2016). In nature, biomolecules self-assemble into supramolecular structures with defined characteristics and functions. These naturally occurring biopolymeric self-associations can be mimicked and/or used as templates to design procedures that allow generating food-grade nanostructures with specific functionality during processing, storage and/or ingestion and digestion (Garg, Ghatmale, Tarwadi, & Chavan, 2017; Zan & Wu, 2016). Assembly of proteins and polysaccharides into nanostructures can be triggered by altering processing parameters (*e.g.*, the continuous phase solvent quality or the electrical properties of the system), by applying a thermal gradient, or by modifying the concentration of select compounds (Joye & McClements, 2013; McClements, 2017; Prosapio, De Marco, & Reverchon, 2018). Specific examples of bottom-up techniques used to produce biopolymer-based nanoparticles are coacervation, inclusion complex formation, liquid antisolvent precipitation, and gelation (Devi, Sarmah, Khatun, & Maji, 2017; Joye & McClements, 2013; Kong, Bhosale, & Ziegler, 2018; Tsuchido, Sasaki, Sawada, & Akiyoshi, 2014). The distinction between liquid antisolvent precipitation and coacervation is not always clear.

Simple coacervation is triggered by changing the conditions, such as salt concentration, pH, temperature of a polymer solution, causing molecular dehydration of the macromolecule, which drives the separation of a concentrated colloidal phase. In contrast, complex coacervation involves more than a single polymer, typically a polycation and a polyanion, and phase separation is driven by anion-cation interactions forming soluble or insoluble complexes between both polymers (Devi et al., 2017). Liquid antisolvent precipitation (LAS) refers to all particle production methods that manipulate solvent quality. In LAS, polymers are dissolved in a "good" solvent (*i.e.*, a solvent capable of solubilizing both polymers) and then the quality of the solvent is reduced by titration or dialysis using an antisolvent that is completely miscible with the solvent (Corradini, Demol, Boeve, Ludescher, & Joye, 2017; Li, Gu, & Gao, 2017). The change in solvent composition causes supersaturation of the biopolymers leading to precipitation (Joye & McClements, 2013). Besides typical solvents (*e.g.*, water, ethanol), alternative LAS procedures use supercritical fluids, such as carbon dioxide, for nanofabrication. The replacement of the antisolvent with supercritical fluids allows for operation at room or lower temperatures and facilitates the separation of the nanoparticles due to the phase change of the antisolvent (Prosapio et al., 2018).

Inclusion complex formation is also a bottom-up technique defined as a host molecule-guest molecule interaction. Prime examples of inclusion complexes are amylose-lipid complexes and cyclodextrin encapsulation of hydrophobic molecules (Al-Nasiri, Cran, Smallridge, & Bigger, 2018; Zhang, Guan, Zhang, Dai, & Hao, 2018). These molecular inclusion complexes assemble to form nanostructures (Kong et al., 2018). Nanoscale structures formed by the gelation of polymers are driven to assemble by a pH change, temperature treatment, and/or addition of ions (Clark, 1996). Physically stabilized gels (*i.e.*, gelation arising from non-covalent interactions) suffer from poor mechanical stability. Conversely, chemically crosslinked gels display enhanced stability, broadening their applications.

The popularity of bottom-up techniques to fabricate biopolymeric nanostructures arises from the fine control of tailor-ability during particle formation, which results in a narrow distribution of particle sizes with customized characteristics and properties. Additionally, low energy requirements and ease of scalability make bottom-up procedures preferred techniques for fabricating food-grade biopolymer-based nanostructures. Bottom-up processes allow the nanoscale properties (*e.g.*, the size (distribution) and charge of the particles) to be designed and engineered; however, controlling and characterizing these structures is difficult (Corradini et al., 2017; Foster et al., 2018; Orecchioni, Duclairoir, Renard, & Nakache, 2006).

Conversely, as it was mentioned in the previous section, top-down techniques involve "breaking-down" a bulk material or composite into nanosized particles. The most commonly employed top-down methods in food-grade nanofabrication using biopolymers include, but are not limited to, shredding, emulsification, homogenization, and grinding. Some techniques are not as simple to categorize as top-down or bottom-up processes. Spray-drying, usually yields particles that fall into the micrometer range, but advances in nano-spray drying and electro-spraying, which atomizes droplets with the assistance of electrical forces, expand its applications to produce food nanoparticles (Arpagaus, Collenberg, Rütti, Assadpour, & Jafari, 2018; Arpagaus, John, Collenberg, & Rütti, 2017; Li, Anton, Arpagaus, Belleteix, & Vandamme, 2010; Pérez-Masiá et al., 2015; Tapia-Hernández et al., 2015). Additive nano-manufacturing techniques can also be classified as top-down or bottom-up based on the process configuration. Among the former, two-photon laser nanofabrication and 3D printing have been used to fabricate high resolution, complex 3D biological nanostructures for medicinal uses (Koumoulos, Gkartzou, & Charitidis, 2017). Their utilization in foods is currently limited by their resolution, compatibility with existing food-grade biomaterials, and cost. However, this rapidly-evolving field is expected to gain momentum in food nanostructure fabrication. The use of food-grade biopolymers as functional materials in nanostructures has received increasing attention due to their physiological innocuity, biocompatibility, versatility, extensive accessibility, functionality, and biodegradability (Rajendran, Udenigwe, & Yada, 2016).

Nanomaterials derived from animal proteins, such as gelatin and serum albumins, have been less commonly studied in recent years than those produced with plant-derived proteins including soy glycinin and beta-conglycinin, corn zein or wheat gliadin. Plant protein matrices are more attractive materials because of greater availability and sustainability (Mohammadinejad, Karimi, Iravani, & Varma, 2016; Wan, Guo, & Yang, 2015). Although protein nanostructures have found applications as colloidal stabilizers and sensing devices, they have been predominantly proposed as delivery systems for bioactive compounds (Pan & Zhong, 2016).

Although *polysaccharides* are hydrophilic molecules, they are often modified, through esterification and other methods, to confer the desired physical properties and amphiphilic characteristics (Sweedman, Tizzotti, Schäfer, & Gilbert, 2013). The size and performance of carbohydrate nanoparticles are affected by the characteristics of the carbohydrates. In general, lower degrees of branching and shorter chains are associated with the formation of smaller nanoparticles. Hence, enzymatic pre-treatment of the carbohydrate sources to reduce chain length or branching tunes the nanoparticle size and functionality (Chang, Yang, Ren, & Zhou, 2018; Qiu et al., 2016). The simultaneous use of proteins and polysaccharides permits large variability of properties, depending on the combination of biopolymers utilized, the manufacturing processes, and conditions employed. The preferred production methods of composite nanoparticles are co-precipitation or post-production nanostructure coatings. Selection of polymers with opposite charges (*i.e.*, negatively charged pectin and positively charged zein) provides a simple pathway to synthesize nanoscale delivery systems through complexation by electrostatic interactions (Yan, Qiu, Wang, & Wu, 2017). Coating of simple or composite particles with proteins, carbohydrates or their combination allows modification and fine-tuning of the properties, interactions, carrier and release capacity of the system (McClements, 2017; Yao, Chen, Song, McClements, & Hu, 2018).

Protein nanostructures are usually stabilized by electrostatic repulsion, where the zeta-potential exceeds 30 mV (in absolute value) to induce colloidal stability. However, changing the pH and/or ionic strength affects the particle charge and aggregation (Davidov-Pardo, Pérez-Ciordia, Marı´n-Arroyo, & McClements, 2015); in turn, affecting bio-accessibility of the encapsulated molecule. Altering the electrochemical characteristics of proteins, for example by cationization, increases the loading capacity of hydrophilic compounds (Fathi, Donsi, & McClements, 2018), but will also alter the aggregation stability of the nanostructure. Most proteins adopt a globular, tight conformation, while polysaccharides exhibit extended stiff or flexible random conformations. The latter makes polysaccharides extremely suitable to stabilize colloidal systems by steric repulsion. Steric stabilization is not exclusively associated with polysaccharides as some proteins also adopt more flexible structures. Colloidal systems stabilized by steric repulsion are less susceptible to changing pH and ionic strength conditions; hence, increasing the stability of these nanoparticles relative to those stabilized by mere electrostatic repulsion. The combined utilization of proteins and polysaccharides for producing nanostructures results in materials with greater stability over a broader range of conditions.

A nanostructure shape affects its interactions with other constituents of a system, as well as its bio-accessibility and bio-specificity. Although most nanoparticles produced with biopolymers exhibit spherical shapes, anisotropic dextran and albumin nanoparticles (*e.g.*, cylindrical) have been obtained using microfluidic methods such as electro-jetting (Meyer & Green, 2015). The ability to design protein, carbohydrate or composite anisotropic nanoparticles, although not fully realized in food applications, could increase their functionality as indicated in medical and pharmaceutical applications (Kinnear, Moore, Rodriguez-Lorenzo, Rothen-Rutishauser, & Petri-Fink, 2017).

Proteins and polysaccharides can also be used to produce nanostructures other than the nanoparticles, the ultimate 0D nanostructure. Nanorods, nanotubes, nanowires, and nanofibers are 1D nanostructures with large surface/volume ratios and have cross-sectional diameters between tens and hundreds of nanometers. Naturally-occurring biological molecules are currently used as nanofiber templates and functionalized as biosensors through mineralization, deposition, or enzyme immobilization (Weiss, Takhistov, & McClements, 2006). Alternatively, electrospun nanofibers are produced using food-grade proteins and carbohydrates (Ghorani & Tucker, 2015; Wen, Wen, Zong, Linhardt, & Wu, 2017). These nanofibers are employed in water treatment and ultrafiltration processes (Kenry & Lim, 2017; Schiffman & Schauer, 2008).

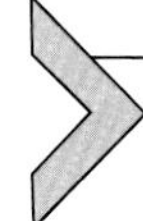

## 3. Applications

### 3.1 Encapsulation and delivery of bioactives

Nanoencapsulation improves the biological efficiency of enzymes, vitamins, and minerals (Pradhan et al., 2015; Thomas & Sayre, 2005). Nanoparticles are designed to enhance biological activity compared to micro- and macro-scale particles of identical chemical composition due to greater surface areas (Sozer & Kokini, 2009). This enhances bioactive solubility, improves digestibility (*via* prolonged residence time in the gastrointestinal tract), and increases biological uptake (Dickinson, 2012; Watanabe, Iwamoto, & Ichikawa, 2005).

The amphiphilicity of proteins allows them to interact with both hydrophilic and hydrophobic molecules through non-covalent and covalent interactions (Fathi et al., 2018; Matalanis, Jones, & McClements, 2011; McClements, 2015; Pathakoti, Manubolu, & Hwang, 2017). Additional benefits of using proteins in nanofabrication include the protection confer to bioactives against oxidation or UV-light degradation because they can scavenge free radicals or chelate pro-oxidants (Charve & Reineccius, 2009; Samaranayaka & Li-Chan, 2011). Specific mucoadhesive properties of proteins allow nanoparticles to target and deliver compounds to specific regions of the gastrointestinal tract (Arangoa et al., 2000). By interacting with the mucosa at specific points along the gastrointestinal tract, the residence time and release profiles of the bioactives are extended. Similarly, several charged polysaccharides, including alginate and chitosan, display mucoadhesive properties for targeted delivery and enhanced bioaccessibility of bioactive compounds encapsulated in carbohydrate-based nanoparticles (Martínez-Ballesta, Gil-Izquierdo, García-Viguera, & Domínguez-Perles, 2018). Protein-polysaccharide composite systems exhibit enhanced loading capacity, additional protection of the encapsulated bioactive, and improved digestibility (Davidov-Pardo et al., 2015; Yan et al., 2017).

## 3.2 Encapsulation and delivery of probiotics

Oral ingestion of probiotics imposes mechanical, chemical and enzymatic stresses that reduce the viability of living organisms throughout the gastrointestinal tract. In contrast to other nutraceutical compounds, orally consumed probiotics usually exert their direct activity in the gut lumen and on the gut epithelium, so there is no requirement for absorption, although it is useful to target release in specific regions of the gut. The four main factors influencing the viability of probiotics are temperature, humidity, acidity and atmospheric conditions (oxygen). Each type of microorganism has optimal conditions for growth, along with minimal and maximal parameters for survival. Sporulating bacteria, such as *Bacillus* spp., can withstand high processing temperatures. Spores are considered a natural encapsulation nanomaterial. However, many of the microbial genera commonly used as probiotics, such as Bifidobacteria and Lactobacilli, do not form spores, so they do not survive thermal treatments. Microbial encapsulation has been studied for many years and applied to the delivery of probiotics, improving survival during gastrointestinal transit, extension of shelf life at room temperature, increased heat and shear stress resistance, as well as improved acid and bile tolerance. Encapsulation may also delay gas exchange by lengthening the path of diffusion, prolonging the shelf life of anaerobic probiotic species.

Methods for encapsulating bacteria traditionally include dispersal through extrusion, emulsification, coacervation (phase separation) or drying (freeze-drying, spray-drying or fluidized bed drying). Extrusion methods for probiotics must be thermally "gentle" and the particle size depends on the nozzle size. Emulsification produces smaller particles, but at a higher cost. During the emulsification process, the solution viscosity and the agitation rate control particle diameter. Drying produces materials with low storage stability (Huq, Khan, Khan, Riedl, & Lacroix, 2013). In addition, the low moisture and high temperature associated with drying reduce probiotic survival. Modifications to spray-drying methods have succeeded in minimizing these effects by incorporating emulsifiers. Although particle size reduction is achieved through micronization, care must be taken to limit cell damage by excessive reduction of particle size. Emerging technologies for producing nanomaterials for encapsulation include electrospinning, impinging aerosol, or using supercritical fluids. The advantages and disadvantages of these technologies have been previously compared (Chen, Wang, Liu, & Gong, 2017). Electrospinning is simple, producing small-size microcapsules without using high temperatures. The disadvantages still needing to be overcome are the scale of production and the availability of materials for food applications. Electrospinning, using whey protein concentrate, resulted in matrices with better cell viability than pullulan matrices (Martín, Lara-Villoslada, Ruiz, & Morales, 2015). Formulating materials requires taking into consideration protecting the bacteria from environmental and gastrointestinal conditions, while providing for their release at the intended site of action. Encapsulation materials can be proteins, polysaccharides or lipids. Each type of biopolymer will have specific properties for protection and release of probiotic cells. Polysaccharides such as dextran mixtures, kappa-carrageenan, gellan gum, and starch have all been studied as encapsulation materials. Plant-based hydrocolloids generally have drawbacks associated with forming low-density gels and larger capsules, due to high viscosity at low concentration, which limits the barrier effect provided to live cells. Proteins can form high-density gels and smaller-sized microcapsules, an important consideration for sensory aspects in foods (Heidebach, Först, & Kulozik, 2009). In addition, the buffering capacity of milk proteins aids in maintaining local pH higher in the protein matrix, which protects probiotic cells from acidic environments. Animal-derived proteins such as hydrolyzed collagen (*i.e.*, gelatin) and casein have been used individually and in combinations with other materials. Whey proteins such as hydrolyzed α-lactalbumin form self-assembled nanotubes with cavities of 8 nm, which would allow binding of vitamins or enzymes (Cushen, Kerry, Morris, Cruz-Romero, & Cummins, 2012). These have applications in coating bacteria, enabling combined formulations of probiotics with other bioactive compounds. Whey proteins have even been conjugated with pectin. Cellulose acetate phthalate (CAP), a cellulose derivative, is a common enteric coating for delivering bioactives to the intestine to bypass the negative effects on taste. Lipids are used in formulations to coat and protect freeze-dried encapsulate probiotics.

Lack of information on structural and compositional properties of materials used to encapsulate probiotics is the main source of inconsistency between studies. Calcium alginate is the most common material used for probiotic encapsulation, yet provides the most contradictory findings (Chen et al., 2017). Early studies using calcium alginate improved the viability of *Lactobacillus acidophilus*, *Lactobacillus casei*, *Lactobacillus rhamnosus* and *Bifidobacterium* spp. in freeze-dried yogurt. Survival was improved by combining calcium alginate with starch as a coating polymer. However, alginate offered no protection for Bifidobacteria in cheese (Gobbetti, Corsetti, Smacchi, Zocchetti, & De Angelis, 1998), or from low pH environments (Huq et al., 2013). Probiotic survival was altered by alginate concentration, porosity and microbial species. Some disadvantages associated with alginates include loss of mechanical integrity in acidic environments, as gels require calcium ions. Spray-coating with starch has been used to microencapsulate Bifidobacteria, which may be sensitive to oxygen (O'Riordan, Andrews, Buckle, & Conway, 2002). Inlet temperature for spray-drying was reduced to 100 °C, as higher temperatures led to reduced viability. Although viable during drying, *Bifidobacterium* survival was not promoted in two spray-dried food products. As spray-drying can lead to damaging microbial cells due to heat and physical injury, temperatures at both the inlet and outlet must be carefully optimized. In the case of freeze drying, cryoprotectants are used to improve microbial survival. Although nanomaterials can be used to encapsulate bacteria, the size of the bacteria themselves dictates the particle size of the final product, regardless of the encapsulation technique used.

Polymeric blends show more promise for wider functionality in protection and targeted delivery of probiotics. The layer-by-layer (LBL) approach (Anselmo, McHugh, Webster, Langer, & Jaklenec, 2016) provides alternate layers of alginate and chitosan to enhance probiotic survival *in vivo*. *Ex vivo* sections of intestine and the EpiIntestinal Model showed enhanced adhesion of encapsulated *Bacillus coagulans* compared to free cells. Adhesion of probiotics prolongs their residence in the gut and exerts potential activity. Other considerations for use of nanomaterials are permeability to nutrient substrates for bacteria and host immune system protection.

Formulation improvements include the addition of antioxidants (protection from oxygen) or prebiotics (see below). The damaging effect of oxygen can be addressed by deoxygenation of solutions for strict anaerobes, but antioxidants such as ascorbic acid, L-cysteine-HCl or tocopherol are common oxygen-neutralizing additives described so far. However, detrimental interactions with the encapsulation material may also occur. Survival of *B. animalis* BB12 improved by adding L-cysteine-HCl to alginate, but it was negatively impacted when mixed with whey proteins (Chen et al., 2017). Possible probiotic strain effects on the antioxidants must be considered, as for the catabolism of ascorbate by *Lactobacillus rhamnosus*. In addition to the chemical oxygen-neutralizing agents, new applications of nanotechnology on the market provide oxygen scavenging or absorbing material, based on food grade enzymes.

A prebiotic is defined as "a selectively fermented ingredient that allows specific changes, both in the composition and/or activity in the gastrointestinal microflora that confers benefits upon host well-being and health" (Roberfroid, 2007). Products combining probiotics with prebiotics are termed synbiotics. As an example, Raftilose P95 (a commercial fructooligosaccharide) improved the viability of Bifidobacteria and Lactobacillus species during storage for 4 weeks and 4 °C. Prebiotics can be considered natural biodegradable nanomaterials that protect probiotics and provide health benefits in the gut. Colon-specific delivery can be achieved with oligosaccharides such as inulin, which is hydrolyzed by microorganisms in the colon. This material will resist passage through the stomach and small intestine. Release of microbes can be accomplished through mechanical stresses (peristaltic motion), pH changes, and time-dependent or enzymatic (enzymes produced by colonic microbiota) structure degradation. Degradation of encapsulation material can be triggered by changes in water content, pH (*e.g.*, CAP, polyvinyl pyrrolidone, vinyl acetate-co-crotonic acid) or enzymes. Novel materials being tested for probiotic encapsulation are yeast cell walls (*Saccharomyces cerevisiae*), which provide bioactive compounds stimulating the immune system (Mokhtari, Jafari, Khomeiri, Maghsoudlou, & Ghorbani, 2017).

Even though probiotics may retain viability through encapsulation, the desired probiotic attribute or activity may be lost. Mycotoxin detoxification by strain LS 100 was lost following encapsulation in calcium alginate. It was hypothesized that the physiology of the encapsulated bacteria changed (Yu et al., 2010). Avoiding microbial contamination requires aseptic conditions and sterile materials and equipment during the encapsulation process to ensure quality control.

## 3.3 Structure building and stabilization

Nanotechnology can also be used to form and stabilize specific micro-, meso- and/or macrostructures in food (Chaudhry et al., 2008; Sekhon, 2014; Weiss et al., 2006). Nano-structuring food proteins can generate particles with nanocavities. These have been applied to produce ingredients with improved functional properties such as water holding capacity, gelation, and emulsification properties (Acosta-Domínguez, Hernández-Sánchez, Gutiérrez-López, Alamilla-Beltrán, & Azuara, 2016). Protein nanoparticles have been recently used to stabilize air/water interfaces in foams and oil/water interfaces in simple and multiple emulsions (Chen et al., 2017; Wen et al., 2017; Zhou et al., 2018). Nanostructures produced with common protein isolates, such as soy and whey proteins, as well as ovalbumin, lactoferrin, gliadin and karafin have successfully stabilized food interfaces. Foam bubbles or emulsion droplets coated with solid nanoparticles (*i.e.*, Pickering foams and emulsions) are less prone to coalesce than their uncoated counterparts and typically exhibit improved functionality (Meshulam & Lesmes, 2014; Wu et al., 2015; Xiao, Lu, & Huang, 2017; Xiao, Wang, Perez Gonzalez, & Huang, 2016). The exact behavior of the particles at the interface, however, remains to be elucidated. Classical Pickering emulsions make use of solid particles that remain in their particulate form at the interface (Berton-Carabin & Schroën, 2015). Previous research, however, suggested that protein particles do not remain in their particulate form at the interface, but instead form an elastic protein film around the dispersed phase (Peng et al., 2017). By making polysaccharides more amphiphilic, their corresponding nanostructures, such as starch nanoparticles, stabilize emulsions and delay phase inversion. The rheological characteristics of carbohydrate nanoparticle-stabilized systems including high viscosity and shear thinning suggest stabilization by the formation of 3D networks and steric repulsion (Ye et al., 2018, 2017).

## 3.4 Sensory attributes

Nanotechnology has the potential to transform the food industry, allowing for the creation of food products with unique sensory properties. Colors and flavors, that are encapsulated, can be manipulated through the bottom-up development of nano-delivery systems whereby the active ingredient (color or flavor) is encapsulated making it stable during processing and storage. Textures of foods have also been manipulated through the top-down development of nanoparticles that are added to the foods, so the mouthfeel can be improved. To date, however, there are few publications where sensory evaluation has been conducted on actual food products that have incorporated nanotechnology.

Natural colorants are generally unstable and degrade in the presence of light, oxygen and/or heat (Ghidouche, Rey, Michel, & Galaffu, 2013). Additionally, natural colorants tend to be insoluble in water, making it a challenge to include them in aqueous media such as beverages. Lycopene, a red colored carotenoid, is insoluble in water and unstable to processing conditions. However, when electrospun with gelatin nanofibers, the thermal stability as well as the water-solubility is greatly enhanced, making it possible to add lycopene as a natural and stable colorant to beverages (İnanç Horuz & Belibağlı, 2018). Nanoemulsification has also been used to stabilize hydrophobic colorants, such as β-carotene (Astete, Sabliov, Watanabe, & Biris, 2009) and capsaicin (Akbas, Soyler, & Oztop, 2018; Choi, Kim, Cho, Hwang, & Kim, 2011) for use in beverages. For each of these compounds, the resulting color was stable over time showing that a top-down approach allows for the protection of colorants when added into food products.

Food products contain both volatile and non-volatile compounds. During consumption, these compounds are released from the product matrix and affect the way we perceive and appreciate food. The volatiles, in the form of odorants and irritants, travel retronasally to olfactory receptors while the non-volatile compounds are mixed with saliva and travel to gustatory receptor cells in the mouth for the perception of tastes (Laing & Jinks, 1996). It is the integration of signals generated by the olfactory and gustatory receptor cells that leads to the perception of the flavor of a food. Often flavors in foods are substantially degraded through chemical or physical changes during processing. Nanotechnology can ensure that consumers are perceiving appropriate flavors and can control the release profile. Nanotechnology also assists with reduction of off-flavors either from added ingredients or that develop during storage. A prime example are oils used as flavorings in beverages. These oils can be unstable and deteriorate in the presence of oxygen and light leading to a reduction of the original flavor and/or the development of off-flavors (Chen et al., 2018). Microemulsions composed of zein nanoparticles have been shown to slow the degradation of unstable oils. These emulsions are comprised of a zein based shell around the unstable oil. In comparison to similar emulsions stabilized with whey protein isolate, flavor release from these microemulsions is gradual, slowing the oxidation rate (Chen et al., 2018). Similar observations have been noted for zein-coated peppermint oil microemulsions (Chen & Zhong, 2015). Starch-flavor volatile complexes have also been suggested as a means of controlled release of volatiles during food consumption. A number of different flavor complexes have been examined including menthol and menthone-starch complexes (Ades, Kesselman, Ungar, & Shimoni, 2012), hexanol-, heptanol-, caprylic alcohol-, nonylalcohol- and naphthal-amylose complexes (Feng et al., 2017), and fenchone-, geraniol- and menthone-amylose complexes. It is speculated that when a food containing such a flavor-starch complex is ingested, the starch will hydrolyze in the presence of salivary α-amylase, releasing the volatiles to produce a flavor. All of this work, however, has been tested *in vitro* using simulated saliva (Ades et al., 2012). There is still a need to conduct *in vivo* testing to track volatile release and resulting flavor perception. While much of the research focused on flavors is to protect volatiles against deterioration and ensure a sustained and controlled flavor release, there is also a need to mask off-flavors produced by compounds added into foods. One of the most well documented commercial examples of nanotechnology being used to mask off-flavors is the commercialization of Up™ bread (Tip Top Bakeries, Australia). This product contains microencapsulated omega-3 fish oils (Chaudhry et al., 2008). The microencapsulation of oil prevents a fishy flavor from being perceived as the bread is consumed. Nanoparticles have also been developed that will deliver taste to consumers without compromising on health. Soda-Lo® marketed by Tate & Lyle as sodium-based salt is composed of hollow spheres with particle sizes of 5–10 μm. The density of this salt is lower than other salts and only a small amount of Soda-Lo® is required for salt taste perception to occur. It is stated that this salt can reduce the sodium content of a food by 25–50% (Tate & Lyle, 2012). Soda-Lo has been incorporated into sushi, leading to a lower amount of sodium consumed with no change to the sensory properties of the food (Đorđević, Buchtová, & Macharáčková, 2017). This product has been added to pizza dough; however, the microcrystalline structure of the salt did not enhance the saltiness of the pizza crust, possibly due to the loss of microcrystalline state due to dissolution during mixing of the dough (Mueller, Koehler, & Scherf, 2016).

Nanotechnology also builds texture in food. Textures in food are perceived when the underlying structural matrix of the product is broken during consumption. Any alterations to product formulations can lead to a change in product structure ultimately influencing the resulting textures of the food (Wilkinson, Dijksterhuis, & Minekus, 2000). Often these new textures are not preferred by consumers which can be a challenge when reformulating foods for health benefits. This can be a particular problem when developing low-fat products as fat reduction is known to alter the mouthfeel of a product (Drewnowski, 1992). Nanoparticles, and cellulose nanoparticles in particular, can be added to these foods to aid in maintenance of appropriate lubricity. Cellulose nanoparticles were first patented by Turbak, Snyder, and Sandberg (1982) for use as a thickener, flavor carrier and stabilizer. Cellulose nanoparticles have been added to low-fat mayonnaise (Golchoobi, Alimi, Shokoohi, & Yousefi, 2016) and low-fat ice cream (Velásquez-Cock et al., 2019) without significantly impacting the sensory properties of these products.

## 3.5 Food preservation

Nanotechnology can preserve and maintain the quality of food products. Examples include the preservation or even enhancement of: (1) product appearance, (2) product function, and (3) product sensory and nutritional attributes. Product appearance is improved through the use of nano-ingredients or packaging technologies. Metallic nanoparticles (such as silver and nanoparticles) have not only been used as antimicrobial additives, they also elicit favorable responses in the preservation of appearance (Bouwmeester et al., 2009). Titanium dioxide ($TiO_2$), for example, can be used as white color enhancer in foods such as milk, cheeses, and other dairy products (Calvano, Jensen, & Zambonin, 2009; Peters et al., 2014; Weir, Westerhoff, Fabricius, Hristovski, & von Goetz, 2012). Nanoemulsions are preferred in comparison to regular emulsions primarily because of their improved stability and texture; however, there are also several examples (salad dressing, flavored oils, sweeteners, and beverages) where improved stability also results in enhanced product appearance and overall consumer acceptability (Sekhon, 2014; Silva, Cerqueira, & Vicente, 2012). Nanocomposites are most notably used in the food packaging industry, where they can indirectly affect food product appearance through their oxygen or ethylene gas scavenging capacity, coating of food products, and/or modification of the atmosphere that food products are stored in (Lagarón, López-Rubio, & Fabra, 2015). Other notable nanostructured materials include nanoencapsulated ingredients, which aid in the functionality of certain ingredients (Assadpour & Jafari, 2018; Fathi et al., 2018; Fathi, Martín, & McClements, 2014), and nanosensors, which detect changes or alterations in appearance (specifically color) of food products during storage (Kumar, Guleria, & Mehta, 2017; Pathakoti et al., 2017). Ongoing research in the field of food nanotechnology is focused on providing insight into the preservation of food function over prolonged storage time (Dickinson, 2012). A few specific areas of interest in this context include the molecular factors controlling nanoparticle function (*i.e.*, wettability, adsorption, dispersion/uniformity and stability) over extended storage times, the structural and mechanical properties of novel nanostructure-stabilized food interfaces, and the stabilization and improved product characteristics of particle-coated ingredients and/or food products. There is potential for exploiting this emerging knowledge in new food product applications with particular emphasis on preservation with nanotechnology.

Perhaps one of the most notable aspects of nanotechnology in food products is its effect on the preservation of sensory attributes and nutritional quality. As it was explained in Section 3.4, from a sensory standpoint, nanotechnology processing techniques and ingredients can improve product flavor, texture, and mouthfeel and also it can be used to preserve those attributes over prolonged storage times (Moraru, Huang, Takhistov, Dogan, & Kokini, 2009). An example of improved preservation of sensory attributes is the use of surface functionalized nanomaterials that increase antimicrobial activity and prevent the absorption of oxygen, thus extending product shelf life (improved sensory attributes) and product safety (less bacteria growth). From a nutritional standpoint, nanotechnology processing techniques and ingredients can improve the availability of bioactive molecules (as outlined above) and overall food digestibility (Sekhon, 2014). An example of improved preservation of nutritional attributes is the use of nanocarrier systems for the delivery of nutrients and supplements in the form of liposomes or biopolymer-based nanocomposites. There is an opportunity to improve initial food product attributes and maintain them over time with various applications of nanotechnology.

## 3.6 Food safety

In Canada, it is estimated that each year there are 1.6 and 2.4 million episodes of domestically-acquired foodborne illness related to 30 known pathogens and unspecified agents, respectively, for a total estimate of 4.0 million episodes of domestically acquired foodborne illnesses. Foodborne illness caused by unknown etiological agents accounted for the majority of the cases (60%), as compared with 40% attributed to known agents (Thomas et al., 2013). These figures highlight the need for additional resources and tools for decreasing foodborne disease in Canada. Nanotechnology holds great promise to provide cutting-edge tools to improve the detection of foodborne pathogens, as well as to provide innovative methods for inactivating them, and thus to improve food safety.

The incentive for better detection of foodborne pathogens and more effective antimicrobial agents will drive the research in the application of nanotechnology in the food and agricultural sector, leading this technology to become more prominent. There is a continual need for more rapid and sensitive real-time methods for detecting foodborne pathogens, and nanotechnology holds great promise in achieving these objectives. For example, the large surface area of nanomaterials allows for the efficient isolation and pre-concentration of analytes from complex matrices, an issue that has always plagued the traditional isolation of pathogens from foods. Some nanoparticles that have been used for the detection of foodborne pathogens include gold nanoparticles, gold/silicon nanorods, gold nanorods, quantum dots, magnetic bead/quantum dots, single walled carbon nanotubes, immunomagnetic liposome nanoparticles, aptamer conjugated gold nanoparticles and liposome nanoparticles (Bajpai et al., 2018).

Surface-enhanced Raman scattering (SERS), in combination with silver nanoparticles, can be used for the quick and efficient detection of microbial pathogens (Bajpai et al., 2018). The simultaneous detection of *Escherichia coli*, *Listeria monocytogenes*, and *Salmonella* Typhimurium was achieved using this method (Duncan, 2011). When used along with gold nanoparticles (AuNPs), the sensitivity of SERS is powerful enough to detect a single Bacillus spore and distinguish Bacillus at the species level (He, Liu, Lin, Mustapha, & Wang, 2008). Researchers are also using fiber optic surface plasmon resonance sensors to detect *E. coli* O157:H7 in juice. The test uses antimicrobial peptides as recognition elements and silver nanoparticles-reduced graphene oxide nanocomposites assisted signal amplification (Zhou et al., 2018).

The detection of Salmonella, *L. monocytogenes*, *E. coli*, and Campylobacter in beef, chicken, pork, and turkey meat has also been done using AuNPs (Belluco et al., 2018). Other than silver and gold nanoparticles, gold-silver core-shell nanoparticles have also been used to simultaneously detect multiple organisms when combined with the use of SERS and UV-Vis spectroscopy (Duncan, 2011). Similarly, a combination of magnetic separation with labeled silica coated magnetic nanoparticles and gold nanoparticles labeled with Raman reporter molecules was used for the multiplexed detection of *Salmonella Typhimurium* and *Staphylococcus aureus* in a peanut butter emulsion and a spinach wash, with a detection limit of $10^3$ CFU/mL (Wang, Ravindranath, & Irudayaraj, 2011). When attached to specific antibodies, magnetic iron oxide nanoparticles can effectively separate target bacteria, which can then be detected and quantified using real-time PCR (Yang, Qu, Wimbrow, Jiang, & Sun, 2007). The main characteristics of nanotechnology applications during animal farming, meat processing and storage, as well as detection/diagnostic tools applicable to the meat chain have been nicely summarized by Belluco et al. (2018). For example, cadmium sulfide and lead sulfide nanoparticles have been used for the selective detection of *E. coli* O157:H7 and *S. aureus* in fresh beef, iron oxide and gold nanoparticles for *E. coli* O157:H7 in ground beef and gold nanoparticles for the detection of *E. coli*, Salmonella or Listeria in raw ground beef, poultry and deli-turkey.

Another nano-method of microbial detection is the use of cantilever bioMEMs (devices or systems constructed using techniques inspired by micro-nanoscale fabrication) which can detect the presence of pathogens such as *E. coli* and *Listeria* spp., as well as DNA and proteins (Powell & Colin, 2008). In one study, Wang et al. developed a multiplexed micro-cantilever biosensor that used phage-derived peptides for the detection of Salmonella (Wang et al., 2014). Multi-walled carbon nanotubes have also been used to increase the sensitivity of biosensors when detecting enzymes (Powell & Colin, 2008). Many bioMEMSs rely on electrical methods such as electrophoresis, which are easy to miniaturize. These electrochemical methods (*e.g.*, amperometric, potentiometric, and conductometric) can be used to detect a wide variety of foodborne pathogens. Similarly, optical biosensors based on detecting changes in optical properties when stimulated, *e.g.*, biosensors that incorporate the bacterial lux gene, can monitor for bioluminescence to detect environmental pollutants (Powell & Colin, 2008). For example, the combination of reflective interferometry and nanotechnology has been used to detect *E. coli* by detecting and measuring light scattered by the mitochondria (Trujillo et al., 2016).

In another methodology application, the use of fluorescence quantum dot labeling using antibodies was shown to result in rapid and sensitive bacterial detection (Sadeghi, Rodriguez, Yao, & Kokini, 2017). In addition, the simultaneous detection of two foodborne pathogens, *E. coli* O157:H7 and *Salmonella Typhimurium*, at a detection limit of $10^3$ CFU/mL, has been observed with the use of magnetic microbeads and quantum dot techniques (Sadeghi et al., 2017). An array-based immunosorbent method using protein G-liposomal nanovesicles has also been used to detect *E. coli* O157:H7, *Salmonella* spp., and *L. monocytogenes* in both pure and mixed cultures (Chen & Durst, 2006). Previous studies using lateral-flow immune test strips with palladium nanoparticles have also shown that the detection of specific bacteria, such as Klebsiella, is possible through specific binding and visualized detection (Tominaga, 2018).

## 3.7 Antimicrobial activity

Silver nanoparticles (AgNP) can exhibit antimicrobial activity against strains resistant to strong chemical antimicrobials such as methicillin-resistant *S. aureus* (MRSA), methicillin-resistant *S. epidermidis* (MRSE), vancomycin-resistant enterococci (VRE), and extended-spectrum β-lactamase (ESBL) producing Klebsiella (Duncan, 2011). In fact, MIC values of 2–4 μg/mL were observed for AgNPs of 45–50 nm in diameter against *E. coli*, *Vibrio cholerae*, *Shigella flexneri*, and at least one strain of *S. aureus* (Sarkar, Jana, Samanta, & Mostafa, 2007). Research done by several groups has shown that AgNPs are effective against more than 650 disease-causing pathogens with 6 min of contact time, as compared to general antibiotics which are only effective against five or six pathogens (Trujillo et al., 2016). While the mode of action of AgNPs is due to Ag+ ions, it appears that they are toxic to microorganisms by mechanisms that are different from Ag+ ions alone. For example, it has been demonstrated that the toxicity of AgNPs increases significantly as the nanoparticle diameter increases, and nanoparticle shape can also influence bactericidal properties (Duncan, 2011).

The use of superparamagnetic iron oxide (ION) has also been proven to be effective against *B. subtilis* biofilm formation, bacterial growth and cell viability (Bajpai et al., 2018). $TiO_2$ nanoparticles are also promising and, unlike AgNPs, were also found to contain antimicrobial properties against *Salmonella choleraesuis*, *Vibrio parahaemolyticus*, and *L. monocytogenes*

(Duncan, 2011). However, $TiO_2$-based antimicrobials have only been found to have antimicrobial activity under UV illumination, *i.e.*, not in the dark (Kim, Kim, Cho, & Cho, 2003).

Natural phytoglycogen nanoparticles have exhibited effective antimicrobial activity against *L. monocytogenes* when nisin was present (Bi, Yang, Narsimhan, Bhunia, & Yao, 2011). Results from a previous study showed that the antimicrobial activity against *L. monocytogenes* of amines anchored to mesoporous silica nanoparticles was 100-fold greater than their free counterparts (Ruiz-Rico et al., 2018). Research is also being conducted to create permanent non-leaching antimicrobial surfaces by linking various nano-based antimicrobials to flexible polymeric chains firmly anchored to surfaces (Powell & Colin, 2008). This is important, because the slow release of antimicrobials into the environment can create conditions which can lead to an increase in antimicrobial resistance.

The use of engineered water nanostructures (EWNS) was shown to be an effective method of inactivating *E. coli*, *S. enterica*, and *L. innocua* on both equipment and food products (Pyrgiotakis et al., 2015). EWNS are small (25 nm), can remain airborne indoors for hours, contain reactive oxygen species and have a very strong surface charge. The study showed that EWNS were capable of causing bacterial reductions of 1 and 2 logs in Gram-positive and Gram-negative bacteria, respectively (Pyrgiotakis et al., 2015). However, the EWNS were not effective against *Bacillus atrophaeus* endospores (Pyrgiotakis, McDevitt, Yamauchi, & Demokritou, 2012). In further studies, the same research group found that the EWNS were able to reduce the microbial load on stainless steel surfaces by 0.7–1.8 logs. Overall, the EWNS shows promise for disinfecting produce and CIP equipment (Pyrgiotakis et al., 2015).

Innovations in the application of nanotechnology in the food industry will continue to expand rapidly as the demand for more sensitive detection methods and more effective antimicrobial agents increases. Nanotechnology has the potential to revolutionize the food quality and safety areas. The incorporation of active nanomaterials, such as antimicrobials and oxygen scavenging reagents, can extend the shelf-life of a food and enhance its quality and safety. The use of "intelligent" packaging containing nano-sensors and nanodevices can give us information regarding freshness and/or potential spoilage in foods, pathogen contamination or monitor changes in package conditions or integrity. Novel antimicrobial nanoparticles that can be used for the food industry will continue to be discovered. However, as the use of nanotechnologies in the agri/feed/food sector increases, public health officials and consumers will demand more information on the human health impacts of this technology. In fact, a more complete assessment of the potential risks in relation to nanotechnology in the food and agriculture sectors will be essential to influence the publics' perception of the potential risks associated with foods derived from nanotechnology.

Most countries do not have defined regulations for nanotechnology-derived food products, and various approaches have been used in regulating nano-based products in foods. As advances are made in the field of nanotechnology, strong regulatory guidance for the food industry will be needed. This will be challenging, as the science and technology of nanomaterials is moving faster than the regulatory frameworks.

## 3.8 Nanosensing

Nanosensing structures present important advantages in comparison to other sensing tools. As it was mentioned in the food safety section, their size provides a large surface area highly populated with recognition elements and confers high sensitivity and low limits of detection (LOD). Their specificity, granted from the possibility to easily dock, entrap or imprint one or more recognition elements in the nanostructure, provides flexibility and multiplexing capabilities at a low cost and high convenience. All these properties justify the vigorous development and continuous employment of these structures as ultimate devices for monitoring quality attributes and food safety in-line and intelligent food packaging. Their use can expand real-time detection of changes in quality attributes and presence of toxins and pathogens during food storage and distribution (Pathakoti et al., 2017).

Nanostructured scaffolds have been predominantly produced with noble metals such as gold and silver due to their biocompatibility, direct interaction with proteins and amino acids, and easy modification with specific functional groups (*e.g.*, carboxylic and thiol) or organic molecules (*e.g.*, citrate, 3-mercaptopropionic acid) that improves sensing and increases specificity (Ahari et al., 2017; Bhamore, Ganguly, & Kailasa, 2016; Mi et al., 2009). Growing interest in biodegradable, nontoxic constituents and green synthesis of nanostructures is currently driving the utilization of carbon-based materials and biomolecules as supports for chemical, biological, and synthetic recognition elements (Cauerhff & Castro Guillermo, 2013; Mohapatra et al., 2018; Prietto et al., 2017). The functionalization of nanostructures with chemical or biological recognition elements can take place through non-covalent (*e.g.*, van der Waals, electrostatic interactions, hydrogen bonds) or covalent binding, adsorption or entrapment (Ansari & Husain, 2012; Basu et al., 2017; Pathakoti et al., 2017). Nanostructures can also be designed to carry specific charges to facilitate interactions with recognition elements. Among the potential recognition elements available, as of today, the most applied nanomaterials for food applications have been biomolecules and the resulting sensing nanostructure is dubbed biosensors. Within biosensors, antigens, antibodies, proteins, enzymes, and DNA are among the most frequently used biorecognition elements. These biomolecules have been incorporated into an array of nanostructures to provide them with specific sensing ability toward food components ranging from small molecules to large macromolecules. For example, as described in previous sections antibodies and DNA probes attached to nanowires and nanoparticles have been used to detect the presence of several pathogens (Papadakis et al., 2018; Vanegas, Gomes, Cavallaro, Giraldo-Escobar, & McLamore, 2017; Zeinhom et al., 2018). Enzymes, *e.g.*, glucose oxidase and alcohol dehydrogenase, were chemisorbed on a core-shell of AuNPs and immobilized onto electrodes to assess wine quality based on sugar and alcohol contents (Samphao et al., 2018). Phenolic compounds have also been detected using nanosensors in which the biorecognition elements were tyrosinase, laccase, or horseradish peroxidase (Rodríguez-Delgado et al., 2015).

Molecular imprinting technology allows the development of nanomaterials and nanostructures bestowed with recognition sites specific for food ingredients, adulterants and/or contaminants so that the nanostructure becomes part of the recognition element (Algieri, Drioli, Guzzo, & Donato, 2014). In molecularly imprinted nanostructures, artificial receptors have been embedded using the target molecule as a template during the production steps. These highly selective, synthetic recognition sites often exhibit higher physical and chemical stability than the biological molecules used as their templates (Ahari et al., 2017). Additionally, coating of nanostructures with molecularly imprinted polymers results in materials with high surface area, imparted by the nanostructure, and high specificity and sensitivity, contributed by the molecularly imprinted polymer (Kozitsina et al., 2018; Zhang et al., 2017). Molecularly imprinted inorganic or organic nanostructures have been prepared to detect small molecules, such as monosaccharides, toxins, antibiotics, or macromolecules without the necessity of adding a layer of a sensing element such as enzymes, or other affinity receptors such as antigens, antibodies, or a DNA probe (Kara, Uzun, Kolayli, & Denizli, 2013).

The inherent and vast array of physical, chemical, optical and electrical properties of nanostructures can contribute to amplifying and translating the identification of the targeted analyte into a signal. Thus, nanostructures can perform as transducers or antennas (Farzin, Shamsipur, Samandari, & Sheibani, 2018). Notably, optical AuNPs, quantum and carbon dots, and nanofibers interfaced to photometric detection systems have been utilized to this end (Debnath et al., 2010; Fang et al., 2016; Long, Li, Zhang, & Yao, 2015; Sivasankaran, Cyriac, Menon, & Kumar, 2017; Sun, Yuan, Liu, & Shen, 2018; Yun et al., 2018). The physical state of nanoparticles has a significant role in its activity as an optical transducer. The aggregation of optical particles, *e.g.*, AuNPs, AgNPs, carbon dots, can elicit a change in color or fluorescence emission or spectral shifts. Nanoparticles that have been functionalized with a biorecognition element that promotes their aggregation will allow the detection of the analyte of interest using optical methods by observing the shade, a quench or increase in fluorescence emission, or a modification of their localized surface plasmon resonance peak or Raman spectra (Chen & Park, 2016; Lakade, Sundar, & Shetty, 2017; Wang, Zhou, & Li, 2017; Zhang et al., 2017). In addition to optical detection, the use of conductive nanostructures as a sensing device, or part of thereof, allows them to perform as electrochemical transducers. In these devices, target analytes can be detected based on changes in resistivity or capacitance. Conductive metal or biopolymer nanoparticles, nanowires and nanotubules assembled onto electrodes have been reported to enhance signal intensity during the electrochemical detection of an array of food-related metabolites such as gallic and ascorbic acid (Ghaani et al., 2016; Karimi-Maleh et al., 2014). For more information on the use of nanostructures as transducers or signal amplification antennas, the review on nanobiosensors complied by Chen and Park provides a detailed section on detection methods based on nanostructures (Chen & Park, 2016).

## 3.9 Nanoreactors

Physical and chemical characteristics that facilitate the use of nanostructures as sensing devices also have promoted their application as biocatalysts supports and nanoreactors. It should be noted that the term nanoreactor is predominantly applied to structures where the catalytic component and consequently the reaction occur in an enclosed nanoscale environment. Therefore, surface attachment of catalytic compounds to nanoparticles is not strictly included within this category and should be considered nano-immobilization techniques.

The large surface area of nanostructures enables the attachment of catalytic biomolecules, which transforms the nanostructures into effective immobilization supports and bioreactors. Not only does the high surface area to volume ratio increase loading of the biocatalyst, but it also can potentially eliminate or minimize diffusion limitations, depending on the geometry of the nanostructure (Khlobystov, 2011; Piazza & Traytak, 2015). Nanostructures, from nanotubules to nanoparticles, have contributed to effective immobilization of enzymes and enhanced performance as bioreactors by increasing enzyme loading, activity and stability (Abbasi, Amiri, Bordbar, Ranjbakhsh, & Khosropour, 2016; Ansari & Husain, 2012). Immobilization of catalytic constituents onto nanostructures has resulted in broadening of their working pH and temperature ranges, as well as enhancing storage stability and reducing their substrate inhibition than free enzymes or catalytic components immobilized on conventional macro- or micro-supports (Li et al., 2014). Additionally, the use of magnetic nanoparticle supports facilitates the recovery of the immobilized biocatalysts, reducing costs due to their potential re-utilization and diminishing safety concerns associated to the presence of nanostructures (Shelby, Sulthana, McAfee, Banerjee, & Santra, 2017; Warriner, Reddy, Namvar, & Neethirajan, 2014). Porous nanospheres, nanotubes, hollow nanostructures, as well as clusters of nanostructures are the most predominant nanostructures used as nanoreactors. The development of intricate 3D nanostructures such as nano-onions or nanoflowers makes co-immobilization of multi-enzymes on these materials feasible. For example, electrostatic interactions can give rise to onion-like multilayer structures where functionality in each layer can be chosen selectively by attaching different enzymes (*e.g.*, glucose oxidase, horseradish peroxidase, and alkaline phosphatase) which gives rise to multiplex catalytic activity and the possibility to perform cascade or multistep enzymatic reactions in a single procedure within the same system (Ahari et al., 2017; Jia, Narasimhan, & Mallapragada, 2013; Sok & Fragoso, 2018).

Moreover, the photocatalytic activity of several nanostructures, *e.g.*, metal and metal oxide nanoparticles such as AuNP or $TiO_2$ and their clusters, has enabled their use in bioremediation applications to degrade aromatic compounds, such as pesticides and colorants in water, waste stream and produce surfaces (Das et al., 2017; Mao et al., 2014). Nanostructures endowed with sensing and photocatalytic activities have provided enhanced specificity to this task. Dual function nanoparticles, nanotubules, nanofibers, nanoflowers, and nanocomposites have been developed for the recognition, degradation, or separation of pesticides (Rawtani, Khatri, Tyagi, & Pandey, 2018). The optical properties of the nanomaterials enhance the photocatalytic oxidization of pesticide attached to the recognition element.

## 3.10 Safety

The ingestion of "engineered" nanostructures may pose a risk to consumer health as the properties of a nanoscale material differ from the bulk form. Increased reactivity, enhanced penetration rates in biological tissues and improved solubility will affect the way in which these materials behave within a food matrix, but also how they will react in the gastrointestinal tract. The use of nanotechnology to encapsulate bioactive compounds has long been recognized as an innovative strategy to: (i) protect encapsulated bioactive compounds from degradation in the gastrointestinal tract; (ii) enable controlled release; and (iii) enhance biodistribution and accumulation in target biological tissues (Sabliov, Chen, & Yada, 2015). Nanodelivery systems for bioactives, such as emulsions, polymeric nanoparticles, liposomes, nanocomplexes, solid lipid nanoparticles, *etc.*, have been primarily focused on developing optimal particle characteristics, bulk physicochemical properties, and bioactive release profiles (Borel & Sabliov, 2014). However, their interactions with biological systems such as cell lines, animals, and humans are understudied. It is estimated that over 1800 products containing nanomaterials are commercially available around the world, among which nano-formulations of dietary supplements are being sold with unsupported claims without any intervention from regulatory bodies (Vance et al., 2015). While regulatory guidelines in the United States (Hamburg, 2012) and Canada (Health Canada, 2011) on nanotechnology based health products are still adjusting, the importance of showing safety and efficacy of nanodelivery systems in pre-clinical and clinical studies is now paramount to gaining full consumer trust (McClements & Xiao, 2017).

Current methods collectively being used to evaluate the biological fate of new nanodelivery systems for bioactives are highlighted in Table 1. *In vitro* digestion is currently the most popular and cost-effective method to determine bioaccessibility as it primarily tests changes in physicochemical properties of a delivery system when sequentially exposed to a series of artificial fluids that simulate conditions of the mouth, stomach, small intestine and colon (McClements & Li, 2010). The execution of such *in vitro* methods is extremely complicated as parameters of these fluids like pH, ionic

**Table 1** Current methods and proposed considerations to determine the fate of nanodelivery systems for bioactives.

| Stage of assessment | Current method(s) | Proposed methods or additional methodical considerations |
|---|---|---|
| *In vitro*—bioaccessibility (passage through GIT) | *In vitro* digestion | TNO Gastro-Intestinal Model (TIM) |
| *In vitro*—absorption | • Physical models<br>• *Ex vivo* permeation methods<br>• Cell culture models: Caco-2 cells | Assessment in additional cell lines like HT29-MTX (colon), HepG2 (liver), human or animal liver microsomes or hepatocytes (liver), HASMC (smooth muscle), J774 (macrophage) |
| *In vivo*—bioavailability (animals) | Pharmacokinetic study of an oral bolus dose of unformulated bioactive or nanodelivery system | Ascertain distribution and excretion of both bioactive and nanoparticle by measuring concentrations in key tissues, urine and feces |
| Safety | Minimally performed | • *In vitro studies:* cytotoxicity, genotoxicity, reactive oxygen species generation, inflammation, and chemical degradation<br>• *In vivo:* acute repeat dose oral toxicity (14 days), chronic repeat dose oral toxicity (28 days and then 90 days) |

composition, enzyme activity, surface active components, flow profile, and mechanical forces are difficult to optimize (McClements & Li, 2010). Furthermore, reported bioaccessible fractions from *in vitro* digestion studies can often be significantly different from results of *in vivo* (animal or human) studies (Kostewicz et al., 2014; Van de Wiele et al., 2007). While static or dynamic *in vitro* digestion methods are still a valuable tool for measuring bioaccessibility, the recent development of the TNO gastrointestinal model (TIM) is predicted to become the gold-standard approach for determining bioaccessibility of delivery systems as it closely simulates the gastrointestinal tract (GIT) processes (Verwei, Minekus, Zeijdner, Schilderink, & Havenaar, 2016). Studies published on pharmaceutical formulations that utilized the TIM system have reported good agreement with results of *in vivo* studies suggesting a far superior *in vitro-in vivo* correlation compared to *in vitro* digestion methods (Barker, Abrahamsson, & Kruusmägi, 2014; Verwei et al., 2016).

Various components of a nanodelivery system (intact particles, digestion products of particles and the bioactive) can be absorbed from GIT fluids but typically absorption occurs at the epithelium cells in the walls of the small intestine. Studies to date have reported three main methods of simulating absorption: (i) physical models; (ii) *ex vivo* permeation methods; and (iii) cell culture models. Physical models assume that solubility of the bioactive in the small intestinal fluid is the rate-limiting step for absorption in the epithelium cells; thus dialysis bags or semipermeable membranes are used to measure the concentration of the bioactive that moves across the membrane over time (Christensen, Schultz, Mollgaard, Kristensen, & Mullertz, 2004). *Ex vivo* permeation methods are very similar in concept to physical models but provide more physiological relevance as they utilize a section of the gastrointestinal tract directly excised from an animal (Dahan & Hoffman, 2006). Despite the cost-effectiveness of physical or *ex vivo* permeation methods, cell culture models that utilize the human epithelium colorectal adenocarcinoma cell line, Caco-2, have become the preferred method of assessing absorption. Typically, a solution containing the nanomaterials, which was previously exposed to a simulated gastrointestinal tract, is added directly to the media where cells are cultured. This allows for the cellular uptake and absorption into the cells to be measured using fluorescent, spectrometric, chromatographic or chemical methods (Dhuique-Mayer et al., 2007). Evaluation of absorption in cell culture models is essential in understanding how nanodelivery systems (once exposed to GIT fluids) change. Caco-2 cells should be used in tandem with the mucus secreting human colon carcinoma cell line, HT29-MTX, to provide more physiologically relevant information on absorption and uptake (Hilgendorf et al., 2000). HepG2 is a human liver hepatocellular carcinoma cell line in which uptake and absorption studies can be used to estimate the extent of biotransformation of bioactives in a nanodelivery system (Sadeghi Ekbatan et al., 2018). Although not readily available, human or animal liver microsomes and whole hepatocytes are the standard models in the pharmaceutical industry to study drug metabolism and should be considered when assessing nanoparticle and bioactive absorption and metabolism. A human arterial smooth muscle cell line, HASMCs, can be utilized to predict transport of bioactive and components of the delivery system (Sahoo, Panyam, Prabha, & Labhasetwar, 2002). Uptake into immune cells thought to play an important role in absorption of bioactives can also be assessed using the mouse peritoneal macrophage cell line, J774 22.

Current methods to assess bioavailability of nanodelivery systems for bioactives are limited to a single pharmacokinetic (PK) assessment where an oral bolus dose of a formulation is given to animals and blood is collected to measure plasma concentration of bioactive over time using areas-under-the-curve (AUC) analysis. This analysis is then used to mathematically calculate the maximum serum concentration the bioactive achieves ($C_{max}$) and at which time ($T_{max}$) as well as the fraction of bioactive that is eliminated per unit of time (elimination rate constant or $K_{el}$). Such pharmacokinetic studies often report nanodelivery systems to possess 5–10-fold greater oral bioavailability compared to unformulated bioactives (Yu & Huang, 2012).

Regulatory bodies have recognized that nanodelivery systems, due to their novel properties, have altered absorption, distribution, metabolism and excretion (ADME) in biological systems. As such, future studies will be required not only to focus on ascertaining PK parameters in plasma, but also to collect information on the distribution and excretion of both the bioactive and a traceable component of the nanodelivery system. This can be done by measuring bioactive and nanoparticle concentrations at all time intervals in several key tissues (liver, heart, brain, spleen, bone marrow, muscle, pancreas, and adipose tissue) as well as urine and feces. Guidelines and best practices for carrying out such extensive assessments in animals are well established and should be used in the planning process of all animal studies.

The debate on the safety of nanomaterials and delivery systems is still ongoing (McClements, 2017). Given the toxicokinetic properties of nanodelivery systems are severely understudied, regulatory bodies around the world have identified a need for studies that assess short-term and long-term toxicity and bioaccumulation. Guidelines from the pharmaceutical industry can easily be used to carry out safety studies that require both *in vitro* and *in vivo* models. *In vitro* studies are required to determine cytotoxicity, genotoxicity, reactive oxygen species generation (or oxidative stress), inflammation and chemical degradation of nanomaterials (Cockburn et al., 2012). *In vivo* studies are recommended to assess acute and chronic toxicity where repeat oral dose studies for 14, 28 and 90 days are typically required (Cockburn et al., 2012). Results from the repeat oral dose 90 day studies are critical in determining carcinogenicity, reproductive toxicity, teratogenicity, and neurotoxicity associated with a given nanodelivery system (Cockburn et al., 2012).

Nanotechnology has begun to transform the way in which the world population consumes food, drugs and bioactive compounds. Several lessons can be learned from the success of nanopharmaceuticals currently approved for the treatment of several debilitating human diseases (Barenholz, 2012). The regulatory requirements of drug approval by the FDA, Health Canada or European Commission are comprehensive and ensure new drugs are safe and efficacious in the global population. As such, the studies and methodical considerations proposed in this section will greatly assist the food industry in obtaining regulatory approval, gaining public trust, and commercializing nanodelivery systems that have great potential for improving human health.

## 4. Future opportunities

Major progress has been made in the formulation of nanostructures for food applications. Nanotechnology is poised to allow the food industry to lower costs of processing while improving the healthiness of foods by increasing bioactive content and reducing fat, salt and calories without compromising on food quality. Future research will focus on testing novel matrix material sources, optimizing the current existing or developing new technologies to produce better performing nanostructures. Very specific research foci are:

- Performant encapsulation systems that will preserve and release the future generation of probiotic species originating from the gut mucosa (*e.g.*, *Akkermansia* and *Faecalibacterium*) in a targeted way. These new generation are strict anaerobes that are sensitive to oxygen. Creating systems which offer these probiotics the ultimate protection against oxygen will be a future opportunity. In addition, only very few studies have paid detailed attention to the molecular structure and composition of the materials used in probiotic encapsulation, even though these parameters influence the functional and release properties of the materials in complex food matrices and upon digestion.
- The development of interactive foods tailored to individual consumer preferences as well as to nutritional and health needs. Companies and academia are already actively researching the capability to change the properties of foods according to desired taste, texture, and color. In addition, *in vitro* stability and release studies are numerous, but follow-up studies on the behavior and functionality of these nanostructures in a complex food matrix and/or during *in vivo* digestion are largely lacking.

- Encapsulation of molecules that can (to some extent) accurately monitor the temperature history of foods to be used in intelligent and active packaging applications (Singh, Gaikwad, Lee, & Lee, 2018) and particles that can change their shape, size and surface characteristics upon environmental changes (Zhang et al., 2016). The latter is only one example of a range of stimuli-responsive nanoparticles, a new research line that offers a vast scope of potential applications in the food industry, from stability tuning of colloidal dispersions to stimuli-induced release of encapsulated molecules and self-assembling micro- or even macrostructures upon specific triggers (Motornov, Roiter, Tokarev, & Minko, 2010).
- New technologies, able to structure liquid oils into gels, containing low saturated and zero trans fats are now able to make solid fat mimics (Rogers, 2011, 2017; Rogers et al., 2014; Wang & Rogers, 2015). These systems rely on food grade fatty acids, monoglycerides, ceramides, phytosterols and oryzanol, waxes and other small molecules to self-assemble into nano-fibers that then form a gel network similar to hydrocolloids which are capable of entraining liquid oils (Rogers, Spagnuolo, & Wang, 2016). Using nanotechnologies, it is now possible to produce a solid-like fat with zero trans, and zero saturated fat and only 2 wt% self-assembling molecular gelator.
- Despite the prevalent desire to use food-grade, biocompatible and biodegradable materials to produce nanostructures such as food lipids, proteins and carbohydrates, the toxicology of these green-engineered food nanostructures needs to be properly assessed before their incorporation into food products. The manipulation of these innocuous biopolymers to produce nanoscale structures comes with an inherent risk for human and environmental health. Novel nanostructures are engineered to introduce new properties and desirable functionality in a food matrix, but may well behave in an unexpected and unwanted way upon entering a biological system such as the human body and/or the environment (Raynes, Carver, Gras, & Gerrard, 2014). Therefore, new methodology needs to be developed to detect, characterize, monitor and even quantify these engineered materials in food.

## 5. Conclusions

Biomaterial-based nanotechnology is a rapidly developing toolbox that opens up a myriad of solutions to long-standing issues in the food industry and fuels the development of new experimental prototype systems and procedures. However, these advances should be accompanied with the fast development of accurate and sensitive techniques to identify, characterize and/or quantify these structures in complex food matrices and to assess the risk for human and environmental health (Raynes et al., 2014; Sadeghi et al., 2017). At the same time, limited knowledge and lack of adequate testing methods to assess the interactions and conformations at play when forming nanostructures hinder the rational development of performant systems with tailor-made functionality. Protein-protein interactions in nature, for example, are known to produce very complex and functional structures. The understanding of these natural structures and how we can manipulate them to engineer particles/assemblies with well-defined functionality is needed to rationalize and economize this process (Sanchez-deAlcazar, Mejias, Erazo, Sot, & Cortajarena, 2018). In addition, open and full information disclosure will be needed to win over the enthusiasm of consumers regarding the application of nanotechnology in food.

## References

Abbasi, M., Amiri, R., Bordbar, A.-K., Ranjbakhsh, E., & Khosropour, A.-R. (2016). Improvement of the stability and activity of immobilized glucose oxidase on modified iron oxide magnetic nanoparticles. *Applied Surface Science*, *364*, 752–757.

Acevedo, N. C., Block, J. M., & Marangoni, A. G. (2012). Critical laminar shear-temperature effects on the nano- and mesoscale structure of a model fat and its relationship to oil binding and rheological properties. *Faraday Discussions*, *158*, 171–194.

Acevedo, N. C., & Marangoni, A. G. (2010). Characterization of the nanoscale in triacylglycerol crystal networks. *Crystal Growth & Design*, *10*, 3327–3333.

Acevedo, N. C., & Marangoni, A. G. (2015). Nanostructured fat crystal systems. *Annual Reviews of Food Science & Technology*, *6*, 71–96.

Acevedo, N. C., Peyronel, M. F., & Marangoni, A. G. (2011). Nanoscale structure intercrystalline interactions in fat crystal networks. *Current Opinion in Colloid and Interface Science*, *16*, 374–383.

Acosta-Domínguez, L., Hernández-Sánchez, H., Gutiérrez-López, G. F., Alamilla-Beltrán, L., & Azuara, E. (2016). Modification of the soy protein isolate surface at nanometric scale and its effect on physicochemical properties. *Journal of Food Engineering*, *168*, 105–112.

Ades, H., Kesselman, E., Ungar, Y., & Shimoni, E. (2012). Complexation with starch for encapsulation and controlled release of menthone and menthol. *LWT—Food Science and Technology*, *45*, 277–288.

Ahari, H., Hedayati, M., Akbari-adergani, B., Kakoolaki, S., Hosseini, H., & Anvar, A. (2017). Staphylococcus aureus exotoxin detection using potentiometric nanobiosensor for microbial electrode approach with the effects of pH and temperature. *International Journal of Food Properties*, *20*, 1578–1587.

Akbas, E., Soyler, B., & Oztop, M. H. (2018). Formation of capsaicin loaded nanoemulsions with high pressure homogenization and ultrasonication. *LWT—Food Science and Technology*, *96*, 266–273.

Algieri, C., Drioli, E., Guzzo, L., & Donato, L. (2014). Bio-mimetic sensors based on molecularly imprinted membranes. *Sensors*, *14*, 13863–13912.

Al-Nasiri, G., Cran, M. J., Smallridge, A. J., & Bigger, S. W. (2018). Optimisation of β-cyclodextrin inclusion complexes with natural antimicrobial agents: Thymol, carvacrol and linalool. *Journal of Microencapsulation*, *35*, 26–35.

Ansari, S. A., & Husain, Q. (2012). Potential applications of enzymes immobilized on/in nano materials: A review. *Biotechnology Advances*, *30*, 512–523.

Anselmo, A. C., McHugh, K. J., Webster, J., Langer, R., & Jaklenec, A. (2016). Layer-by-layer encapsulation of probiotics for delivery to the microbiome. *Advanced Materials*, *28*, 9486–9490.

Arangoa, M. A., Ponchel, G., Orecchioni, A. M., Renedo, M. J., Duchêne, D., & Irache, J. M. (2000). Bioadhesive potential of gliadin nanoparticulate systems. *European Journal of Pharmaceutical Sciences*, *11*, 333–341.

Arpagaus, C., Collenberg, A., Rütti, D., Assadpour, E., & Jafari, S. M. (2018). Nano spray drying for encapsulation of pharmaceuticals. *International Journal of Pharmaceutics*, *546*, 194–214.

Arpagaus, C., John, P., Collenberg, A., & Rütti, D. (2017). 10—Nanocapsules formation by nano spray drying. In S. M. Jafari (Ed.), *Nanoencapsulation technologies for the food and nutraceutical industries* (pp. 346–401). Academic Press.

Assadpour, E., & Jafari, S. M. (2018). A systematic review on nanoencapsulation of food bioactive ingredients and nutraceuticals by various nanocarriers. *Critical Reviews in Food Science and Nutrition*, *8*, 1–47.

Astete, C. E., Sabliov, C. M., Watanabe, F., & Biris, A. (2009). Ca2+ cross-linked alginic acid nanoparticles for solubilization of lipophilic natural colorants. *Journal of Agricultural and Food Chemistry*, *57*, 7505–7512.

Bajpai, V. K., Kamle, M., Shukla, S., Mahato, D. K., Chandra, P., Hwang, S. K., et al. (2018). Prospects of using nanotechnology for food preservation, safety, and security. *Journal of Food and Drug Analysis*, *26*, 1201–1214.

Barenholz, Y. (2012). Doxil®—The first FDA-approved nano-drug: Lessons learned. *Journal of Controlled Release*, *160*, 117–134.

Barker, R., Abrahamsson, B., & Kruusmägi, M. (2014). Application and validation of an advanced gastrointestinal in vitro model for the evaluation of drug product performance in pharmaceutical development. *Journal of Pharmaceutical Sciences*, *103*, 3704–3712.

Basu, A., Kundu, S., Sana, S., Halder, A., Abdullah, M. F., Datta, S., et al. (2017). Edible nano-bio-composite film cargo device for food packaging applications. *Food Packaging and Shelf Life*, *11*, 98–105.

Belluco, S., Gallocchio, F., Losasso, C., & Ricci, A. (2018). State of art of nanotechnology applications in the meat chain: A qualitative synthesis. *Critical Reviews in Food Science and Nutrition*, *58*, 1084–1096.

Berton-Carabin, C. C., & Schroën, K. (2015). Pickering emulsions for food applications: Background, trends, and challenges. *Annual Review of Food Science and Technology*, *6*, 263–297.

Bhamore, J. R., Ganguly, P., & Kailasa, S. K. (2016). Molecular assembly of 3-mercaptopropinonic acid and guanidine acetic acid on silver nanoparticles for selective colorimetric detection of triazophos in water and food samples. *Sensors and Actuators B: Chemical*, *233*, 486–495.

Bi, L., Yang, L., Narsimhan, G., Bhunia, A. K., & Yao, Y. (2011). Designing carbohydrate nanoparticles for prolonged efficacy of antimicrobial peptide. *Journal of Controlled Release*, *150*, 150–156.

Borel, T., & Sabliov, C. M. (2014). Nanodelivery of bioactive components for food applications: Types of delivery systems, properties, and their effect on ADME profiles and toxicity of nanoparticles. *Annual Review of Food Science and Technology*, *5*, 197–213.

Bouwmeester, H., Dekkers, S., Noordam, M. Y., Hagens, W. I., Bulder, A. S., de Heer, C., et al. (2009). Review of health safety aspects of nanotechnologies in food production. *Regulatory Toxicology and Pharmacology, 53*, 52–62.

Calvano, C. D., Jensen, O. N., & Zambonin, C. G. (2009). Selective extraction of phospholipids from dairy products by micro-solid phase extraction based on titanium dioxide microcolumns followed by MALDI-TOF-MS analysis. *Analytical and Bioanalytical Chemistry, 394*, 1453–1461.

Cauerhff, A., & Castro Guillermo, R. (2013). Bionanoparticles, a green nanochemistry approach. *Electronic Journal of Biotechnology, 16*(3), 2013.

Chang, Y., Yang, J., Ren, L., & Zhou, J. (2018). Characterization of amylose nanoparticles prepared via nanoprecipitation: Influence of chain length distribution. *Carbohydrate Polymers, 194*, 154–160.

Charve, J., & Reineccius, G. A. (2009). Encapsulation performance of proteins and traditional materials for spray dried flavors. *Journal of Agricultural and Food Chemistry, 57*, 2486–2492.

Chaudhry, Q., Scotter, M., Blackburn, J., Ross, B., Boxall, A., Castle, L., et al. (2008). Applications and implications of nanotechnologies for the food sector. *Food Additives & Contaminants. Part A, Chemistry, Analysis, Control, Exposure & Risk Assessment, 25*, 241–258.

Chen, C.-S., & Durst, R. A. (2006). Simultaneous detection of Escherichia coli O157:H7, Salmonella spp. and *Listeria monocytogenes* with an array-based immunosorbent assay using universal protein G-liposomal nanovesicles. *Talanta, 69*, 232–238.

Chen, J., & Park, B. (2016). Recent advancements in nanobioassays and nanobiosensors for foodborne pathogenic bacteria detection. *Journal of Food Protection, 79*, 1055–1069.

Chen, Y., Shu, M., Yao, X., Wu, K., Zhang, K., He, Y., et al. (2018). Effect of zein-based microencapsules on the release and oxidation of loaded limonene. *Food Hydrocolloids, 84*, 330–336.

Chen, J., Wang, Q., Liu, C.-M., & Gong, J. (2017). Issues deserve attention in encapsulating probiotics: Critical review of existing literature. *Critical Reviews in Food Science and Nutrition, 57*, 1228–1238.

Chen, H., & Zhong, Q. (2015). A novel method of preparing stable zein nanoparticle dispersions for encapsulation of peppermint oil. *Food Hydrocolloids, 43*, 593–602.

Choi, A.-J., Kim, C.-J., Cho, Y.-J., Hwang, J.-K., & Kim, C.-T. (2011). Characterization of capsaicin-loaded nanoemulsions stabilized with alginate and chitosan by self-assembly. *Food and Bioprocess Technology, 4*, 1119–1126.

Christensen, J. Ø., Schultz, K., Mollgaard, B., Kristensen, H. G., & Mullertz, A. (2004). Solubilisation of poorly water-soluble drugs during in vitro lipolysis of medium- and long-chain triacylglycerols. *European Journal of Pharmaceutical Sciences, 23*, 287–296.

Clark, A. H. (1996). Biopolymer gels. *Current Opinion in Colloid & Interface Science, 1*, 712–717.

Cockburn, A., Bradford, R., Buck, N., Constable, A., Edwards, G., Haber, B., et al. (2012). Approaches to the safety assessment of engineered nanomaterials (ENM) in food. *Food and Chemical Toxicology, 50*, 2224–2242.

Corradini, M. G., Demol, M., Boeve, J., Ludescher, R. D., & Joye, I. J. (2017). Fluorescence spectroscopy as a tool to unravel the dynamics of protein nanoparticle formation by liquid antisolvent precipitation. *Food Biophysics, 12*, 211–221.

Cushen, M., Kerry, J., Morris, M., Cruz-Romero, M., & Cummins, E. (2012). Nanotechnologies in the food industry—Recent developments, risks and regulation. *Trends in Food Science & Technology, 24*, 30–46.

Dahan, A., & Hoffman, A. (2006). Use of a dynamic in vitro lipolysis model to rationalize oral formulation development for poor water soluble drugs: Correlation with in vivo data and the relationship to intra-enterocyte processes in rats. *Pharmaceutical Research, 23*, 2165–2174.

Das, D., Datta, A. K., Kumbhakar, D. V., Ghosh, B., Pramanik, A., Gupta, S., et al. (2017). Assessment of photocatalytic potentiality and determination of ecotoxicity (using plant model for better environmental applicability) of synthesized copper, copper oxide and copper-doped zinc oxide nanoparticles. *PLoS One*, *12*, e0182823.

Davidov-Pardo, G., Pérez-Ciordia, S., Marı´n-Arroyo, M. R., & McClements, D. J. (2015). Improving resveratrol bioaccessibility using biopolymer nanoparticles and complexes: Impact of protein–carbohydrate maillard conjugation. *Journal of Agricultural and Food Chemistry*, *63*, 3915–3923.

Debnath, M., Prasad, G., Bisen, P. S., Debnath, M., Prasad, G., & Bisen, P. S. (2010). DNA biosensors. In *Molecular diagnostics: Promises and possibilities* (pp. 209–226). Dordrech, Heidelberg, London: Springer.

de Souza, S. L., Madalena, D. A., Pinheiro, A. C., Teixeira, J. A., Vicente, A. A., & Ramos, Ó. L. (2017). Micro- and nano bio-based delivery systems for food applications: In vitro behavior. *Advances in Colloid and Interface Science*, *243*, 23–45.

Devi, N., Sarmah, M., Khatun, B., & Maji, T. K. (2017). Encapsulation of active ingredients in polysaccharide–protein complex coacervates. *Advances in Colloid and Interface Science*, *239*, 136–145.

Dhuique-Mayer, C., Borel, P., Reboul, E., Caporiccio, B., Besancon, P., & Amiot, M.-J. (2007). β-Cryptoxanthin from citrus juices: Assessment of bioaccessibility using an in vitro digestion/Caco-2 cell culture model. *British Journal of Nutrition*, *97*, 883–890.

Dickinson, E. (2012). Use of nanoparticles and microparticles in the formation and stabilization of food emulsions. *Trends in Food Science & Technology*, *24*, 4–12.

Đorđević, Đ., Buchtová, H., & Macharáčková, B. (2017). Salt microspheres and potassium chloride usage for sodium reduction: Case study with sushi. *Food Science and Technology International*, *24*, 3–14.

Drewnowski, A. (1992). Sensory properties of fats and fat replacements. *Nutrition Reviews*, *50*, 17–20.

Duncan, T. V. (2011). Applications of nanotechnology in food packaging and food safety: Barrier materials, antimicrobials and sensors. *Journal of Colloid and Interface Science*, *363*, 1–24.

Fang, A. J., Long, Q., Wu, Q. Q., Li, H. T., Zhang, Y. Y., & Yao, S. Z. (2016). Upconversion nanosensor for sensitive fluorescence detection of Sudan I-IV based on inner filter effect. *Talanta*, *148*, 129–134.

Farzin, L., Shamsipur, M., Samandari, L., & Sheibani, S. (2018). Advances in the design of nanomaterial-based electrochemical affinity and enzymatic biosensors for metabolic biomarkers: A review. *Microchimica Acta*, *185*, 276.

Fathi, M., Donsi, F., & McClements, D. J. (2018). Protein-based delivery systems for the nanoencapsulation of food ingredients. *Comprehensive Reviews in Food Science and Food Safety*, *17*, 920–936.

Fathi, M., Martín, Á., & McClements, D. J. (2014). Nanoencapsulation of food ingredients using carbohydrate based delivery systems. *Trends in Food Science & Technology*, *39*, 18–39.

Feng, T., Wang, H., Wang, K., Liu, Y., Rong, Z., Ye, R., et al. (2017). Preparation and structural characterization of different amylose–flavor molecular inclusion complexes. *Starch—Stärke*, *70*, 1700101.

Foster, E. J., Moon, R. J., Agarwal, U. P., Bortner, M. J., Bras, J., Camarero-Espinosa, S., et al. (2018). Current characterization methods for cellulose nanomaterials. *Chemical Society Reviews*, *47*, 2609–2679.

Garg, P., Ghatmale, P., Tarwadi, K., & Chavan, S. (2017). Influence of nanotechnology and the role of nanostructures in biomimetic studies and their potential applications. *Biomimetics*, *2*, 7.

Ghaani, M., Nasirizadeh, N., Ardakani, S. A. Y., Mehrjardi, F. Z., Scampicchio, M., & Farris, S. (2016). Development of an electrochemical nanosensor for the determination of gallic acid in food. *Analytical Methods*, *8*, 1103–1110.

Ghidouche, S., Rey, B., Michel, M., & Galaffu, N. (2013). A rapid tool for the stability assessment of natural food colours. *Food Chemistry*, *139*, 978–985.

Ghorani, B., & Tucker, N. (2015). Fundamentals of electrospinning as a novel delivery vehicle for bioactive compounds in food nanotechnology. *Food Hydrocolloids*, *51*, 227–240.

Gobbetti, M., Corsetti, A., Smacchi, E., Zocchetti, A., & De Angelis, M. (1998). Production of crescenza cheese by incorporation of bifidobacteria. *Journal of Dairy Science*, *81*, 37–47.

Golchoobi, L., Alimi, M., Shokoohi, S., & Yousefi, H. (2016). Interaction between nanofibrillated cellulose with guar gum and carboxy methyl cellulose in low-fat mayonnaise. *Journal of Texture Studies*, *47*, 403–412.

Gutiérrez, J. M., González, C., Maestro, A., Solè, I., Pey, C. M., & Nolla, J. (2008). Nano-emulsions: New applications and optimization of their preparation. *Current Opinion in Colloid & Interface Science*, *13*, 245–251.

Hamburg, M. A. (2012). FDA's approach to regulation of products of nanotechnology. *Science*, *336*, 299.

He, L., Liu, Y., Lin, M., Mustapha, A., & Wang, Y. (2008). Detecting single Bacillus spores by surface enhanced Raman spectroscopy. *Sensing and Instrumentation for Food Quality and Safety*, *2*, 247.

Health Canada. (2011). *Nanotechnology-based health products and food*. .

Heidebach, T., Först, P., & Kulozik, U. (2009). Microencapsulation of probiotic cells by means of rennet-gelation of milk proteins. *Food Hydrocolloids*, *23*, 1670–1677.

Hilgendorf, C., Spahn-Langguth, H., Regårdh, C. G., Lipka, E., Amidon, G. L., & Langguth, P. (2000). Caco-2 versus Caco-2/HT29-MTX Co-cultured cell lines: Permeabilities via diffusion, inside- and outside-directed carrier-mediated transport. *Journal of Pharmaceutical Sciences*, *89*, 63–75.

Huq, T., Khan, A., Khan, R. A., Riedl, B., & Lacroix, M. (2013). Encapsulation of probiotic bacteria in biopolymeric system. *Critical Reviews in Food Science and Nutrition*, *53*, 909–916.

İnanç Horuz, T., & Belibağlı, K. B. (2018). Nanoencapsulation by electrospinning to improve stability and water solubility of carotenoids extracted from tomato peels. *Food Chemistry*, *268*, 86–93.

Israelachvili, J. N. (1991). *Intermolecular and surface forces*. SanDiego: Academic Press.

Jia, F., Narasimhan, B., & Mallapragada, S. (2013). Materials-based strategies for multi-enzyme immobilization and co-localization: A review. *Biotechnology and Bioengineering*, *111*, 209–222.

Joye, I. J., & McClements, D. J. (2013). Production of nanoparticles by anti-solvent precipitation for use in food systems. *Trends in Food Science & Technology*, *34*, 109–123.

Joye, I. J., & McClements, D. J. (2016). Biopolymer-based delivery systems: Challenges and opportunities. *Current Topics in Medicinal Chemistry*, *16*, 1026–1039.

Kara, M., Uzun, L., Kolayli, S., & Denizli, A. (2013). Combining molecular imprinted nanoparticles with surface plasmon resonance nanosensor for chloramphenicol detection in honey. *Journal of Applied Polymer Science*, *129*, 2273–2279.

Karimi-Maleh, H., Moazampour, M., Yoosefian, M., Sanati, A. L., Tahernejad-Javazmi, F., & Mahani, M. (2014). An electrochemical nanosensor for simultaneous voltammetric determination of ascorbic acid and Sudan I in food samples. *Food Analytical Methods*, *7*, 2169–2176.

Kenry, & Lim, C. T. (2017). Nanofiber technology: Current status and emerging developments. *Progress in Polymer Science*, *70*, 1–17.

Khlobystov, A. N. (2011). Carbon nanotubes: From nano test tube to nano-reactor. *ACS Nano*, *5*, 9306–9312.

Kim, B., Kim, D., Cho, D., & Cho, S. (2003). Bactericidal effect of TiO2 photocatalyst on selected food-borne pathogenic bacteria. *Chemosphere*, *52*, 277–281.

Kinnear, C., Moore, T. L., Rodriguez-Lorenzo, L., Rothen-Rutishauser, B., & Petri-Fink, A. (2017). Form follows function: Nanoparticle shape and its implications for nanomedicine. *Chemical Reviews*, *117*, 11476–11521.
Kong, L., Bhosale, R., & Ziegler, G. R. (2018). Encapsulation and stabilization of β-carotene by amylose inclusion complexes. *Food Research International*, *105*, 446–452.
Kostewicz, E. S., Abrahamsson, B., Brewster, M., Brouwers, J., Butler, J., Carlert, S., et al. (2014). In vitro models for the prediction of in vivo performance of oral dosage forms. *European Journal of Pharmaceutical Sciences*, *57*, 342–366.
Koumoulos, E. P., Gkartzou, E., & Charitidis, C. A. (2017). Additive (nano)manufacturing perspectives: The use of nanofillers and tailored materials. *Manufacturing Review*, *4*, 1–9.
Kozitsina, A., Svalova, T., Malysheva, N., Okhokhonin, A., Vidrevich, M., & Brainina, K. (2018). Sensors based on bio and biomimetic receptors in medical diagnostic, environment, and food analysis. *Biosensors*, *8*, 35.
Kumar, V., Guleria, P., & Mehta, S. K. (2017). Nanosensors for food quality and safety assessment. *Environmental Chemistry Letters*, *15*, 165–177.
Lagarón, J. M., López-Rubio, A., & Fabra, M. J. (2015). Bio-based packaging. *Applied Polymer*, *133*(2). https://doi.org/10.1002/app.42971.
Laing, D. G., & Jinks, A. (1996). Flavour perception mechanisms. *Trends in Food Science & Technology*, *7*, 387–389.
Lakade, A. J., Sundar, K., & Shetty, P. H. (2017). Nanomaterial-based sensor for the detection of milk spoilage. *LWT—Food Science and Technology*, *75*, 702–709.
Li, X., Anton, N., Arpagaus, C., Belleteix, F., & Vandamme, T. F. (2010). Nanoparticles by spray drying using innovative new technology: The Büchi Nano Spray Dryer B-90. *Journal of Controlled Release*, *147*, 304–310.
Li, S., Gu, F., & Gao, Q. (2017). Preparation of rutin-loaded starch nanospheres. *Starch*, *70*(3–4). https://doi.org/10.1002/star.201700116.
Li, D., Li, Q., Hao, X., Zhang, Y., Zhang, Z., & Li, C. (2014). Assembled core-shell nanostructures of gold nanoparticles with biocompatible polymers toward biology. *Current Topics in Medicinal Chemistry*, *14*, 595–616.
Loisel, C., Keller, G., Bourgaux, C., & Ollivon, M. (1998). Phase transitions and polymorphism of cocoa butter. *Journal of American Oil Chemistry Society*, *75*, 425–439.
Long, Q., Li, H. T., Zhang, Y. Y., & Yao, S. Z. (2015). Upconversion nanoparticle-based fluorescence resonance energy transfer assay for organophosphorus pesticides. *Biosensors & Bioelectronics*, *68*, 168–174.
MacMillan, S. D., Roberts, K. J., Rossi, A., Wells, M. A., Polgreen, M. C., & Smith, I. H. (2002). In situ small angle X-ray scattering (SAXS) studies of polymorphism with the associated crystallization of cocoa butter fat using shearing conditions. *Crystal Growth & Design*, *2*, 221–226.
Mao, K., Chen, Y., Wu, Z., Zhou, X., Shen, A., & Hu, J. (2014). Catalytic strategy for efficient degradation of nitroaromatic pesticides by using gold nanoflower. *Journal of Agricultural and Food Chemistry*, *62*, 10638–10645.
Martín, M. J., Lara-Villoslada, F., Ruiz, M. A., & Morales, M. E. (2015). Microencapsulation of bacteria: A review of different technologies and their impact on the probiotic effects. *Innovative Food Science & Emerging Technologies*, *27*, 15–25.
Martínez-Ballesta, M., Gil-Izquierdo, Á., García-Viguera, C., & Domínguez-Perles, R. (2018). Nanoparticles and controlled delivery for bioactive compounds: Outlining challenges for new "smart-foods" for health. *Foods*, *7*, 72.
Matalanis, A., Jones, O. G., & McClements, D. J. (2011). Structured biopolymer-based delivery systems for encapsulation, protection, and release of lipophilic compounds. *Food Hydrocolloids*, *25*, 1865–1880.
Mayer, S., Weiss, J., & McClements, D. J. (2013). Vitamin E-enriched nanoemulsions formed by emulsion phase inversion: Factors influencing droplet size and stability. *Journal of Colloid and Interface Science*, *402*, 122–130.

Mazzanti, G., Li, M., Marangoni, A. G., & Idziak, S. H. K. (2011). Effects of shear rate variation on the nanostructure of crystallizing triglycerides. *Crystal Growth & Design*, *11*, 4544–4550.

McClements, D. J. (2015). Encapsulation, protection, and release of hydrophilic active components: Potential and limitations of colloidal delivery systems. *Advances in Colloid and Interface Science*, *219*, 27–53.

McClements, D. J. (2017). Delivery by design (DbD): A standardized approach to the development of efficacious nanoparticle- and microparticle-based delivery systems. *Comprehensive Reviews in Food Science and Food Safety*, *17*, 200–219.

McClements, D. J., & Li, Y. (2010). Review of in vitro digestion models for rapid screening of emulsion-based systems. *Food & Function*, *1*, 32–59.

McClements, D. J., & Rao, J. (2011). Food-grade nanoemulsions: Formulation, fabrication, properties, performance, biological fate, and potential toxicity. *Critical Reviews in Food Science and Nutrition*, *51*, 285–330.

McClements, D. J., & Xiao, H. (2017). Is nano safe in foods? Establishing the factors impacting the gastrointestinal fate and toxicity of organic and inorganic food-grade nanoparticles. *npj Science of Food*, *1*, 6. https://doi.org/10.1038/s41538-017-0005-1.

Mendes, M., Soares, H. T., Arnaut, L. G., Sousa, J. J., Pais, A. A., & Vitorino, C. (2016). Can lipid nanoparticles improve intestinal absorption? *International Journal of Pharmaceutics*, *515*, 69–83.

Meshulam, D., & Lesmes, U. (2014). Responsiveness of emulsions stabilized by lactoferrin nano-particles to simulated intestinal conditions. *Food & Function*, *5*, 65–73.

Meyer, R. A., & Green, J. J. (2015). Shaping the future of nanomedicine: Anisotropy in polymeric nanoparticle design. *Wiley Interdisciplinary Reviews: Nanomedicine and Nanobiotechnology*, *8*, 191–207.

Mi, L., Zhang, X., Yang, W., Wang, L., Huang, Q., Fan, C., et al. (2009). Artificial nano-bio-complexes: Effects of nanomaterials on biomolecular reactions and applications in biosensing and detection. *Journal of Nanoscience and Nanotechnology*, *9*, 2247–2255.

Mohammadinejad, R., Karimi, S., Iravani, S., & Varma, R. S. (2016). Plant-derived nanostructures: Types and applications. *Green Chemistry*, *18*, 20–52.

Mohapatra, J., Ananthoju, B., Nair, V., Mitra, A., Bahadur, D., Medhekar, N. V., et al. (2018). Enzymatic and non-enzymatic electrochemical glucose sensor based on carbon nano-onions. *Applied Surface Science*, *442*, 332–341.

Mokhtari, S., Jafari, S. M., Khomeiri, M., Maghsoudlou, Y., & Ghorbani, M. (2017). The cell wall compound of Saccharomyces cerevisiae as a novel wall material for encapsulation of probiotics. *Food Research International*, *96*, 19–26.

Moraru, C., Huang, Q., Takhistov, P., Dogan, H., & Kokini, J. (2009). Chapter 21—Food nanotechnology: Current developments and future prospects. In G. Barbosa-Cánovas, A. Mortimer, D. Lineback, W. Spiess, K. Buckle, & P. Colonna (Eds.), *Global issues in food science and technology* (pp. 369–399). San Diego: Academic Press.

Motornov, M., Roiter, Y., Tokarev, I., & Minko, S. (2010). Stimuli-responsive nanoparticles, nanogels and capsules for integrated multifunctional intelligent systems. *Progress in Polymer Science*, *35*, 174–211.

Mueller, E., Koehler, P., & Scherf, K. A. (2016). Applicability of salt reduction strategies in pizza crust. *Food Chemistry*, *192*, 1116–1123.

Orecchioni, A.-M., Duclairoir, C., Renard, D., & Nakache, E. (2006). Gliadin characterization by sans and gliadin nanoparticle growth modelization. *Journal of Nanoscience and Nanotechnology*, *6*, 3171–3178.

O'Riordan, K., Andrews, D., Buckle, K., & Conway, P. (2002). Evaluation of microencapsulation of a Bifidobacterium strain with starch as an approach to prolonging viability during storage. *Journal of Applied Microbiology*, *91*, 1059–1066.

Ostertag, F., Weiss, J., & McClements, D. J. (2012). Low-energy formation of edible nanoemulsions: Factors influencing droplet size produced by emulsion phase inversion. *Journal of Colloid and Interface Science*, *388*, 95–102.

Pan, K., & Zhong, Q. (2016). Organic nanoparticles in foods: Fabrication, characterization, and utilization. *Annual Review of Food Science and Technology*, 7, 245–266.

Papadakis, G., Murasova, P., Hamiot, A., Tsougeni, K., Kaprou, G., Eck, M., et al. (2018). Micro-nano-bio acoustic system for the detection of foodborne pathogens in real samples. *Biosensors & Bioelectronics*, *111*, 52–58.

Pathakoti, K., Manubolu, M., & Hwang, H.-M. (2017). Nanostructures: Current uses and future applications in food science. *Journal of Food and Drug Analysis*, *25*, 245–253.

Peng, D., Jin, W., Li, J., Xiong, W., Pei, Y., Wang, Y., et al. (2017). Adsorption and distribution of edible gliadin nanoparticles at the air/water interface. *Journal of Agricultural and Food Chemistry*, *65*, 2454–2460.

Pérez-Masiá, R., López-Nicolás, R., Periago, M. J., Ros, G., Lagaron, J. M., & López-Rubio, A. (2015). Encapsulation of folic acid in food hydrocolloids through nanospray drying and electrospraying for nutraceutical applications. *Food Chemistry*, *168*, 124–133.

Peters, R. J. B., van Bemmel, G., Herrera-Rivera, Z., Helsper, H. P. F. G., Marvin, H. J. P., Weigel, S., et al. (2014). Characterization of titanium dioxide nanoparticles in food products: Analytical methods to define nanoparticles. *Journal of Agricultural and Food Chemistry*, *62*, 6285–6293.

Piazza, F., & Traytak, S. D. (2015). Diffusion-influenced reactions in a hollow nano-reactor with a circular hole. *Physical Chemistry Chemical Physics*, *17*, 10417–10425.

Powell, M., & Colin, M. (2008). Nanotechnology and food safety: Potential benefits, possible risks? *CAB Reviews: Perspectives in Agriculture, Veterinary Science, Nutrition and Natural Resources*, *3*, 16.

Pradhan, N., Singh, S., Ojha, N., Shrivastava, A., Barla, A., Rai, V., et al. (2015). Facets of nanotechnology as seen in food processing, packaging, and preservation industry. *BioMed Research International*, *2015*, 17.

Prietto, L., Pinto, V. Z., El Halal, S. L. M., de Morais, M. G., Costa, J. A. V., Lim, L.-T., et al. (2017). Ultrafine fibers of zein and anthocyanins as natural pH indicator. *Journal of the Science of Food and Agriculture*, *98*, 2735–2741.

Prosapio, V., De Marco, I., & Reverchon, E. (2018). Supercritical antisolvent coprecipitation mechanisms. *The Journal of Supercritical Fluids*, *138*, 247–258.

Pyrgiotakis, G., McDevitt, J., Yamauchi, T., & Demokritou, P. (2012). A novel method for bacterial inactivation using electrosprayed water nanostructures. *Journal of Nanoparticle Research*, *14*, 1027.

Pyrgiotakis, G., Vasanthakumar, A., Gao, Y., Eleftheriadou, M., Toledo, E., DeAraujo, A., et al. (2015). Inactivation of foodborne microorganisms using engineered water nanostructures (EWNS). *Environmental Science & Technology*, *49*, 3737–3745.

Qiu, C., Yang, J., Ge, S., Chang, R., Xiong, L., & Sun, Q. (2016). Preparation and characterization of size-controlled starch nanoparticles based on short linear chains from debranched waxy corn starch. *LWT—Food Science and Technology*, *74*, 303–310.

Rajendran, S. R. C. K., Udenigwe, C. C., & Yada, R. Y. (2016). Nanochemistry of protein-based delivery agents. *Frontiers in Chemistry*, *4*, 31.

Rawtani, D., Khatri, N., Tyagi, S., & Pandey, G. (2018). Nanotechnology-based recent approaches for sensing and remediation of pesticides. *Journal of Environmental Management*, *206*, 749–762.

Raynes, J. K., Carver, J. A., Gras, S. L., & Gerrard, J. A. (2014). Protein nanostructures in food—Should we be worried? *Trends in Food Science & Technology*, *37*, 42–50.

Roberfroid, M. (2007). Prebiotics: The concept revisited. *The Journal of Nutrition*, *137*, 830S–837S.

Rodríguez-Delgado, M. M., Alemán-Nava, G. S., Rodríguez-Delgado, J. M., Dieck-Assad, G., Martínez-Chapa, S. O., Barceló, D., et al. (2015). Laccase-based biosensors for detection of phenolic compounds. *TrAC Trends in Analytical Chemistry*, *74*, 21–45.

Rogers, M. A. (2011). Co-operative self-assembly of cholesterol and γ-oryzanol composite crystals. *CrystEngComm*, *13*, 7049. https://doi.org/10.1039/C1CE05818E.

Rogers, M. A. (2016). Naturally occurring nanoparticles in food. *Current Opinion in Food Science*, *7*, 14–19.

Rogers, M. A. (2017). Advances in edible oleogel technologies—A decade in review. *Food Research International*, *97*, 307–317.

Rogers, M. A., Spagnuolo, P. A., & Wang, T.-M. (2016). A potential bioactive hard-stock fat replacer comprised of a molecular gel. *Food Science & Nutrition*, *5*, 579–587. https://doi.org/10.1002/fsn3.433.

Rogers, M. A., Strober, T., Bot, A., Toro-Vazques, J. F., Stortz, T., & Marangoni, A. G. (2014). Edible oleogels in molecular gastronomy. *International Journal of Gastronomy and Food Science*, *2*, 22–31.

Ruiz-Rico, M., Pérez-Esteve, É., de la Torre, C., Jiménez-Belenguer, A. I., Quiles, A., Marcos, M. D., et al. (2018). Improving the antimicrobial power of low-effective antimicrobial molecules through nanotechnology. *Journal of Food Science*, *83*, 2140–2147.

Sabliov, C. M., Chen, H., & Yada, R. Y. (2015). Nanotechnology and functional foods: Effective delivery of bioactive ingredients. In *Nanotechnology and functional foods: Effective delivery of bioactive ingredients*. John Wiley & Sons.

Sadeghi Ekbatan, S., Iskandar, M. M., Sleno, L., Sabally, K., Khairallah, J., Prakash, S., et al. (2018). Absorption and metabolism of phenolics from digests of polyphenol-rich potato extracts using the Caco-2/HepG2 co-culture system. *Foods*, *7*, 8.

Sadeghi, R., Rodriguez, R. J., Yao, Y., & Kokini, J. L. (2017). Advances in nanotechnology as they pertain to food and agriculture: Benefits and risks. *Annual Review of Food Science and Technology*, *8*, 467–492.

Sahoo, S. K., Panyam, J., Prabha, S., & Labhasetwar, V. (2002). Residual polyvinyl alcohol associated with poly (d,l-lactide-co-glycolide) nanoparticles affects their physical properties and cellular uptake. *Journal of Controlled Release*, *82*, 105–114.

Samaranayaka, A. G. P., & Li-Chan, E. C. Y. (2011). Food-derived peptidic antioxidants: A review of their production, assessment, and potential applications. *Journal of Functional Foods*, *3*, 229–254.

Samphao, A., Butmee, P., Saejueng, P., Pukahuta, C., Svorc, L., & Kalcher, K. (2018). Monitoring of glucose and ethanol during wine fermentation by bienzymatic biosensor. *Journal of Electroanalytical Chemistry*, *816*, 179–188.

Sanchez-deAlcazar, D., Mejias, S. H., Erazo, K., Sot, B., & Cortajarena, A. L. (2018). Self-assembly of repeat proteins: Concepts and design of new interfaces. *Journal of Structural Biology*, *201*, 118–129.

Sarkar, S., Jana, A. D., Samanta, S. K., & Mostafa, G. (2007). Facile synthesis of silver nano particles with highly efficient anti-microbial property. *Polyhedron*, *26*, 4419–4426.

Sato, H., & Ueno, S. (2005). Polymorphism in fats and oils. In F. Shahidi (Ed.), *Bailey's industrial oil and fat products* (6th ed., pp. 77–120). New York, USA: John Wiley & Sons, Inc.

Schiffman, J. D., & Schauer, C. L. (2008). A review: Electrospinning of biopolymer nanofibers and their applications. *Polymer Reviews*, *48*, 317–352.

Schubert, H., & Engel, R. (2004). Product and formulation engineering of emulsions. *Chemical Engineering Research and Design*, *82*, 1137–1143.

Sekhon, B. S. (2014). Nanotechnology in agri-food production: An overview. *Nanotechnology, Science and Applications*, *7*, 31–53.

Shelby, T., Sulthana, S., McAfee, J., Banerjee, T., & Santra, S. (2017). Foodborne pathogen screening using magneto-fluorescent nanosensor: Rapid detection of *E. coli* O157:H7. *Journal of Visualized Experiments*, *127*, e55821.

Silva, H. D., Cerqueira, M. Â., & Vicente, A. A. (2012). Nanoemulsions for food applications: Development and characterization. *Food and Bioprocess Technology*, *5*, 854–867.
Singh, S., Gaikwad, K. K., Lee, M., & Lee, Y. S. (2018). Microwave-assisted microencapsulation of phase change material using zein for smart food packaging applications. *Journal of Thermal Analysis and Calorimetry*, *131*, 2187–2195.
Sivasankaran, U., Cyriac, S. T., Menon, S., & Kumar, K. G. (2017). Fluorescence turn off sensor for brilliant blue FCF—An approach based on inner filter effect. *Journal of Fluorescence*, *27*, 69–77.
Sok, V., & Fragoso, A. (2018). Preparation and characterization of alkaline phosphatase, horseradish peroxidase, and glucose oxidase conjugates with carboxylated carbon nano-onions. *Preparative Biochemistry & Biotechnology*, *48*, 136–143.
Sozer, N., & Kokini, J. L. (2009). Nanotechnology and its applications in the food sector. *Trends in Biotechnology*, *27*, 82–89.
Sun, X. Y., Yuan, M. J., Liu, B., & Shen, J. S. (2018). Carbon dots as fluorescent probes for detection of VB12 based on the inner filter effect. *RSC Advances*, *8*, 19786–19790.
Sweedman, M. C., Tizzotti, M. J., Schäfer, C., & Gilbert, R. G. (2013). Structure and physicochemical properties of octenyl succinic anhydride modified starches: A review. *Carbohydrate Polymers*, *92*, 905–920.
Tapia-Hernández, J. A., Torres-Chávez, P. I., Ramírez-Wong, B., Rascón-Chu, A., Plascencia-Jatomea, M., Barreras-Urbina, C. G., et al. (2015). Micro- and nanoparticles by electrospray: Advances and applications in foods. *Journal of Agricultural and Food Chemistry*, *63*, 4699–4707.
Tate & Lyle. October 2012, https://www.tateandlyle.com/news/tate-lyle-launches-soda-lo-salt-microspheres-globally (Accessed September 1, 2018).
Thomas, M. K., Murray, R., Flockhart, L., Pintar, K., Pollari, F., Fazil, A., et al. (2013). Estimates of the burden of foodborne illness in Canada for 30 specified pathogens and unspecified agents, circa 2006. *Foodborne Pathogens and Disease*, *10*, 639–648.
Thomas, K., & Sayre, P. (2005). Research strategies for safety evaluation of nanomaterials, part I: Evaluating the human health implications of exposure to nanoscale materials. *Toxicological Sciences*, *87*, 316–321.
Tominaga, T. (2018). Rapid detection of Klebsiella pneumoniae, Klebsiella oxytoca, Raoultella ornithinolytica and other related bacteria in food by lateral-flow test strip immunoassays. *Journal of Microbiological Methods*, *147*, 43–49.
Trujillo, L., Avalos, R., Granda, S., Guerra, L., & Pais-Chanfrau, J. (2016). Nanotechnology application for food and bioprocessing industries. *Biology and Medicine*, *8*, 1–6.
Tsuchido, Y., Sasaki, Y., Sawada, S., & Akiyoshi, K. (2014). Protein nanogelation using vitamin B6-bearing pullulan: Effect of zinc ions. *Polymer Journal*, *47*, 201.
Turbak, A., Snyder, F., & Sandberg, K. (1982). *Food products containing microfibrillated cellulose.* US Patent 43,418,071,982.
Vance, M. E., Kuiken, T., Vejerano, E. P., McGinnis, S. P., Hochella, M. F., Rejeski, D., et al. (2015). Nanotechnology in the real world: Redeveloping the nanomaterial consumer products inventory. *Beilstein Journal of Nanotechnology*, *6*, 1769–1780.
Van de Wiele, T. R., Oomen, A. G., Wragg, J., Cave, M., Minekus, M., Hack, A., et al. (2007). Comparison of five in vitro digestion models to in vivo experimental results: Lead bioaccessibility in the human gastrointestinal tract. *Journal of Environmental Science and Health, Part A*, *42*, 1203–1211.
Vanegas, D. C., Gomes, C. L., Cavallaro, N. D., Giraldo-Escobar, D., & McLamore, E. S. (2017). Emerging biorecognition and transduction schemes for rapid detection of pathogenic bacteria in food. *Comprehensive Reviews in Food Science and Food Safety*, *16*, 1188–1205.

Velásquez-Cock, J., Serpa, A., Vélez, L., Gañán, P., Gómez Hoyos, C., Castro, C., et al. (2019). Influence of cellulose nanofibrils on the structural elements of ice cream. *Food Hydrocolloids*, *87*, 204–213.

Verwei, M., Minekus, M., Zeijdner, E., Schilderink, R., & Havenaar, R. (2016). Evaluation of two dynamic in vitro models simulating fasted and fed state conditions in the upper gastrointestinal tract (TIM-1 and tiny-TIM) for investigating the bioaccessibility of pharmaceutical compounds from oral dosage forms. *International Journal of Pharmaceutics*, *498*, 178–186.

Wan, Z.-L., Guo, J., & Yang, X.-Q. (2015). Plant protein-based delivery systems for bioactive ingredients in foods. *Food & Function*, *6*, 2876–2889.

Wang, J., Morton, M. J., Elliott, C. T., Karoonuthaisiri, N., Segatori, L., & Biswal, S. L. (2014). Rapid detection of pathogenic bacteria and screening of phage-derived peptides using microcantilevers. *Analytical Chemistry*, *86*, 1671–1678.

Wang, Y., Ravindranath, S., & Irudayaraj, J. (2011). Separation and detection of multiple pathogens in a food matrix by magnetic SERS nanoprobes. *Analytical and Bioanalytical Chemistry*, *399*, 1271–1278.

Wang, T.-M., & Rogers, M. A. (2015). Biomimicry—An approach to engineering oils into solid fats. *Lipid Technology*, *27*, 175–178.

Wang, L., Sun, Y., Li, Z., Wu, A., & Wei, G. (2016). Bottom-up synthesis and sensor applications of biomimetic nanostructures. *Materials*, *9*, 53.

Wang, Y. Y., Zhou, J. H., & Li, J. H. (2017). Construction of plasmonic nano-biosensor-based devices for point-of-care testing. *Small Methods*, *1*, 1–13.

Warriner, K., Reddy, S. M., Namvar, A., & Neethirajan, S. (2014). Developments in nanoparticles for use in biosensors to assess food safety and quality. *Trends in Food Science & Technology*, *40*, 183–199.

Watanabe, J., Iwamoto, S., & Ichikawa, S. (2005). Entrapment of some compounds into biocompatible nano-sized particles and their releasing properties. *Colloids and Surfaces B: Biointerfaces*, *42*, 141–146.

Weir, A., Westerhoff, P., Fabricius, L., Hristovski, K., & von Goetz, N. (2012). Titanium dioxide nanoparticles in food and personal care products. *Environmental Science & Technology*, *46*, 2242–2250.

Weiss, J., Takhistov, P., & McClements, D. J. (2006). Functional materials in food nanotechnology. *Journal of Food Science*, *71*, R107–R116.

Wen, P., Wen, Y., Zong, M.-H., Linhardt, R. J., & Wu, H. (2017). Encapsulation of bioactive compound in electrospun fibers and its potential application. *Journal of Agricultural and Food Chemistry*, *65*, 9161–9179.

Wilczewska, A. Z., Niemirowicz, K., & Markiewicz, K. H. (2012). Nanoparticles as drug delivery systems. *Pharmacological Reports*, *64*, 1020–1037.

Wilkinson, C., Dijksterhuis, G. B., & Minekus, M. (2000). From food structure to texture. *Trends in Food Science & Technology*, *11*, 442–450.

Wille, R. L., & Lutton, E. S. (1966). Polymorphism of cocoa butter. *Journal of American Oil Chemistry Society*, *43*, 491–496.

Wu, J., Shi, M., Li, W., Zhao, L., Wang, Z., Yan, X., et al. (2015). Pickering emulsions stabilized by whey protein nanoparticles prepared by thermal cross-linking. *Colloids and Surfaces B: Biointerfaces*, *127*, 96–104.

Xiao, J., Lu, X., & Huang, Q. (2017). Double emulsion derived from kafirin nanoparticles stabilized Pickering emulsion: Fabrication, microstructure, stability and in vitro digestion profile. *Food Hydrocolloids*, *62*, 230–238.

Xiao, J., Wang, X., Perez Gonzalez, A. J., & Huang, Q. (2016). Kafirin nanoparticles-stabilized Pickering emulsions: Microstructure and rheological behavior. *Food Hydrocolloids*, *54*, 30–39.

Yan, J.-K., Qiu, W.-Y., Wang, Y.-Y., & Wu, J.-Y. (2017). Biocompatible polyelectrolyte complex nanoparticles from lactoferrin and pectin as potential vehicles for antioxidative curcumin. *Journal of Agricultural and Food Chemistry*, *65*, 5720–5730.

Yang, H., Qu, L., Wimbrow, A. N., Jiang, X., & Sun, Y. (2007). Rapid detection of Listeria monocytogenes by nanoparticle-based immunomagnetic separation and real-time PCR. *International Journal of Food Microbiology*, *118*, 132–138.

Yao, K., Chen, W., Song, F., McClements, D. J., & Hu, K. (2018). Tailoring zein nanoparticle functionality using biopolymer coatings: Impact on curcumin bioaccessibility and antioxidant capacity under simulated gastrointestinal conditions. *Food Hydrocolloids*, *79*, 262–272.

Ye, F., Miao, M., Cui, S. W., Jiang, B., Jin, Z., & Li, X. (2018). Characterisations of oil-in-water Pickering emulsion stabilized hydrophobic phytoglycogen nanoparticles. *Food Hydrocolloids*, *76*, 78–87.

Ye, F., Miao, M., Jiang, B., Campanella, O. H., Jin, Z., & Zhang, T. (2017). Elucidation of stabilizing oil-in-water Pickering emulsion with different modified maize starch-based nanoparticles. *Food Chemistry*, *229*, 152–158.

Yin, L.-J., Chu, B.-S., Kobayashi, I., & Nakajima, M. (2009). Performance of selected emulsifiers and their combinations in the preparation of β-carotene nanodispersions. *Food Hydrocolloids*, *23*, 1617–1622.

Yu, H., & Huang, Q. (2012). Improving the oral bioavailability of curcumin using novel organogel-based nanoemulsions. *Journal of Agricultural and Food Chemistry*, *60*, 5373–5379.

Yu, L., Li, C., Xu, J., Hao, J., & Sun, D. (2012). Highly stable concentrated nanoemulsions by the phase inversion composition method at elevated temperature. *Langmuir*, *28*, 14547–14552.

Yu, H., Zhou, T., Gong, J., Young, C., Su, X., Li, X.-Z., et al. (2010). Isolation of deoxynivalenol-transforming bacteria from the chicken intestines using the approach of PCR-DGGE guided microbial selection. *BMC Microbiology*, *10*, 182.

Yun, J. R., Duan, F., Liu, L. M., Chen, X. L., Liu, J., Luo, Q. L., et al. (2018). A selective and sensitive nanosensor for fluorescent detection of specific IgEs to purified allergens in human serum. *RSC Advances*, *8*, 3547–3555.

Zan, G., & Wu, Q. (2016). Biomimetic and bioinspired synthesis of nanomaterials/ nanostructures. *Advanced Materials*, *28*, 2099–2147.

Zeinhom, M. M. A., Wang, Y. J., Sheng, L. N., Du, D., Li, L., Zhu, M. J., et al. (2018). Smart phone based immunosensor coupled with nanoflower signal amplification for rapid detection of Salmonella Enteritidis in milk, cheese and water. *Sensors and Actuators B: Chemical*, *261*, 75–82.

Zhang, L., Guan, P., Zhang, Z., Dai, Y., & Hao, L. (2018). Physicochemical characteristics of complexes between amylose and garlic bioactive components generated by milling activating method. *Food Research International*, *105*, 499–506.

Zhang, W. G., Han, Y., Chen, X. M., Luo, X. L., Wang, J. L., Yue, T. L., et al. (2017). Surface molecularly imprinted polymer capped Mn-doped ZnS quantum dots as a phosphorescent nanosensor for detecting patulin in apple juice. *Food Chemistry*, *232*, 145–154.

Zhang, H., Li, G., Xie, F. Y., Zhang, W. H., Chen, Q. L., & Chen, J. (2017). Surface enhanced spectroscopy based on lab-on-a-chip technology and its applications in analytical science. *Spectroscopy and Spectral Analysis*, *37*, 350–355.

Zhang, Y., Zhang, J., Xing, C., Zhang, M., Wang, L., & Zhao, H. (2016). Protein nanogels with temperature-induced reversible structures and redox responsiveness. *ACS Biomaterials Science & Engineering*, *2*, 2266–2275.

Zhou, C., Zou, H., Li, M., Sun, C., Ren, D., & Li, Y. (2018). Fiber optic surface plasmon resonance sensor for detection of E. coli O157:H7 based on antimicrobial peptides and AgNPs-rGO. *Biosensors and Bioelectronics*, *117*, 347–353.

CHAPTER TWO

# Assembled protein nanoparticles in food or nutrition applications

**Young-Hee Cho, Owen Griffith Jones***
Department of Food Science, Purdue University, West Lafayette, IN, United States
*Corresponding author: e-mail address: joneso@purdue.edu

## Contents



## Abstract

Proteins are one of the essential components of nutritional food materials and an excellent source for food-grade nanomaterials. This review focuses on select examples of nanoparticles assembled naturally, found in food-relevant materials, major approaches in assembling nanoscale structure from proteins, and general applications of protein nanoparticles in food or nutrition. Animal-sourced casein and non-animal grain storage proteins and legume storage proteins are discussed in terms of their structural assemblies. Protein solubility is a key factor in assembling protein nanoparticles with desired functional properties. Desolvation is the most common technique to prepare protein nanoparticles for insoluble proteins. Well-hydrated protein assemblies have been extensively studied through electrostatic complexes, assembled with fatty acid and starch, reassembled protein structure, and nanogels. These protein-based nanoparticles have been utilized for filler materials of films, encapsulation of bioactive molecules, and stabilization of emulsions. Most studies exploiting protein-based nanoparticles have focused on developing technologies in extraction of proteins from sources and assembly of nanoparticles in different environmental conditions.

*Advances in Food and Nutrition Research*, Volume 88
ISSN 1043-4526
https://doi.org/10.1016/bs.afnr.2019.01.002

## 1. Introduction

Proteins are an essential component of nutritional food materials and an excellent source for food-grade nanomaterials. The nanoscale size and compact conformational states of many proteins leads to some confusion as to whether they are already "nanoparticles," and so it is important to create a suitable definition before generating a useful summary. Literature provides a somewhat ambiguous definition for nanoparticles and nanomaterials, leading to recent discussions on a suitable definition that is commonly understood, functional, and agreeable to all involved parties (Boholm & Arvidsson, 2016). There are strong arguments to define nanoparticles as having one dimension of 1–100 nm or a surface area per volume of $60\,m^2/cm^3$ or less. These two definitions are relevant to the ability of particles to bypass standard cell internalization routes or to increases in surface reactivity with active molecules/enzymes, respectively. A relatively simple working definition, which will be utilized for the remainder of this chapter, defines "nanoparticles" as objects with all dimensions on the nanometer-length scale (i.e., 1–999 nm) (Boholm et al., 2016).

A separate yet practically important question is whether there is a *desire* to label assembled proteins or polysaccharides as nanoparticles, particularly if one wishes to distinguish these from potentially toxic inorganic materials. Further, the public may have a distrust of nanotechnology stemming from its newness (i.e., the public's unfamiliarity and therefore their skepticism of its necessity or safety) and/or from unwarranted associations with specific nanotechnology linked with toxicity. For example, there was concern that ingestion of protein fibrils, having two dimensions in the nanometer-length scale but another that is typically on the micrometer-length scale (and therefore not discussed further in this chapter), would be toxic if ingested because the presence of fibrillar protein structures have been linked with pathogenic states in living creatures (Riek, 1996). However, there is evidence showing that particles of proteins and polysaccharides including protein fibrils are quickly digested by the conditions and enzymes present in the digestive tract (Bateman, Ye, & Singh, 2010). With this in mind, a case could be made for avoiding the use of "nano-" terms when describing assembled structures of proteins and polysaccharides, and instead refer to them as a subclass of colloids or colloidal assemblies. In this way, it is possible to discuss a class of assembled structures that are below the threshold of tactile sensory perception and possess a greater reactivity (via increased surface area per volume) when compared to macro-scale assemblies while avoiding confusion with very different inorganic materials.

Basic information on proteins as biopolymers has been well reviewed in existing literature (Damodaran, 1996; Nakai & Modler, 1996). In brief, proteins are polymers synthesized from amino acids within living cell tissues. Because of the diverse amino acid structures, including positive/negative-charges, polar-aprotic, sulfhydryl-/disulfide-, and hydrophobic functional groups, proteins undergo highly ordered arrangement both intramolecularly and intermolecularly. Depending on which amino acids are exposed to the solvent in the final structure, proteins are soluble in very different media: globulin proteins are soluble in water, albumins in high ionic strength media, prolamins in aqueous alcohols, and glutelins are soluble in acidic, basic, or detergent-based aqueous solutions. Much of the functional behaviors of proteins can be described by their excluded volume based on solvent conditions and intramolecular interactions (e.g., conformational states and assemblies).

In this chapter, a description is given on the utilization and identity of assembled structures of proteins with all dimensions in the nanometer-scale. In the first section, a description is given on nanometer-scale assembled structures naturally found in food-relevant materials. Then, a review is provided of major approaches in assembling nanometer-scale structures from proteins and polysaccharides in order to create desired functional properties in foods or food-relevant materials. In the final section, functions and applications of such structures are discussed from theoretical and practical standpoints.

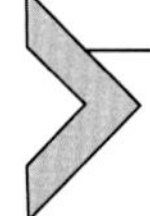

## 2. Nanoparticles assembled naturally within biological materials

The large majority of proteins are prepared in cells to have specified molecular weights and peptide sequences. Sequence ordering, as well as folding enzymes, contributes to very specific folding arrangements so highly reproducible structural features develop. As we recall from biology and biochemistry, many proteins also assemble into supramolecular structures for enzymatic, transport, or structural functions. Natural, nano-scale protein assemblies created by cells or viruses have been recently reviewed. The major classes are described as linear (e.g., collagen, actin), rings, tubes, and catenanes (Pieters, van Eldijk, Nolte, & Mecinovic, 2016). Some of these structures have been suggested as inspiration for next-generation of nanostructure designs in biotechnology applications (Howorka, 2011). It is sobering to consider the frequency that we consume "nanoparticles" within food products that originate from these natural cellular structures. Of course, the relative weight fraction of such specific assemblies that remain as nano-structures after processing could be considered negligible, and much of the non-meat protein consumed in our diet is in the form of larger aggregates meant for nutrition. Examples of this include grain storage proteins and many of the egg and dairy proteins. Some assemblies of such proteins do fall somewhat within the nanometer-scale range, and the following section describes some select examples from major commodity protein-sources.

## 2.1 Casein

Casein is a collection of the dominant protein fractions found in milk, which include α-, β-, γ-, and κ-caseins. These protein subunits typically form a supramolecular assembly in milk, termed a casein "micelle," with an average diameter of 100–200 nm (de Kruif, Huppertz, Urban, & Petukhov, 2012). Forces holding these spheroidal assemblies together include hydrogen bonding, ionic bridging, and hydrophobic forces. Calcium phosphate is particularly important for forming salt bridges between phosphoseryl residues on the β- and α-caseins to maintain the micelle structure, making up ~6% of the total casein dry weight (O'Mahoney & Fox, 2013). When preparing acidic dairy foods, such as certain cheeses, the calcium phosphate dissolves and leads to structural changes that can have macroscopic impacts to texture (Lucey, Johnson, & Horne, 2003).

Casein assemblies have been very well studied by a variety of structural analyses, yet, like the starch granule, its scale and potential for variability makes it a challenge to build an accurate universal model. Neutron and X-ray scattering indicate that the structure is comprised of 4-nm-sized calcium phosphate nanoclusters dispersed throughout a loose matrix of protein with nanometer-scale density fluctuations (de Kruif et al., 2012). Such a depiction agrees fairly well with fractal models of the casein micelle, with calcium phosphate clusters as one of the building blocks (Fig. 1). Recent evidence further indicates that the salt bridges and hydrophobic interactions responsible for the internal networks are interchangeable and in a pseudo-equilibrium state, as indicated by the shifting of casein micelles to sub-micellar assembles using only extra sodium caseinate to sequester some of the calcium phosphate (Thomar & Nicolai, 2015).

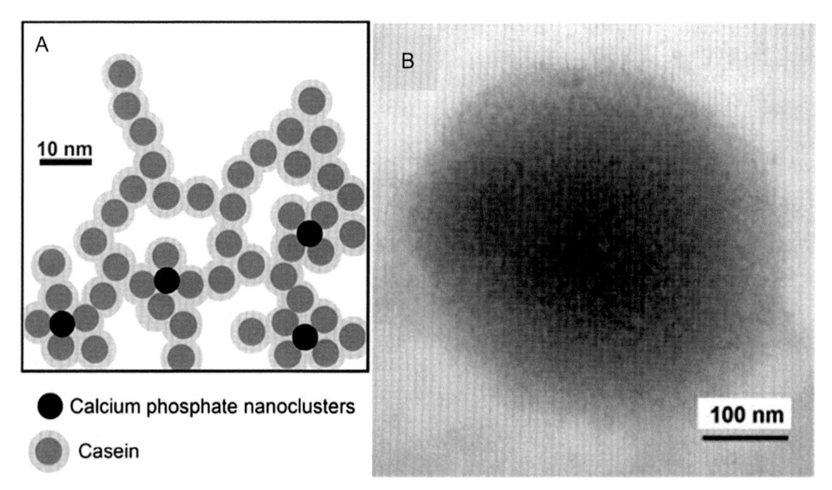


**Fig. 1** Structure of the casein micelle (A) with a proposed interlocking lattice model based on electron microscopy or (B) direction from a transmission electron micrograph. *Reprinted from McMahon, D. J., & Oommen, B. S. (2008). Supramolecular structure of the casein micelle.* Journal of Dairy Science, *91(5), 1709–1721, Copyright 2008, with permission from Elsevier.*

Assembled casein is compact yet remains adequately stable in milk as a suspension, allowing the milk to be both an excellent protein source and of low viscosity for easy delivery, yet there have been attempts to improve the functionality of the assembly by enzymatic modification. Covalent cross-links have been successfully formed within the casein micelle using the enzyme Transglutaminase without inducing inter-micelle interactions (Huppertz & de Kruif, 2008; Mounsey, O'Kennedy, & Kelly, 2005). Such enzymatic cross-linking improves gelation properties of casein (Schorsch, Carrie, & Norton, 2000), increasing the gel strength of skim milk gels and yogurt (Anema, Lauber, Lee, Henle, & Klostermeyer, 2005; Lauber, Henle, & Klostermeyer, 2000). Controlled delivery applications of casein micelles have so far not been shown because of the challenge in perfusing the active material within the micelle structure. However, this can be solved by disassembling the micelle and then reassembling it with the active agent, as will be discussed later (Section 3.2.3).

## 2.2 Grain storage proteins

Grain storage proteins include those from wheat, corn, sorghum, rye, and other commodity grains. The most recognized assembled proteins among these grains include gluten, primarily from wheat but also found in many of the grains. "Gluten" proteins are a collection of prolamins and glutelins found in a variety of grains with high content of glutamine, glycine, and nonpolar amino acids (Day, 2011). These proteins have low solubility in neutral or acidic water; they are typically found as partially hydrated materials within food products. Prolamin assemblies are also found in significant supply among corn and sorghum.

In baked products, such as wheat-based breads, the glutelins and gliadins participate significantly in the network formation, contributing characteristic viscoelastic properties for gaseous expansion and final product texture (Delcour et al., 2012). It is perhaps instructional to recognize that glutelins and gliadins are found as multimeric assemblies of proteins. In food, alcohol-based solvents, gliadin fractions ($\alpha$-, $\gamma$-, and $\omega$-) were found to be present as oblate ellipsoids with dimensions of $\sim$3.25 nm $\times$ $\sim$15–20 nm, while high molecular weight glutenin was also in the form of ellipsoids or rods with a diameter of 6.4 nm and a length of 69 nm (Thomson et al., 1999).

Wheat glutelin assemblies can be very large, and interactions between these assemblies are believed to contribute significantly to the viscoelastic network of baked products. These glutelins can be divided into high molecular weight and low molecular weight fractions. High-molecular weight glutenin is a large, linear protein that participates in significant disulfide bonding with other high-molecular weight glutenin chains (Wang, Jin, & Xu, 2015). Recent differentiation of wheat gluten proteins indicated that proteins above $45 \times 10^3$ Da molecular weight experience less interpolymeric interactions and behave as weakly hydrated polymers, with the protein becoming completely insoluble above a molecular weight of $6.5 \times 10^6$ Da (Boire, Menut, Morel, & Sanchez, 2013). Although present as only 20–40 mg/g of flour, the largest assemblies of fibrous glutenin, possessing molecular weights in excess of 10 million Da, have been found to most strongly impact the viscoelasticity of dough (Wang, Zhao, & Zhao, 2007). Low-molecular weight glutenin and the alcohol-soluble gliadin fractions are globular proteins and are not believed to contribute significantly to the formation of elastic networks, but instead contribute toward the cohesiveness and extensibility of gluten-based doughs (Wieser & Kieffer, 2001). Gliadins also decrease the transition temperature of gluten and lead to the formation of more open networks (Khatkar, Barak, & Mudgil, 2013). Looking at flours of different wheat strains, stronger doughs are formed if there are larger protein assemblies (>10 μm) comprised of a greater complement of protein subunits (Don, Lichtendonk, Plijter, & Hamer, 2003a), and it is feasible that the large glutenin can provide structure while smaller subunits can contribute to adhesion between the assemblies for a relatively continuous network. Such assemblies are likely weakly associated, as intrinsic viscosity of these suspensions diminishes with the duration and intensity of mixing (and should be correlated with performance of wheat during dough mixing) (Don, Lichtendonk, Plijter, & Hamer, 2003b).

Since gliadin and glutenin are poorly soluble in pure water, they are naturally present as aggregates in aqueous dispersions. Among the large molecular weight fraction of glutenin, simple shear was shown to break particles into smaller size fractions, yet these comminuted particles were all well above several microns in diameter (Peighambardoust, van der Goot, Hamer, & Boom, 2005). Other natural structures comprised of glutenin are also well into the micrometer range and will not be considered as nanoparticles.

In corn, there are many storage proteins that are poorly digestible and removed during processing. Zein is a major component of this fraction. Zein solubility is fairly poor in most solvents, and even the typically chosen solvent of aqueous ethanol mixtures does not always provide sufficient solubilization for full dispersion of individual protein chains. A recent X-ray scattering study of 30–60% (w/v) zein suspensions in 70–90% (v/v) ethanol revealed a minor, initial concentration of 50–150 nm discs and spheroidal shapes with diameters on the scale of $10^2$–$10^3$ that marginally increased in concentration with aging (Uzun, Ilavsky, & Padua, 2017). The authors concluded that zein, comprised of proteins with gyrational radii of ~2 nm, was highly prone to the formation of sheets via hydrogen-bonding between beta-sheet rich regions, and these sheets interacted with each other and other free protein to form larger spheroids. Thus, even in relatively good solvents, prolamins have a pronounced tendency toward self-assembly.

Kafirin is a prolamin from sorghum with similar properties to maize zein and is often found in assembled structures. Structural studies performed with X-ray scattering in different solvents and increasing concentrations showed that kafirin, comprised of polypeptides ($\alpha_1$, $\alpha_2$, $\beta$, and $\gamma$) with molecular weight between 20 and 30 kDa, exists in small ellipsoidal assemblies even in good alcohol-water solvents yet will readily assemble into larger aggregate assemblies with increasing solvent polarity (Xiao et al., 2015). Kafirin, similar to many grain storage proteins, exists in large, spheroidal "protein bodies" throughout the starch endosperm and in larger concentration near the sorghum germ layer, and these protein bodies are highly resistant to digestion unless reducing agents are also present (Oria, Hamaker, & Shull, 1995).

## 2.3 Legume storage proteins

Legume storage proteins possess two major groupings defined by their sedimentation coefficients: the 7S fraction (vicilins), which are trimeric assemblies, and 11S fraction (legumins), which are hexameric assemblies. In many beans, there is a major vicilin fraction with the name of phaseolin possessing a molecular weight of ~150,000 Da. The structure of the trimer is a planar (triangular) assembly with a diameter of 9 nm and a thickness of 3.5 nm yet further coordinates with three other trimers near pH 4.5, forming an 18S dodecamer (Lawrence et al., 1990).

A very interesting natural nano-assembly of soy proteins and other fractions was recently found. Specifically, a fraction of soy protein was isolated after the removal of glycinin- and conglycinin-rich fractions that possessed a collection of conglycinin, glycinin, and oleosins (among others), as well as a variety of lipid components, such as phospholipids (Gao et al., 2013). These fractions naturally form assemblies when dispersed in water with varied polydisperse size ranges, yet sonication was found to induce surprisingly monomodal distributions of assemblies with particle diameter ~250 nm. Because of the phospholipid and oleosin content, surface activity was superior to β-conglycinin, with observations of improved oil-in-water emulsion stability and reduced interfacial displacement by uncharged small-molecule surfactants.

Other vegetal crops are emerging as potential sources of protein ingredients with useful properties, largely driven by the need to increase full utilization of biomass and to capitalize on consumer interest in novel, natural food sources. A great example of such a protein source that is present in natural assemblies is quinoa. Quinoa protein is rather nutritious (amino acid score 60–94) and contains a collection of proteins most soluble in mildly basic conditions and partially soluble in neutral to acidic conditions with molecular weights of 50,000, 38,000, 32,000, 25,000, and 20,000 Da, comprising 11S and 2S fractions (Steffolani et al., 2016). A study was performed on the use of quinoa protein isolate to stabilize emulsions, finding that aqueous dispersions containing assemblies of 300–500 nm could be reduced to ~200 nm after 120 kJ of ultrasonic treatment (100 W, 20 min, bath unit) or dispersed further with a strong surfactant (Qin, Luo, & Peng, 2018).

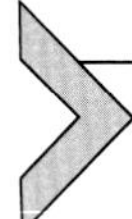

# 3. Assembled protein nanoparticles

## 3.1 Insoluble, suspended assemblies

As mentioned above (Section 2.2), glutelins and prolamins are present as water-insoluble assemblies within grains and legumes, yet still form multimeric assemblies even within better solvent conditions. Their relatively poor solubility in most food-grade solvents, particularly water, makes them well suited for production of environmentally robust films, fibers, or particles. Further, the preparation of aggregate particles from water-insoluble materials is relatively straight-forward, as the process is dependent upon well-studied nucleation and growth mechanisms.

Joye and McClements have prepared a detailed review on the formation of such water-insoluble protein particles, where they summarize several examples of desolvated protein nanoparticles and relate the major factors leading to small, dispersed particles (Joye & McClements, 2013). In brief, smaller particles will be formed by encouraging rapid nucleation of particles from solution by sudden immersion into poor solvent conditions. In the absence of heterogeneous nucleation sites (e.g., bubbles, dust, etc.), nuclei generation rate is enhanced with a higher degree of supersaturation (concentration in relation to maximum solubility in given solvent) and reduced surface tension between the solvent and newly formed particles. Growth of particles occurs as the concentration falls below the critical supersaturation ratio until it is no longer supersaturated (maximum solubility), although particles can undergo further aggregation by collision with each other. Beyond concentration and surface tension, factors promoting smaller particle sizes by desolvation include faster mixing techniques, application of ultrasound of higher intensities, and reduction in temperature.

Kafirin is a prolamin with many structural similarities to zein and can be readily desolvated within aqueous solutions to form suspended particles. An early study prepared particles well above 1 μm in diameter by bulk desolvation, and the resulting particles were shown to be effective encapsulants for catechins and sorghum tannins (Taylor, Taylor, Belton, & Minnaar, 2009b). The authors proposed that the porous surface of the particle structure provided improved performance for gradual release applications. Only a small fraction of kafirin particles desolvated from acetic acid solutions below 1 μm and a more significant fraction of sub-micrometer particles were prepared using aqueous ethanol as the initial solvent. However, the porosity of the particles was significantly diminished with decreasing particle size (Taylor, Taylor, Belton, & Minnaar, 2009a). In a recent study by Xiao et al., kafirin particles of 200–250 nm diameter were prepared by desolvation from 80% aqueous ethanol (Xiao, Nian, & Huang, 2015). These particles were subsequently coated by a modified chitosan (carboxy-methylation) via pH-induced association with only a modest increase in diameter.

Gliadin, prolamin from wheat, has been formed into nanoparticles by desolvating aqueous alcohol solutions into aqueous salt solutions. An early study on gliadin nanoparticles obtained a conversion factor close to 90% and used them to encapsulate retinoic acid (Ezpeleta et al., 1996). These particles were readily digested by trypsin, indicating their quick digestibility, yet digestion was drastically reduced after cross-linking the particles with glutaraldehyde. Other gliadin particles were electrostatically interacted with chitosan (50% of gliadin content) at neutral pH to form a secondary coating, with particle size increasing from ~150 nm up to 600 nm as the concentration of sodium chloride in the continuous phase was increased (Yuan et al., 2017).

## 3.2 Well-hydrated assemblies

### *3.2.1 Soluble electrostatic complexes*

Polymers are much more likely to interact with other polymers of the same kind, forming a relatively homogenous demixed phase, a simple coacervate, or a precipitate, rather than to interact with a different polymer. The exception to this occurs if there are a number of coordinated associative interactions between functional groups of two different polymers. In such systems, a significant yet limited number of associations leads to soluble complexes where much of the polymers are freely surrounded by solvent, while a greater number of associations leads to the formation of complex coacervates where the associative polymers form a separate phase from the bulk solution. Excellent reviews exist on the theory and factors involved in segregative (Norton & Frith, 2001) and associative polymer interactions (Cooper, Dubin, Kayitmazer, & Turksen, 2005; Schmitt & Turgeon, 2011), and it will not be further covered here. Surveys have also been performed on the variety of biopolymers studied in the formation of associative systems (Moschakis & Biliaderis, 2017). The following sub-section describes discrete formation of small (nanometer-scale) soluble complexes formed by electrostatic interactions between proteins and polysaccharides, which could be used for general nanoassembly applications.

Since many proteins utilized by the food industry have an isoelectric pH between 4 and 6 (largely due to feasible isolation strategies), many electrostatic complexes with protein have involved complexes with anionic polysaccharides in mildly acidic conditions or cationic polysaccharides (usually chitosan) in neutral to mildly alkaline conditions. Stable and soluble proteins such as whey proteins have been shown to form soluble complexes with anionic polysaccharides, such as gum Arabic, in the pH range of ~5.5 and 4.75 while forming complex coacervates at lower pH values (Weinbreck, de Vries, Schrooyen, & de Kruif, 2003). Observed hydrodynamic diameters

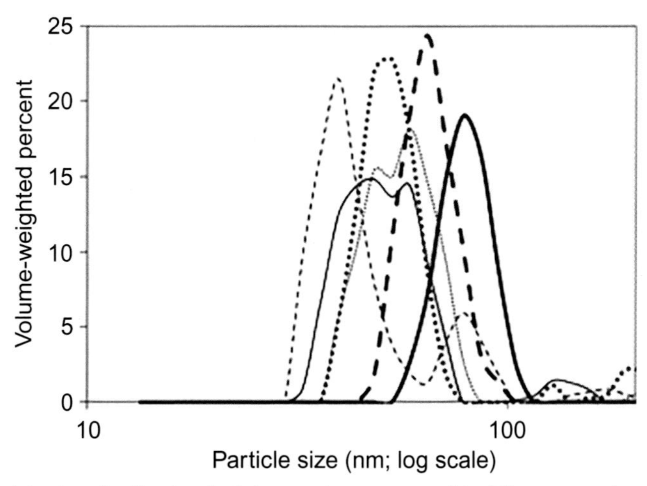


**Fig. 2** Particle size distribution in β-Lg-pectin systems with different pectin concentrations (······, 0.025%wt; ---, 0.05%wt; ——, 0.075%wt; •••, 0.1%wt; — —, 0.125%wt; —, 0.15%wt). β-Lg concentration was 0.05%wt and the pH was 4.25. Particle size distribution was measured after an overnight equilibration at room temperature. *Reprinted from Ron, N., Zimet, P., Bargarum, J., & Livney, Y. D. (2010). Beta-lactoglobulin–polysaccharide complexes as nanovehicles for hydrophobic nutraceuticals in non-fat foods and clear beverages.* International Dairy Journal, 20*(10), 686–693, Copyright 2010, with permission from Elsevier.*

of these weak and well-hydrated complexes were ~20–40 nm, as would be expected for coordinated complexes between a few proteins and a corresponding number of oppositely charged polyelectrolytes. Similarly small assemblies (hydrodynamic diameter 50–100 nm) were also observed among β-lactoglobulin and pectin mixtures at lower pH values (pH 4–5) and correspondingly smaller protein-to-polysaccharide ratios (1:1–1:3; Fig. 2) to ensure soluble complexes rather than coacervates (Ron, Zimet, Bargarum, & Livney, 2010). Larger structures than this may be observed at lower pH or higher protein-to-polysaccharide ratios, although this is typically indicative of complex coacervate phases that will merge and enlarge over time.

Associative electrostatic complexes, which are equilibrium structures with significant solvation, can be readily disassembled with further changes in pH or concentrations. This phenomenon can be arrested if one or both of the biopolymers form irreversible interactions, through controlled aggregation processes (as discussed later) or when the biopolymers are in close proximity. For example, soluble complexes of sodium caseinate and gum Arabic formed spherical structures of 100–150 nm diameter in the pH ranges of pH 5.6–4 at protein-to-polysaccharide ratios of 1:1 or

1:5 (Ye, Flanagan, & Singh, 2006). The authors speculated that the relatively spherical and dense structures observed by electron microscopy could be due to strong associations between caseinate chains within the complex, a supposition that is logical given the nature of caseinate but not proven by analysis. Spherical assemblies of slightly larger diameters (~200–300 nm) were formed between caseinate and chitosan at pH 5–6 if the protein concentration was less than 0.05 wt%, which could indicate that the protein aggregation was just as important a factor as interaction with polyelectrolyte (Anal, Tobiassen, Flanagan, & Singh, 2008). Protein aggregation combined with interactivity with polyelectrolyte may also be responsible for the formation of 200–400 nm spherical assemblies observed in mixtures of succinylated-cruciferin (from rapeseed) and chitosan at mildly acidic pH with diameter of 200–400 nm (Wang, Yang, Ju, Udenigwe, & He, 2018). These nanoparticle suspensions were quite stable to aggregation for up to a month in refrigerated storage and showed promise as an encapsulant for curcumin.

Apart from associative complexes based on electrostatic interactions, complexes also form between charged (good hydrogen-donor) and nonionic (hydrogen-acceptor) polymers by hydrogen bonding (Such, Johnston, & Caruso, 2011). This is readily observed in easily controlled polymer systems, such as strong donor/acceptor pairs (e.g., polymethacrylic acid and poly(vinyl pyrrolidone); interact at near-neutral pH values) and weaker donor/acceptor pairs (e.g., poly(acrylic acid) and poly(ethylene oxide); interact only below pH 3.6) (Sukhishvili & Granick, 2002). Increased ionic strength and proximity to critical solution temperatures facilitate such associations (Kharlampieva & Sukhishvili, 2006). Similar hydrogen-bond associations may lead to *intra*-polymeric complexes among block co-polymers possessing an acidic polyelectrolyte block and a nonionic block, such as poly(acrylic acid)-block-poly(ethylene oxide) (Holappa, Karesoja, Shan, & Tenhu, 2002). It is not clear to what degree these relatively weaker charge-dipole interactions contribute to complex formation among associative proteins and polysaccharides, yet there has been convincing evidence that hydrogen-bonding strengthens the associative interactions between whey proteins and high-methoxy-pectin (Girard, Turgeon, & Gauthier, 2002).

### 3.2.2 Whey protein/fatty-acid/starch assemblies

There is evidence indicating strong affinity between fatty acids and amylose, which have been ascribed to the coordination of the fatty acid tail within the helix. Taking this a step further, it was found that when amylose was heated with whey protein and fatty acids (linoleic, palmitic, or oleic), there was an increased peak viscosity attributable to significant interactions between the components. This phenomenon was not observed in tests omitting one of the components (Zhang & Hamaker, 2003). Size-exclusion chromatography showed the elution of a product larger than amylose, with a dimension of ~150 nm as determined from light scattering. The product would not form if the linoleic acid was replaced with an alcohol or monoglyceride, indicating the role of the free carboxylic acid group in engendering the structure (Zhang, Maladen, Campanella, & Hamaker, 2010). Isolated product was verified to contain each of the components, including protein that was significantly aggregated because of the thermal preparation. A later study with iodine-binding and molecular modeling indicated that the linoleic acid did bind within the amylose helix and that the protein interacted electrostatically with the negatively charged carboxylate residue that was left emerged at the end of the helix (Bhopatkar et al., 2015).

As the amylose is typically much longer than the length of fatty-acid tails, there should be significant vacancy remaining in the helix to accommodate a small hydrophobic molecule for use as a delivery vehicle. This was demonstrated for the model compound naphthol in the original system (Bhopatkar et al., 2015). The ternary complex (β-lactoglobulin/fatty-acid/starch/naringin) was recently utilized to encapsulate naringin from citrus, obtaining encapsulation efficiency of ~79% with a loading content of ~15% (Feng et al., 2017). This latter study also obtained some of the most definitive images of these nanoparticles (Fig. 3).

### 3.2.3 Reassembled proteins

As stated above, casein micelles are natural assemblies of proteins found in milk. Techniques to temporarily disrupt this structure and then promote reassembly are useful to create novel particles for controlled delivery purposes. Studies performed several decades ago demonstrated the feasibility to reassemble micelles from sodium caseinate by reincorporating calcium, phosphate, and citrate ions at milk-relevant contents (Knoop, Knoop, & Wiechen, 1979). Such structures were used to encapsulate vitamin $D_2$ with remarkable morphological similarity to the native micelles (Fig. 4) (Semo et al., 2007). More recently, high pressure was shown to increase micelle size (lower pressures) or decrease size (>250 MPa) (Huppertz, Fox, & Kelly, 2004). It was found that 40 min of 250 MPa pressurization caused disassembly of micelles, particularly with lower casein concentration, added sodium

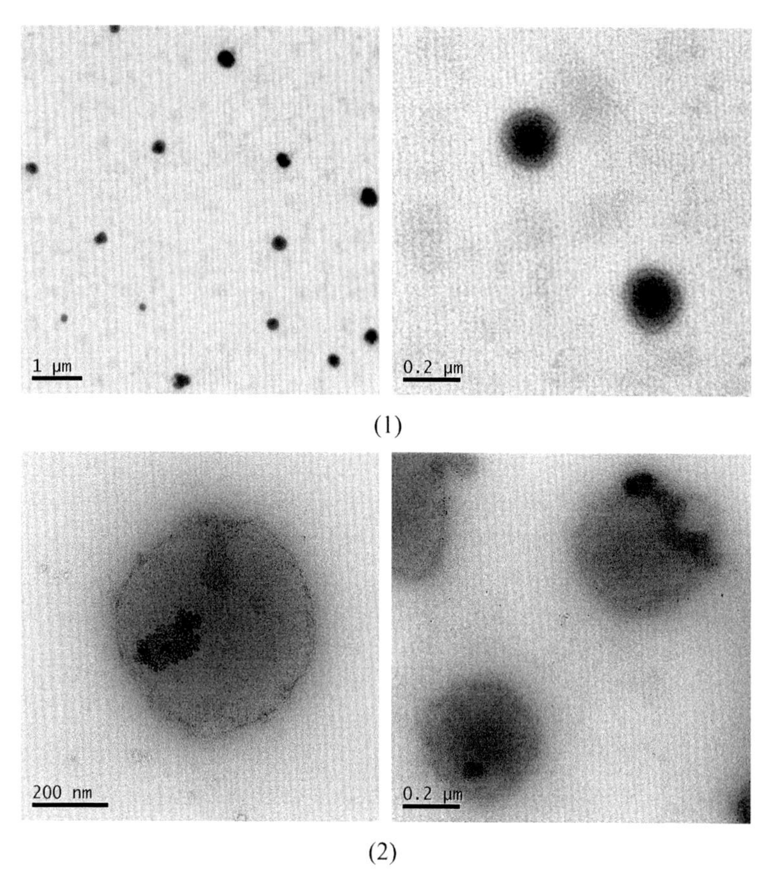


**Fig. 3** TEM micrographs of ternary complex particle comprised of amylose, linoleic acid, and whey protein. *Reprinted from Feng, T., Wang, K., Liu, F., Ye, R., Zhu, X., Zhuang, H. (2017). Structural characterization and bioavailability of ternary nanoparticles consisting of amylose, α-linoleic acid and β-lactoglobulin complexed with naringin.* International Journal of Biological Macromolecules, 99, *365–374, Copyright 2017, with permission from Elsevier.*

phosphate, and higher pH but less so with added ethanol or calcium/sodium chloride (Huppertz & De Kruif, 2006). Reassembly was more likely if the micelle was significantly disassembled without degradation of the subunits and was further promoted by adding sodium phosphate or ethanol.

Since pressurization and depressurization seem to induce temporary dissociation and increased dispersion of casein subunits, it is logical that changes in solvent conditions, particularly pH of aqueous solutions, can also temporarily disassemble casein micelles. Caseinates are relatively more soluble in strong acidic or alkali conditions. It is possible to solubilize caseinate in such

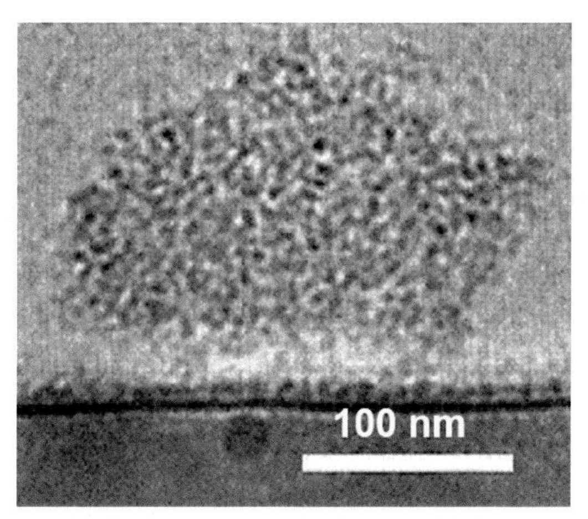


**Fig. 4** Cryo-transmission electron micrograph of reassembled casein micelle containing vitamin $D_2$ (Semo, Kesselman, Danino, & Livney, 2007). *Adapted from Semo, E., Kesselman, E., Danino, D., & Livney, Y. D. (2007). Casein micelle as a natural nano-capsular vehicle for nutraceuticals.* Food Hydrocolloids, 21*(5–6), 936-942, Copyright 2007, with permission from Elsevier.*

conditions and then assemble them under more neutral pH conditions. As a means to form delivery vehicles for the hydrophobic bioactive curcumin, both sodium caseinate and curcumin were dissolved in alkaline water (pH 12) and then neutralized to form 100–210 nm particles with a 1–7% curcumin loading (Pan, Luo, Gan, Baek, & Zhong, 2014). An interesting variant on this strategy was to disperse sodium caseinate at high pH with zein, which is mildly soluble in alkaline water, and subsequent pH neutralization caused the assembly and desolvation of the two components into ~100 nm particles (Pan & Zhong, 2016). These particles were found to be significantly smaller and more monodisperse when compared to zein nanoparticles desolvated in sodium caseinate aqueous solution (Fig. 5).

Assembled structures of soybean proteins also have the potential to be disrupted and later reassembled, although this is largely underexplored. Using pieces of information gleaned from years of studying the assemblies of soybean glycinin, reversible shifts between larger and smaller assembled states of glycinin were achieved by changes in solution conditions (Pizones Ruiz-Henestrosa, Martinez, Patino, & Pilosof, 2012). Specifically, the 11S assembly that is dominant at neutral pH with salt could be reduced to 7S or 3S forms with acidification ($pH < 4$) or dilution. Mild increases in temperature (30–50 °C), likely disrupting intermolecular hydrogen bonds, also led to disassembly, unlike higher temperature treatments that induced

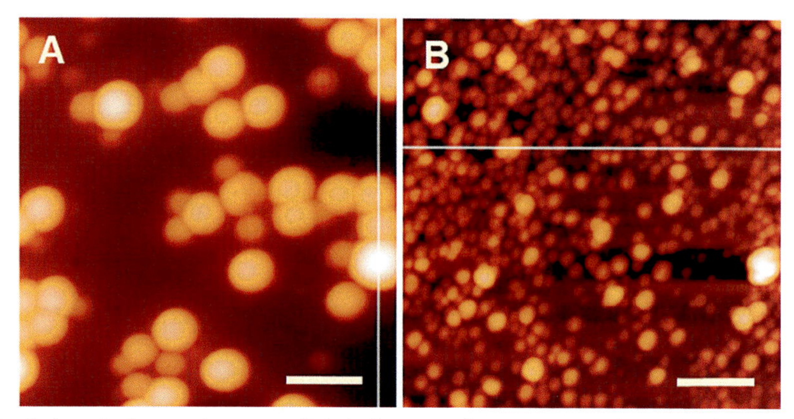


**Fig. 5** Atomic force microscopy height images of zein-caseinate particles formed by (A) desolvating zein from 80% aq. ethanol into sodium caseinate solution (1:1 final protein ratio) or (B) solubilizing both proteins in pH 11.5 buffer and neutralizing the pH to induce particle formation; scale bars = 400 nm. *Adapted from Pan, K., & Zhong, Q. (2016). Low energy, organic solvent-free co-assembly of zein and caseinate to prepare stable dispersions.* Food Hydrocolloids, 52, *600–606, Copyright 2016, with permission from Elsevier.*

greater aggregation. However, utilization of such reversible assemblies can be challenging, since even the largest assemblies are fairly small (~20 nm), making it difficult to trap significant amounts of material for a controlled delivery application.

Desolvation is also an approach to prepare nanoparticles from water-soluble proteins by adding non-aqueous solvents to induce aggregation followed by optional cross-linking of the aggregates and removal of solvent. For example, the formation of 200–300 nm gelatin nanoparticles by partial desolvation in water with added acetone followed by cross-linking by glutaraldehyde, where the formed particles showed promise as particle-based stabilizers in concentrated oil-in-water emulsions (Tan, Tu, Jia, Gou, & Ngai, 2018). Soybean β-conglycinin was desolvated from aqueous solution by the addition of ethanol during continuous vortexing, and the resulting particles possessed diameters between 50 and 500 nm depending on the ethanol fraction during desolvation (Levinson, Israeli-Lev, & Livney, 2014). Cross-linking was not required in these particles because of the significant hydrophobicity of the β-conglycinin.

### 3.2.4 Nanogels

Certain protein molecules will aggregate in solution but will not expel much of the hydrating water, forming a percolated, gel-like structure. If the concentration is relatively low and the protein possesses some residual, exposed charge via ionization, aggregation effectively ceases once a general particle size is reached. Such particles in nanometer- to micron-scale with well-hydrated internal structure are often termed as "microgels" or "nanogels." While microgels have been well studied among synthetic polymer systems (Lyon & Fernandez-Nieves, 2012), the preparation of protein-based nanogels has received significant attention only in the past decade. Building on findings from their previous studies, Schmitt and others systematically described the formation of 200 nm nanogels after heat treatment of whey protein isolate. At pH values above the isoelectric pH, smaller worm-like aggregates and larger amorphous aggregates were formed at pH 7 and 5, respectively (Schmitt et al., 2009). Such nanogels were well characterized by scattering experiments, indicting a low density percolated structure of sub-aggregates with a fractal dimension of 2.1 (Schmitt et al., 2010). Further exploration of these aggregates demonstrated the importance of pH (5.7–6.1), concentration (~1–3 g/L), and ionic strength on defining the resulting size of nanogel aggregates (Phan-Xuan et al., 2011).

A similar process of controlled heat treatment was used to prepare nanogels from the bovine dairy protein lactoferrin—a transferrin with a relatively high isoelectric pH (Bengoechea, Peinado, & McClements, 2011). Because the isoelectric pH is somewhat alkaline, heat treatment of at least 5 min at 75 °C could be carried out at pH 7, and the resulting 100–150 nm particles were highly stable against further pH changes or 200 mM or greater of sodium chloride.

Soybean particles with diameter of ~100 nm have been formed by thermal treatment in aqueous solution at pH 7 followed by aggregation induced by ionic shielding (300 mM sodium chloride solution), which was confirmed by light scattering and atomic force microscopy (showing fair amounts of polydispersity) (Liu & Tang, 2013). Similarly sized particles were also formed by heating soybean glycinin (2% w/v, 95 °C, 15 min) at pH 7 with 100–500 mM sodium chloride, although a fraction of larger particles (~1 μm) were also formed during the heat treatment (Zhu et al., 2017). Particle formation by thermal treatment of soy protein isolate led to a significant fraction of particles with >1 μm diameter, with the formation of smaller particles more probable as the temperature was increased from 80 to 100 °C and concentration was maintained below 5% (w/v) (Guo et al., 2015). These studies reiterated the complex nature of protein aggregation and difficulty in controlling the formation of monodisperse distributions of nanometer-sized aggregates, particularly for a protein with multiple domains and significant hydrophobicity, such as glycinin and β-conglycinin.

Rather than utilizing solution conditions to selectively ionize proteins and limit aggregation, polysaccharides can be added to interact with the protein to achieve a similar purpose and form nanogels of a narrow size distribution. Over the past 15 years, a number of studies demonstrating the feasibility of forming nanogels by heat treatment of electrostatic complexes between polysaccharides and β-lactoglobulin (Jones & McClements, 2011). Many studies indicated that the addition of interactive polysaccharide can mitigate aggregation (Fig. 6). The polysaccharide interacts partially via hydrogen bonding, resulting in at least some dissociation at higher temperatures and promotes the aggregation into larger particles. Highly charged polysaccharides, such as carrageenan gums, produce smaller nanogels because of extensive electrostatic interactions and do not permit significant dissociation, unlike polysaccharides with fewer ionized functional groups (e.g., pectin). This approach will work with other water-soluble protein systems that are prone to temperature-induced aggregation. For instance, 200–400 nm particles were produced from pectin and sodium caseinate when heated (85 °C/30 min) at pH 3.7 (Luo, Pan, & Zhong, 2015). Because of the thermal treatment, there are limited applications for such particles as

**Type 1** Particles–Protein particles are formed first by heating, then coated wtih polysaccharide

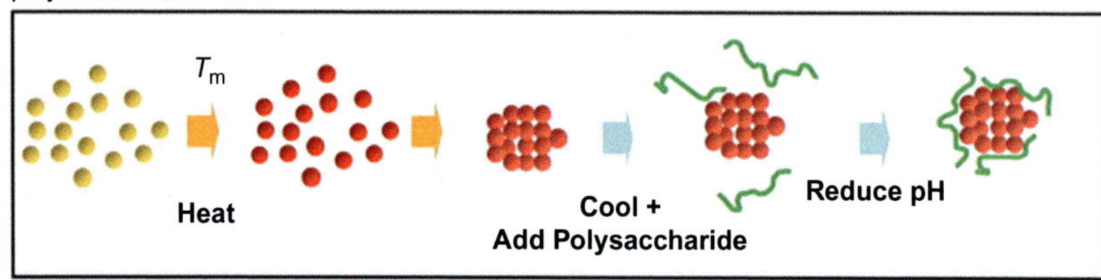


**Type 2** Particles–Protein-polysaccharide complexes are heated together to form particles

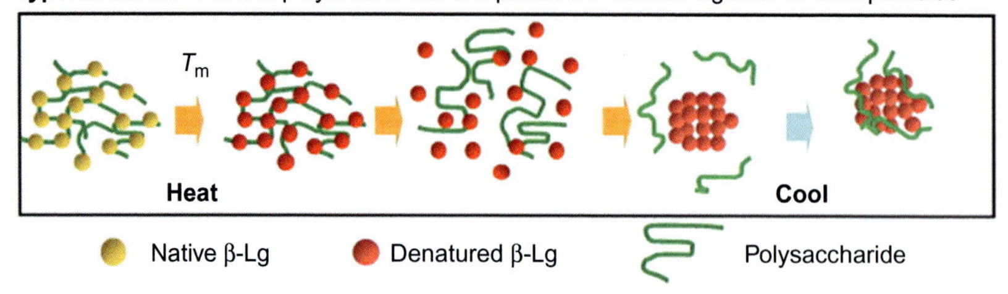


**Fig. 6** Schematic diagram describing the formation of polysaccharide-coated nanogels by heating protein (A) at specific pH values and adding interactive polysaccharide after cooling or (B) at pH values close to the isoelectric pH in the presence of interactive polysaccharides. *Reprinted from Jones, O. G., & McClements, D. J. (2011). Recent progress in biopolymer nanoparticle and microparticle formation by heat-treating electrostatic protein–polysaccharide complexes.* Advances in Colloid and Interface Science, 167*(1), 49–62, Copyright 2011, with permission from Elsevier.*

delivery vehicles, although they may be useful for certain active ingredients that are more heat stable. For example, heated lysozyme/sodium carboxymethyl cellulose complexes (neutral pH, 4:1–1:1 protein-polysaccharide ratio) formed 400–700 nm nanogel as a delivery vehicle for 5-fluorouracil (Zhu et al., 2013).

A similar, yet slightly modified approach, is to covalently attach polysaccharide chains to proteins, reducing the extent of nonspecific agglomeration among proteins and guiding the formation of nanogels during thermal aggregation. As chemically induced modification of protein is typically not desired, covalent attachment of polysaccharide chains in the food industry can be achieved using enzymes. Yet the most preferred, cost-effective approach is to use Maillard reactions. For instance, lysozyme-dextran conjugates (2:1 M ratio) prepared by heating dry mixtures at 60 °C, pH 7–8 with 79% relative humidity were subsequently dissolved in water (~3–30 g/L) and heated at 80 °C for 30 min at pH 10.7 (Li, Yu, Yao, & Jiang, 2008). The resulting aggregates possessed a spherical morphology with average size of 150–250 nm as concentration increased (Fig. 7). Particles of smaller sizes but greater polydispersity were observed when the pH was decreased during heat treatment (Li et al., 2008). Nanogels of 60–90 nm were prepared by heating dextran conjugates of the soybean protein β-conglycinin at 80 °C for 30 min with pH values between 3.5 and 6 (Feng et al., 2015). These nanogels were stable as aqueous suspensions for several months and could even be lyophilized and rehydrated with no significant irreversible agglomeration. Further examples of dextran glycosylation and heat-induced nanogel formation can be found for a variety of proteins, indicating the universal promise of this approach in nanoparticle formation.

Lastly, limiting aggregation to form nanoparticles can be carried out by a combination of polysaccharide-protein interaction and covalent modification approaches, which is truly just a combination of the two approaches discussed above. An example of this is the formation of complexes between the negatively charged biopolysaccharide chondroitin sulfate with whey protein-dextran conjugates, formed via Maillard reaction at pH 5.2 and 85 °C for 15 min, forming ~150 nm nanogels (Dai et al., 2015). Such nanogels were stable against pH changes (pH 1–8) and up to 200 mM sodium chloride, therefore were promising encapsulating vehicle for lutein. A similar study prepared 130–230 nm nanoparticles by heating pH 5.6 complexes of chitosan and bovine serum albumin-dextran Maillard conjugates at 80 °C for 1 h (Qi, Yao, He, Yu, & Huang, 2010). These nanoparticles successfully delivered a cancer drug to reduce viability of tumor cells within in vitro studies.

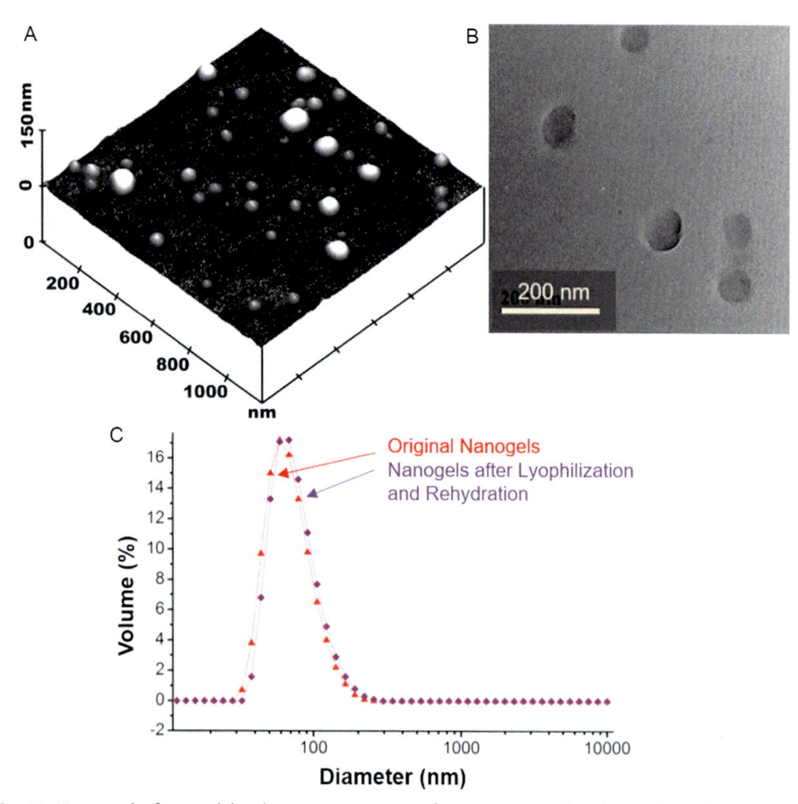


**Fig. 7** Nanogels formed by heat treatment of proteins and polysaccharides (Maillard conjugates), showing (A) atomic force micrograph (topography) of lysozyme-dextran conjugates (1:2 lysozyme:dextran molar ratio; pH 10.7; ~3 g/L concentration for 30 min/80 °C), (B) transmission electron micrograph of whey protein isolate/dextran conjugates (9.7% degree of glycosylation) with chondroitin sulfate, (C) Hydrodynamic diameter distribution of β-conglycinin-dextran conjugates that were stable to lyophilization and rehydration processes. *Panel A: Reprinted with permission from Li, J., Yu, S., Yao, P., & Jiang, M. (2008). Lysozyme–dextran core–shell nanogels prepared via a green process.* Langmuir, 24*(7), 3486–3492, Copyright 2008, American Chemical Society. Panel B: Adapted with permission from Dai, Q., Zhu, X., Abbas, S., Karangwa, E., Zhang, X., Xia, S. (2015). Stable nanoparticles prepared by heating electrostatic complexes of whey protein isolate–dextran conjugate and chondroitin sulfate.* Journal of Agricultural and Food Chemistry, 63*(16), 4179–4189, Copyright 2015, American Chemical Society. Panel C: Adapted with permission from Feng, J.-L., Qi, J.-R., Yin, S.-W., Wang, J.-M., Guo, J., Weng, J.-Y. (2015). Fabrication and characterization of stable soy β-conglycinin–dextran core–shell nanogels prepared via a self-assembly approach at the isoelectric point.* Journal of Agricultural and Food Chemistry, 63*(26), 6075–6083, Copyright 2015, American Chemical Society.*

### 3.2.5 Assemblies of co-polymers

Polysaccharides used as thickeners or stabilizers in commercial products are inherently prone to intra-polymeric assembly in aqueous solution. Mechanical or chemical treatment is often required to achieve their full dispersion (Kravtchenko, Renoir, Parker, & Brigand, 1999; Laneuville, Turgeon, & Paquin, 2013; Ma & Pawlik, 2007). One remaining unknown in understanding average behavior of polysaccharides in solution is the influence of side-chains, many of which are uncharged yet still hydrophilic. This is in contrast to hydrophobic side chains, which polymer scientists have already utilized to prepare unique delivery vehicles with polysaccharides as a natural component, such as dextran-poly(styrene) (Houga et al., 2009) and cholesteryl-pullulan (Akiyoshi, Deguchi, Tajima, Nishikawa, & Sunamoto, 1997). For attached hydrophilic chains, research has begun to systematically investigate how side-chains and grouped "blocks" of ionized saccharides impact interactivity of polysaccharides (Sperber, Stuart, Schols, Voragen, & Norde, 2010; Xu, Melton, Jameson, Williams, & McGillivray, 2015). However, there is still much to learn about the impact of attached oligosaccharide branches on the intramolecular or intermolecular assembly behavior of polysaccharides.

Complex coacervate core micelles (C3M) are nano-assemblies formed by electrostatic interactions between ionic segments of block copolymers and oppositely charged polyelectrolytes (Pergushov, Müller, & Schacher, 2012; Voets, de Keizer, & Cohen Stuart, 2009). Similar to complex coacervates/soluble complexes, the combined ionic segments form a neutralized complex that forms a phase-separated "core." Unlike complex coacervates, the hydrophilic segments of the block copolymer remain as a solubilized outer "shell" (Pergushov et al., 2012). The core-shell structure is reminiscent of surfactant-based micelles, yet C3M assembly is not driven by hydrophobic interactions and is therefore less toxic to cellular life, in vivo, as surfactants potentially are (Swenson & Curatolo, 1992). Increased length of the nonionic block on copolymers, in some cases $\sim 3\times$ the length of the ionic block (van der Burgh, de Keizer, & Cohen Stuart, 2004), has been found to improve stability of C3Ms by increasing solvent interactions (Novoa-Carballal, Pergushov, & Müller, 2013; Novoa-Carballal, Pfaff, & Müller, 2013). The size of the C3Ms formed has also been correlated with the length of the nonionic block, as demonstrated for C3Ms of polystyrene and dextran/poly(2-(dimethylamino)ethyl methacrylate) (Novoa-Carballal, Pfaff, et al., 2013).

While polymer scientists have produced detailed information on complex micelle formation from synthetic copolymers, there have been relatively fewer studies of micelle formation using proteins or polysaccharides. Two decades ago, a copolymer of PEG and poly(aspartic acid) was associated with hen egg white lysozyme in neutral aqueous media to form complex coacervate core micelles of ~50 nm diameter, involving approximately 36 and 42 lysozyme and copolymer molecules, respectively (Fig. 8) (Harada & Kataoka, 1998). Lysozyme also interacts with a copolymer involving bovine serum albumin and a decorated PEG chain, possessing average diameter ~20 nm so that it could be internalized within cells (Jiang et al., 2016). PEG was also covalently attached to the end of a negatively charged hyaluronic acid chain to form a 90–240 nm complex micelle with poly(L-lysine). Blocks of 9 and 5 kDa of the hyaluronic acid and PEG chains, respectively, were necessary to allow for successful complex micelle

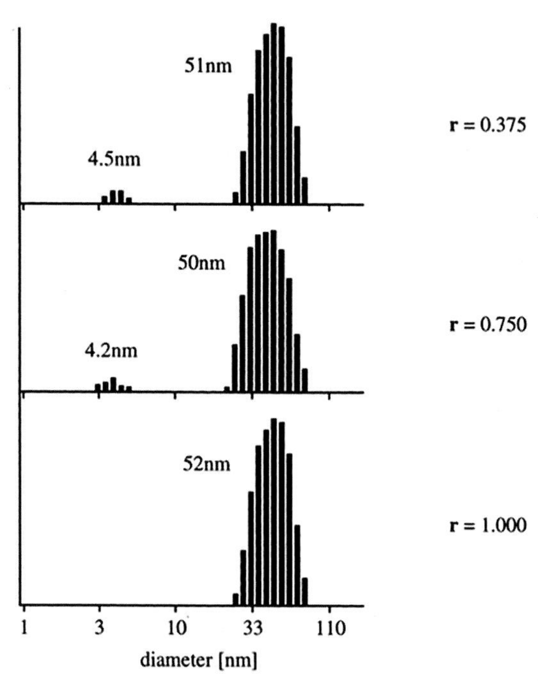


**Fig. 8** Change in z-weighted size distribution obtained from the histogram analysis of DLS with the lysozyme:PEG-p(Asp) mixing ratio, *r*. (25.0 ± 0.2 °C; total concentration 2.5 mg/mL; detection angle 90°; size range of analysis 1–200 nm). *Reprinted with permission from Harada, A., & Kataoka, K. (1998). Novel polyion complex micelles entrapping enzyme molecules in the core: Preparation of narrowly-distributed micelles from lysozyme and poly(ethylene glycol)–poly(aspartic acid) block copolymer in aqueous medium.* Macromolecules, 31*(2), 288–294, Copyright 1998, American Chemical Society.*

formation without further aggregation (Novoa-Carballal, Pergushov, et al., 2013). The polysaccharide dextran has also been utilized as an alternative to the hydrophilic PEG chains by creating a copolymer of dextran with poly (2-(dimethylamino)ethylmethacrylate) and forming a complex micelle with poly(styrene-sulfonate) (Novoa-Carballal, Pfaff, et al., 2013). Thus, proteins and polysaccharides are both very functional as components of copolymer-based assemblies.

In an effort to assemble a complex micelle using both a polysaccharide and protein component, a block copolymer was prepared by attaching a PEG chain ($M_n = 5$ kDa) to carboxymethyl-dextran ($M_n \sim 27$ kDa) (Du, Reuhs, & Jones, 2016). This polysaccharide-based copolymer was then mixed with α-lactalbumin to determine whether the PEG chain would impede or promote association between negatively charged carboxymethyl-dextran and positively charged residues on the protein in acidic pH conditions. ζ-potential values of the mixtures were more negative than pure α-lactalbumin, implying that the two components interacted (Fig. 9A). Interestingly, ζ-potential values of complexes were largely insensitive to the relative biopolymer ratio, which may be due to the formation of an assembled structure with a discrete size limitation, such as a complex coacervate core micelle. The formation of initial complexes and larger, phase-separating structures from these mixtures was observed by the increase in scattered light intensity and turbidity as the pH was reduced from neutral to acidic (Fig. 9B). The critical pH for complex phase separation was higher in these mixtures when compared to mixtures of α-lactalbumin and carboxymethyldextran without the attached PEG chain, indicating that the PEG chain promoted a greater interactivity with the protein. A collection of small complexes and larger spherical assemblies were observed in electron micrographs (Fig. 9C) with corresponding hydrodynamic diameters of 22–80 nm (Du et al., 2016). The proposed structure of these spherical assemblies, based upon the physical evidence, is schematically detailed in Fig. 9D.

Conjugates between the soybean protein β-conglycinin and dextran have been investigated as a means of preparing more stable and functional protein-based ingredients. These conjugates were prepared by Maillard reaction at 60 °C over several days in humid conditions. Such conjugates form particulate aggregates of ~100 nm diameter after dispersion in water because of the preparatory thermal treatment, with improved pH stability of suspensions as compared with the non-glycated protein (glycinin aggregates significantly near the isoelectric pH of 4.8) (Zhang et al., 2012). Desolvation in ethanol and cross-linking by glutaraldehyde, which marginally increased the particle size of such conjugates, was found to increase the

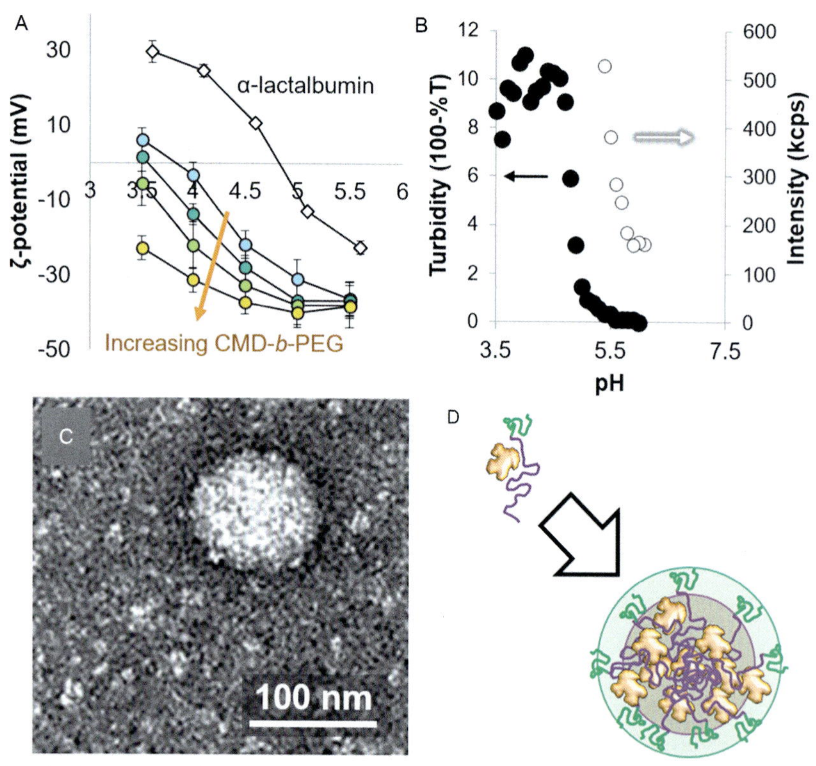


**Fig. 9** Formation of self-assembled complex-coacervate core micelles from a covalently-modified dextran molecule (carboxymethyldextran-block-poly(ethylene glycol)) and the protein α-lactalbumin, showing (A): colloidal charge based on electrophoretic mobility of the protein with or without increasing amounts of modified dextran, (B) development of turbidity and scattered light intensity during acidification of a protein/modified dextran mixture, (C) transmission electron micrographs of mixtures with protein-to-polysaccharide ratio of 3.5 at pH 5.2; (D) Schematic of proposed assembly, showing internalization of the electrostatically-interacting protein and anionic dextran block and externalization of the uncharged poly(ethylene glycol) block. *Panel C: Adapted from Du, J., Reuhs, B. L., & Jones, O. G. (2016). Influence of PEGylation on the ability of carboxymethyl-dextran to form complexes with α-lactalbumin.* Food Chemistry, 196, *853–859, Copyright 2016, with permission from Elsevier.*

development of surface pressure and interfacial elasticity in relation to the untreated assemblies, although similar relative gains were also observed among similarly-treated, non-glycated protein (Wu et al., 2014).

Heat treatment of complexes with a micelle-like structure can arrest further growth or destabilization due to pH or ionic strength changes. For instance, Maillard conjugates of bovine serum albumin and dextran were prepared by dry heating and interacted with the analgesic drug ibuprofen via electrostatic and hydrophobic interactions with the protein component (Li & Yao, 2009). The authors stated that the resultant complexes were prone to precipitation unless a heat treatment was applied, in which case stable suspensions were formed with detected particles possessing an average hydrodynamic radius of 60–130 nm (Li & Yao, 2009). Nanoparticles of ~200 nm diameter were similarly prepared by heating suspensions of complex coacervate core micelles of α-lactalbumin and a chitosan-PEG copolymer. However, it was found that the attached PEG chain had little impact on changing the size in relation to heated α-lactalbumin/chitosan mixtures except when the chitosan chain possessed a molecular weight of greater than 76 kDa (Du, Cho, Murphy, & Jones, 2017).

Ionic cross-linking may also prevent dissolution of polysaccharide-based complex micelles. An example is the use of tripolyphosphate to cross-link hyaluronic acid after it had formed a complex coacervate core micelle with chitosan copolymer (grafted with PEG chains at a fraction of the C-2 amines). The resulting nanoparticles possessed a hydrodynamic diameter of 100–150 nm and were either positively- and negatively charged if the ratio between hyaluronic acid and copolymer was 1:1 or 2:1, respectively (Fig. 10) (Raviña et al., 2010).

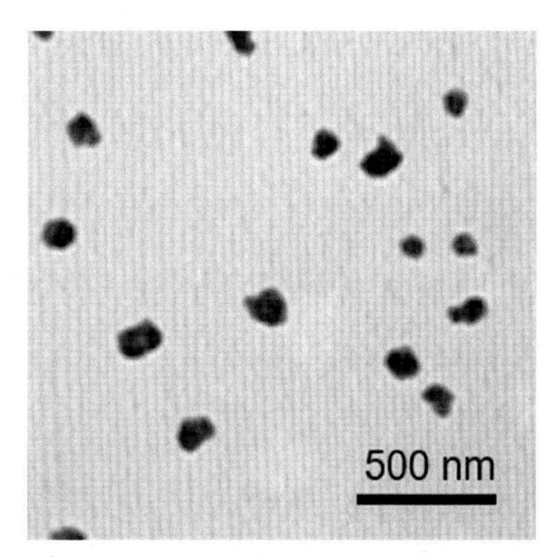


**Fig. 10** Morphology of HA/CS-g-PEG-loaded nanoparticles as visualized in transmission electron microscopy (TEM): 1/1 HA/CS-g-PEG 10% pDNA-loaded nanoparticles (a). *Adapted from Raviña, M., Cubillo, E., Olmeda, D., Novoa-Carballal, R., Fernandez-Megia, E., Riguera, R. (2010). Hyaluronic acid/chitosan-g-poly(ethylene glycol) nanoparticles for gene therapy: An application for pDNA and siRNA delivery.* Pharmaceutical Research, *27(12), 2544–2555, Copyright 2010, Springer Nature.*

# 4. General applications of protein nanoparticles for food or nutrition

Taking the definition that nanoparticles comprised of proteins can be considered as assembled structures with all length scales of nanometer dimensions, we can discuss the application of any protein-based colloidal structures used in foods as long as they are smaller than a micrometer. It is perhaps also wise to limit the minimum dimension of the nanoparticle to 10 nm, so as to disambiguate between nanoparticles and many single-domain proteins.

A major advantage of incorporating nanoscale particles into concentrated or semi-concentrated materials is the enhanced dispersibility of small, numerous particles into viscous/gelling media, which might otherwise remain as concentrated, poorly distributed agglomerates.

## 4.1 Filler materials

Filler materials are frequently used to improve the mechanical and barrier properties of films. Utilizing nanoscale filler materials is older than the cognizant pursuit of nanomaterial preparation, itself, and there exists excellent literature on the basic concepts of creating enhanced composite materials of polymers and nanoparticles (Balazs, Emrick, & Russell, 2006). In brief, decreasing the size of nanoparticle fillers increases the ratio between particle surface area and volume, translating to an increased contact between the dominant film phase and the filler particle. If there is strong compatibility between the continuous film phase and filler particle, increased contact contributes to increased film elasticity/strength or extensibility of the film. However, if the compatibility is poor the mechanical properties of the composite film will diminish with increased contact (i.e., smaller size, increased concentration). For instance, grafting poly(caprolactone) chains onto cellulose or starch nanoparticles (5–10 nm width × 20–250 nm length) significantly increased compatibility with the dominant poly(caprolactone) phase, thereby increasing the strain at which the films broke during extension (Habibi & Dufresne, 2008). Thus, the preparation of enhanced nanoparticle-based composites must focus on improving surface interactions with the dominant phase and ensuring their total dispersion.

Compatibility with the dominant film phase can be evidenced by studies involving different types of filler particles. An example of a compatible system is cellulose nanoparticles (~50–100 nm) dispersed in wheat starch films, where up to 5% content (w/w dry basis) led to consistent increases in tensile strength while decreasing extensibility and water vapor transmissibility, although changes to these parameters were relatively less above filler 2% content (Chang, Jian, Zheng, Yu, & Ma, 2010). In a recent study, the incorporation of zein nanoparticles into modified cellulose films was shown to increase the film elasticity at low concentrations yet decreased elasticity at higher concentrations (Fig. 11) (Gilbert, Cheng, & Jones, 2018). This observation was attributed to the limited compatibility of zein with the continuous film, where zein nanoparticles were able to distribute quite homogeneously throughout the film at low concentrations but readily agglomerated at higher concentrations during drying. Similar behaviors have been observed for titanium dioxide nanoparticles within whey protein films, where initial increases in modulus and tensile strength shifted to overall decreases as the nanoparticles agglomerated into phase instable aggregates within the film (Zhou, Wang, Gunasekaran, 2009). For zein nanoparticles, compatibility within a whey protein-based film was achieved by utilizing sodium caseinate to interact with and stabilize the nanoparticles. The resulting composite film exhibited consistent increases in tensile strength even at ~50% filler content (Oymaci & Altinkaya, 2016). Other than these

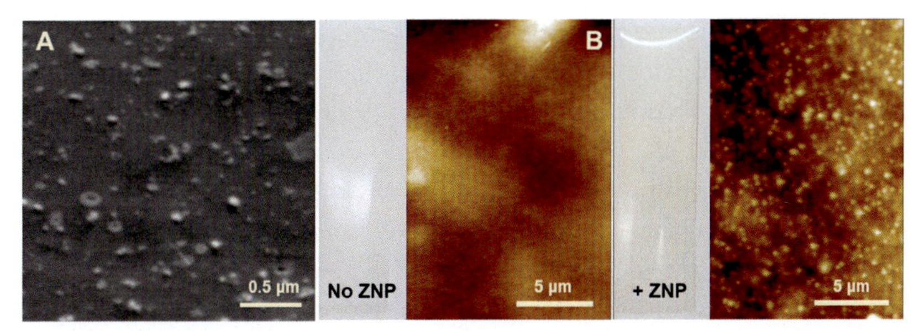


**Fig. 11** Images of composite films prepared with zein protein nanoparticles: (A) scanning electron micrograph of the cross-section of a WPI film with caseinate-stabilized zein nanoparticles (ZNP-WPI ratio of 1.2). (B) Topographical atomic force micrographs and macroscopic photographs of modified cellulose films with or without zein nanoparticles (3.6% by vol.). *Panel A: Adapted from Oymaci, P. and Altinkaya, S.A. (2016). Improvement of barrier and mechanical properties of whey protein isolate based food packaging films by incorporation of zein nanoparticles as a novel bionanocomposite.* Food Hydrocolloids, 54, *1–9, Copyright 2016, with permission from Elsevier. Panel B: Adapted from Gilbert, J., Cheng, C.J., Jones, O.G. (2018). Vapor barrier properties and mechanical behaviors of composite hydroxypropyl methylcelluose/zein nanoparticle films.* Food Biophysics, 13, *25–36, Copyright 2018, Springer Nature.*

studies, surprisingly research is limited in terms of improving film compatibility of biopolymer-based nanoparticles by surface modification.

## 4.2 Hydrogels

Despite the growing interest in preparing hydrogel nanocomposite for biomedical implementation, there has been limited interest in preparing food-relevant hydrogels with protein-based nanoparticles. Adding micrometer-scale particles to hydrogels has more immediate relevance, as such larger particles can act as fat droplet mimetics for reduced-fat product alternatives. An example of this is microparticulated whey protein used to partial replace milk fat in cheese, where the protein particles provided similar lubricant functions as the fat droplets despite failing to produce similar melting behaviors (Philipp, Ravi, & Jörg, 2013). Utilization of protein-based nanoparticles in hydrogels may pick up once there are significant crossover applications from biomedical research. For instance, the incorporation of microgel particles with pH-related swelling behaviors into the structure of the hydrogel network would be highly beneficial toward controlled release applications or even to hydrate-and-serve products in future markets.

## 4.3 Use in suspensions

There is an interest in using protein-based nanoparticles to encapsulate and deliver bioactive molecules to specific sites of action. Delivery vehicles must achieve a satisfactory encapsulation efficiency and yield for the particular vehicle-bioactive molecule pair in order to be considered viable. Further, the particle must possess efficacy with minimal toxicity at a given dosage. For the incorporation in foods, the bioactive molecules of interest would typically be nutraceuticals normally found in foods that may have positive health benefits if consumed in higher quantities or if protected from deteriorative activities after encapsulation. Further, since there is no control over the content of food materials consumed by a single person, the nanoparticle and its encapsulated bioactive molecule must be non-toxic.

Protein-based nanoparticles are often considered for the preparation of delivery vehicles due to the presence of amino acids with hydrophobic moieties that gives some affinity for bioactive molecules with reduced water solubility. Protein nanoparticles composed of casein were assembled with a diameter of ~250 nm for the delivery of cisplatin, a therapeutic drug targeting tumor cells (Zhen, Wang, Xie, Wu, & Jiang, 2013). It was found that the particles were highly effective, being able to permeate the widened cellular junctions found in tumor cells. Such permeation would be considered an undesirable risk in consumed foods, but it must be kept in mind that proteins such as casein are readily digested by gastro-intestinal enzymes.

Desolvation processes of proteins with poor water-solubility are ideal for the formation of controlled delivery vehicles, when the solvents (e.g., alcohols, weak acids) for the proteins are also good solvents for the bioactive molecules that are poorly soluble in water. After desolvation in water, the nanoparticles formed have a strong tendency to enclose the desolvated bioactive molecules, resulting in high encapsulation efficiencies. Prolamin proteins are often utilized for this approach because they are soluble in aqueous ethanol solutions, which are also good solvents for many bioactive molecules. For example, kafirin desolvated from 80% ethanol was utilized as a vehicle for curcumin with 55–85% encapsulation efficiency and a 5% loading capacity. The resulting capsules resulted in an improved curcumin stability against UV-exposure and an increased dispersibility/release during in vitro digestion assays (Xiao, Nian, & Huang, 2015).

Assembly of nanoparticles via electrostatic interactions has also been frequently used among proteins and polysaccharides to create delivery vehicles. Satisfying the increasing interest in the utilization of grain- or vegetable-based proteins, rapeseed-based cruciferin was succinylated and interacted with the positively charged chitosan to form 200–500 nm particles for the encapsulation of curcumin, with an encapsulation efficiency of ~70% (Wang et al., 2018).

## 4.4 Stabilization of emulsions or foams

Gliadin nanoparticles prepared by desolvation from 70% aqueous ethanol and coated with chitosan at neutral pH (~100–200 nm in low ionic strength media) were utilized to prepare concentrated oil-in-water emulsions (Yuan et al., 2017). These emulsions were very stable against coalescence, attributable to the formation of particles bridges between the emulsion droplets, creating a stable concentrated emulsion network, as evidenced from microscopy. Soy protein particles prepared by thermal- and ionic-induced aggregation (~100 nm average diameter) were used to prepare emulsions, and it was also found that increasing the oil phase from 20 to 60% content, as well as increasing the particle concentration up to 6% w/v, led to improved stability against creaming (Liu & Tang, 2013). Similarly, the authors attributed this increased stabilization to network formation between droplets via the adsorbed, interactive protein particles. This is a consistent issue in particle-stabilized emulsion systems, where the particle-laden interface provides excellent resistance to coalescence yet the large particles necessitate the formation of larger emulsion droplets that are, in turn, more prone to gravitational separation. Thus, emulsions stabilized by particles well above 10 nm would typically provide poor stability against creaming unless the oil phase content is very high (leading to an arrested emulsion network) or additional stabilization is supplied. The validity of the former approach can be seen from the large increase in recent studies on "high-internal-phase" emulsions stabilized by particles. Examples of such stable emulsions include those stabilized by gelatin nanoparticles (Fig. 12) (Tan et al., 2018).

Kafirin nanoparticles prepared by desolvation (~90 nm diameter with some clusters ~400 nm) were utilized to stabilize concentrated emulsions (oil phase between 60 and 80%) (Xiao, Wang, Perez Gonzalez, & Huang, 2016). As with most particle-stabilized emulsions, the significant size of the particle necessitates a larger emulsion droplet (based on contact angle-limitations of surface occupancy) so emulsions prepared at lower oil contents are susceptible to rapid gravitational separation. The authors found that the added ions enhanced surface adsorption by shielding surface charges on the protein-particles, although this also caused an increase in particle aggregation and a resultant decrease in effective particles for surface coverage (Xiao, et al., 2016).

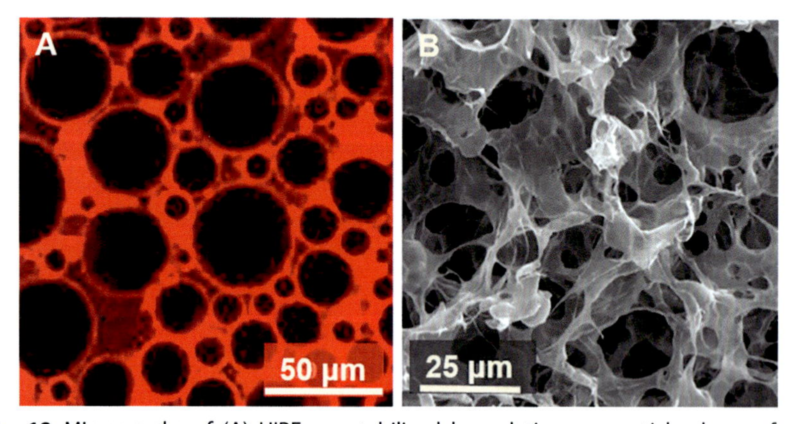


**Fig. 12** Micrographs of (A) HIPEs co-stabilized by gelatin nanoparticles by confocal microscopy, (B) freeze-dried porous scaffolds template from HIPEs of gelatin nanoparticles and unstructured gelatin (3:2 ratio) by scanning electron microscopy. *Adapted with permission from Tan, H., Tu, Z., Jia, H., Gou, X., & Ngai, T. (2018). Hierarchical porous protein scaffold templated from high internal phase emulsion costabilized by gelatin and gelatin nanoparticles.* Langmuir, 34*(16), 4820–4829, Copyright 2018, American Chemical Society.*

There is an ongoing discussion on the factors of particles that contribute to improved adsorption, potential spreading, and functionality at air-liquid and liquid-liquid interfaces. As many biopolymer-based particles, particularly those formed from protein with residual ionized charges on their surface, there is a possibility that electrostatic repulsions between particles could be inhibiting their adsorption and packing at these interfaces. Increasing the ionic strength to 500 mM was shown to increase the adsorption rate and surface pressure for soybean-based glycinin particles at oil-water interfaces (Liu & Tang, 2016).

Oil-in-water emulsions stabilized by particles may be resistant enough to allow for the emulsions to be dried effectively into powders or into scaffolds. Concentrated emulsions stabilized by chitosan-coated gliadin nanoparticles were shown to produce very fine emulsion powders after lyophilization with minimal evidence of oil separation (Yuan et al., 2017). Scaffolds are formed when high-internal-phase emulsions are prepared among numerous small droplets, followed by the gentle removal of the internal phase. Using glutaraldehyde-cross-linked gelatin nanoparticles, an 80% hexane-in-water emulsion was further cross-linked with genipin, and the hexane phase was removed by alcohol extraction to obtain a porous scaffold network.

## 5. Concluding remarks

To meet consumers' increasing demand for protein-rich ingredients, extensive investigations on the functionality and sensory properties of proteins and their assemblies have been performed in the food sector. Protein, due to its versatility as a building material in foods, is an excellent potential source of assembled nanomaterials. Many studies involving protein-based assemblies have focused on animal-sourced protein such as milk proteins. However, there has been a growing interest in exploring sustainable, non-animal-sourced proteins. In contrast to milk proteins, the application of plant-based proteins as building blocks of nanoparticles or nanoassemblies is limited due to their structural complexity and low solubility, as well as difficulties in their isolation for producing useful food ingredients. The incorporation of protein nanoparticles in food or package systems have a lot of potential, such as edible coating containing antioxidant/antimicrobial ingredients, encapsulation and controlled-release of probiotic, enzyme, or essential oils, and so on.

## References

Akiyoshi, K., Deguchi, S., Tajima, H., Nishikawa, T., & Sunamoto, J. (1997). Microscopic structure and thermoresponsiveness of a hydrogel nanoparticle by self-assembly of a hydrophobized polysaccharide. *Macromolecules*, *30*(4), 857–861.

Anal, A. K., Tobiassen, A., Flanagan, J., & Singh, H. (2008). Preparation and characterization of nanoparticles formed by chitosan–caseinate interactions. *Colloids and Surfaces B: Biointerfaces*, *64*(1), 104–110.

Anema, S. G., Lauber, S., Lee, S. K., Henle, T., & Klostermeyer, H. (2005). Rheological properties of acid gels prepared from pressure- and transglutaminase-treated skim milk. *Food Hydrocolloids*, *19*(5), 879–887.

Balazs, A. C., Emrick, T., & Russell, T. P. (2006). Nanoparticle polymer composites: Where two small worlds meet. *Science*, *314*(5802), 1107–1110.

Bateman, L., Ye, A., & Singh, H. (2010). *In vitro* digestion of ß-lactoglobulin fibrils formed by heat treatment at low pH. *Journal of Agricultural and Food Chemistry*, *58*(10), 9800–9808.

Bengoechea, C., Peinado, I., & McClements, D. J. (2011). Formation of protein nanoparticles by controlled heat treatment of lactoferrin: Factors affecting particle characteristics. *Food Hydrocolloids*, *25*(5), 1354–1360.

Bhopatkar, D., Feng, T., Chen, F., Zhang, G., Carignano, M., Park, S. H., et al. (2015). Self-assembled nanoparticle of common food constituents that carries a sparingly soluble small molecule. *Journal of Agricultural and Food Chemistry*, *63*(17), 4312–4319.

Boholm, M., & Arvidsson, R. (2016). A definition framework for the terms nanomaterial and nanoparticle. *NanoEthics*, *10*(1), 25–40.

Boire, A., Menut, P., Morel, M. H., & Sanchez, C. (2013). Phase behaviour of a wheat protein isolate. *Soft Matter*, *9*(47), 11417–11426.

Chang, P. R., Jian, R., Zheng, P., Yu, J., & Ma, X. (2010). Preparation and properties of glycerol plasticized starch (GPS)/cellulose nanoparticle (CN) composites. *Carbohydrate Polymers*, *79*(2), 301–305.

Cooper, C. L., Dubin, P. L., Kayitmazer, A. B., & Turksen, S. (2005). Polyelectrolyte–protein complexes. *Current Opinion in Colloid & Interface Science*, *10*(1), 52–78.

Dai, Q., Zhu, X., Abbas, S., Karangwa, E., Zhang, X., Xia, S., et al. (2015). Stable nanoparticles prepared by heating electrostatic complexes of whey protein isolate–dextran conjugate and chondroitin sulfate. *Journal of Agricultural and Food Chemistry*, *63*(16), 4179–4189.

Damodaran, S. (1996). Amino acids, peptides, and proteins. In O. R. Fennema (Ed.), *Food Chemistry*. (3rd ed.). New York, NY: Marcel Dekker, pp. 321.

Day, L. (2011). Wheat gluten: Production, properties, and application. In G. O. Phillips & P. A. Williams (Eds.), *Handbook of food proteins* (pp. 267–288). Philadelphia, PA: Woodhead Publishing.

de Kruif, C. G., Huppertz, T., Urban, V. S., & Petukhov, A. V. (2012). Casein micelles and their internal structure. *Advances in Colloid and Interface Science*, *171–172*, 36–52.

Delcour, J. A., Joye, I. J., Pareyt, B., Wilderjans, E., Brijs, K., & Lagrain, B. (2012). Wheat gluten functionality as a quality determinant in cereal-based food products. *Annual Review of Food Science and Technology*, *3*(3), 469–492.

Don, C., Lichtendonk, W., Plijter, J. J., & Hamer, R. J. (2003a). Glutenin macropolymer: A gel formed by glutenin particles. *Journal of Cereal Science*, *37*(1), 1–7.

Don, C., Lichtendonk, W. J., Plijter, J. J., & Hamer, R. J. (2003b). Understanding the link between GMP and dough: From glutenin particles in flour towards developed dough. *Journal of Cereal Science*, *38*(2), 157–165.

Du, J., Cho, Y.-H., Murphy, R., & Jones, O. (2017). Impact of chitosan molecular weight and attached non-interactive chains on the formation of α-lactalbumin nanogel particles. *Gels*, *3*(2), 14.

Du, J., Reuhs, B. L., & Jones, O. G. (2016). Influence of PEGylation on the ability of carboxymethyl-dextran to form complexes with α-lactalbumin. *Food Chemistry*, *196*, 853–859.

Ezpeleta, I., Irache, J. M., Stainmesse, S., Chabenat, C., Gueguen, J., Popineau, Y., et al. (1996). Gliadin nanoparticles for the controlled release of all-trans-retinoic acid. *International Journal of Pharmaceutics*, *131*(2), 191–200.

Feng, J.-L., Qi, J.-R., Yin, S.-W., Wang, J.-M., Guo, J., Weng, J.-Y., et al. (2015). Fabrication and characterization of stable soy β-conglycinin–dextran core–shell nanogels prepared via a self-assembly approach at the isoelectric point. *Journal of Agricultural and Food Chemistry*, *63*(26), 6075–6083.

Feng, T., Wang, K., Liu, F., Ye, R., Zhu, X., Zhuang, H., et al. (2017). Structural characterization and bioavailability of ternary nanoparticles consisting of amylose, α-linoleic acid and β-lactoglobulin complexed with naringin. *International Journal of Biological Macromolecules*, *99*, 365–374.

Gao, Z.-M., Wang, J.-M., Wu, N.-N., Wan, Z.-l., Guo, J., Yang, X.-Q., et al. (2013). Formation of complex interface and stability of oil-in-water (O/W) emulsion prepared by soy lipophilic protein nanoparticles. *Journal of Agricultural and Food Chemistry*, *61*(32), 7838–7847.

Gilbert, J., Cheng, C. J., & Jones, O. G. (2018). Vapor barrier properties and mechanical behaviors of composite hydroxypropyl methylcelluose/zein nanoparticle films. *Food Biophysics*, *13*, 25–36.

Girard, M., Turgeon, S. L., & Gauthier, S. F. (2002). Interbiopolymer complexing between β-lactoglobulin and low- and high-methylated pectin measured by potentiometric titration and ultrafiltration. *Food Hydrocolloids*, *16*(6), 585–591.

Guo, F., Xiong, Y. L., Qin, F., Jian, H., Huang, X., & Chen, J. (2015). Surface properties of heat-induced soluble soy protein aggregates of different molecular masses. *Journal of Food Science*, *80*(2), C279–C287.

Habibi, Y., & Dufresne, A. (2008). Highly filled bionanocomposites from functionalized polysaccharide nanocrystals. *Biomacromolecules*, ***9***(7), N50–N56.

Harada, A., & Kataoka, K. (1998). Novel polyion complex micelles entrapping enzyme molecules in the core: Preparation of narrowly-distributed micelles from lysozyme and poly(ethylene glycol)–poly(aspartic acid) block copolymer in aqueous medium. *Macromolecules*, *31*(2), 288–294.

Holappa, S., Karesoja, M., Shan, J., & Tenhu, H. (2002). Solution properties of linear and branched block copolymers consisting of acidic and PEO blocks. *Macromolecules*, *35*(12), 4733–4738.

Houga, C., Giermanska, J., Lecommandoux, S., Borsali, R., Taton, D., Gnanou, Y., et al. (2009). Micelles and polymersomes obtained by self-assembly of dextran and polystyrene based block copolymers. *Biomacromolecules*, *10*(1), 32–40.

Howorka, S. (2011). Rationally engineering natural protein assemblies in nanobiotechnology. *Current Opinion in Biotechnology*, *22*(4), 485–491.

Huppertz, T., & De Kruif, C. G. (2006). Disruption and reassociation of casein micelles under high pressure: Influence of milk serum composition and casein micelle concentration. *Journal of Agricultural and Food Chemistry*, *54*(16), 5903–5909.

Huppertz, T., & de Kruif, C. G. (2008). Structure and stability of nanogel particles prepared by internal cross-linking of casein micelles. *International Dairy Journal*, *18*(5), 556–565.

Huppertz, T., Fox, P. F., & Kelly, A. L. (2004). High pressure treatment of bovine milk: Effects on casein micelles and whey proteins. *Journal of Dairy Research*, *71*(1), 97–106.

Jiang, Y., Lu, H., Chen, F., Callari, M., Pourgholami, M., Morris, D. L., et al. (2016). PEGylated albumin-based polyion complex micelles for protein delivery. *Biomacromolecules*, *17*(3), 808–817.

Jones, O. G., & McClements, D. J. (2011). Recent progress in biopolymer nanoparticle and microparticle formation by heat-treating electrostatic protein–polysaccharide complexes. *Advances in Colloid and Interface Science, 167*(1), 49–62.

Joye, I. J., & McClements, D. J. (2013). Production of nanoparticles by anti-solvent precipitation for use in food systems. *Trends in Food Science & Technology, 34*(2), 109–123.

Kharlampieva, E., & Sukhishvili, S. A. (2006). Hydrogen-bonded layer-by-layer polymer films. *Journal of Macromolecular Science, Part C, 46*(4), 377–395.

Khatkar, B. S., Barak, S., & Mudgil, D. (2013). Effects of gliadin addition on the rheological, microscopic and thermal characteristics of wheat gluten. *International Journal of Biological Macromolecules, 53*, 38–41.

Knoop, A.-M., Knoop, E., & Wiechen, A. (1979). Sub-structure of synthetic casein micelles. *Journal of Dairy Research, 46*(2), 347–350.

Kravtchenko, T. P., Renoir, J., Parker, A., & Brigand, G. (1999). A novel method for determining the dissolution kinetics of hydrocolloid powders. *Food Hydrocolloids, 13*(3), 219–225.

Laneuville, S. I., Turgeon, S. L., & Paquin, P. (2013). Changes in the physical properties of xanthan gum induced by a dynamic high-pressure treatment. *Carbohydrate Polymers, 92*(2), 2327–2336.

Lauber, S., Henle, T., & Klostermeyer, H. (2000). Relationship between the crosslinking of caseins by transglutaminase and the gel strength of yoghurt. *European Food Research and Technology, 210*(5), 305–309.

Lawrence, M. C., Suzuki, E., Varghese, J. N., Davis, P. C., Van Donkelaar, A., Tulloch, P. A., et al. (1990). The three-dimensional structure of the seed storage protein phaseolin at 3 Å resolution. *The EMBO Journal, 9*(1), 9–15.

Levinson, Y., Israeli-Lev, G., & Livney, Y. D. (2014). Soybean β-conglycinin nanoparticles for delivery of hydrophobic nutraceuticals. *Food Biophysics, 9*(4), 332–340.

Li, J., & Yao, P. (2009). Self-assembly of ibuprofen and bovine serum albumin – dextran conjugates leading to effective loading of the drug. *Langmuir, 25*(11), 6385–6391.

Li, J., Yu, S., Yao, P., & Jiang, M. (2008). Lysozyme–dextran core–shell nanogels prepared via a green process. *Langmuir, 24*(7), 3486–3492.

Liu, F., & Tang, C.-H. (2013). Soy protein nanoparticle aggregates as pickering stabilizers for oil-in-water emulsions. *Journal of Agricultural and Food Chemistry, 61*(37), 8888–8898.

Liu, F., & Tang, C. H. (2016). Soy glycinin as food-grade Pickering stabilizers: Part II improvement of emulsification and interfacial adsorption by electrostatic screening. *Food Hydrocolloids, 60*, 620–630.

Lucey, J. A., Johnson, M. E., & Horne, D. S. (2003). Invited review: Perspectives on the basis of the rheology and texture properties of cheese. *Journal of Dairy Science, 86*(9), 2725–2743.

Luo, Y., Pan, K., & Zhong, Q. (2015). Casein/pectin nanocomplexes as potential oral delivery vehicles. *International Journal of Pharmaceutics, 486*(1), 59–68.

Lyon, L. A., & Fernandez-Nieves, A. (2012). The polymer/colloid duality of microgel suspensions. *Annual Review of Physical Chemistry, 63*(1), 25–43.

Ma, X., & Pawlik, M. (2007). Intrinsic viscosities and Huggins constants of guar gum in alkali metal chloride solutions. *Carbohydrate Polymers, 70*(1), 15–24.

Moschakis, T., & Biliaderis, C. G. (2017). Biopolymer-based coacervates: Structures, functionality and applications in food products. *Current Opinion in Colloid & Interface Science, 28*, 96–109.

Mounsey, J. S., O'Kennedy, B. T., & Kelly, P. M. (2005). Influence of transglutaminase treatment on properties of micellar casein and products made therefrom. *Lait, 85*(4–5), 405–418.

Nakai, S., & Modler, H. W. (1996). *Food proteins: Properties and characterization.* Wiley VCH, pp. 560.

Norton, I. T., & Frith, W. J. (2001). Microstructure design in mixed biopolymer composites. *Food Hydrocolloids*, *15*(4), 543–553.

Novoa-Carballal, R., Pergushov, D. V., & Müller, A. H. (2013). Interpolyelectrolyte complexes based on hyaluronic acid-block-poly (ethylene glycol) and poly-l-lysine. *Soft Matter*, *9*(16), 4297–4303.

Novoa-Carballal, R., Pfaff, A., & Müller, A. H. (2013). Interpolyelectrolyte complexes with a polysaccharide corona from dextran-block-PDMAEMA diblock copolymers. *Polymer Chemistry*, *4*(7), 2278–2285.

O'Mahoney, J. A., & Fox, P. F. (2013). Milk proteins: Introduction and historical aspects. In P. F. Fox & P. L. H. McSweeney (Eds.), *Advanced dairy chemistry: Vol. 1* (4th ed., pp. 43–86). New York: Springer.

Oria, M. P., Hamaker, B. R., & Shull, J. M. (1995). Resistance of sorghum alpha-, beta-, and gamma-kafirins to pepsin digestion. *Journal of Agricultural and Food Chemistry*, *43*(8), 2148–2153.

Oymaci, P., & Altinkaya, S. A. (2016). Improvement of barrier and mechanical properties of whey protein isolate based food packaging films by incorporation of zein nanoparticles as a novel bionanocomposite. *Food Hydrocolloids*, *54*, 1–9.

Pan, K., Luo, Y., Gan, Y., Baek, S. J., & Zhong, Q. (2014). pH-driven encapsulation of curcumin in self-assembled casein nanoparticles for enhanced dispersibility and bioactivity. *Soft Matter*, *10*(35), 6820–6830.

Pan, K., & Zhong, Q. (2016). Low energy, organic solvent-free co-assembly of zein and caseinate to prepare stable dispersions. *Food Hydrocolloids*, *52*, 600–606.

Peighambardoust, S. H., van der Goot, A. J., Hamer, R. J., & Boom, R. M. (2005). Effect of simple shear on the physical properties of glutenin macro polymer (GMP). *Journal of Cereal Science*, *42*(1), 59–68.

Pergushov, D. V., Müller, A. H., & Schacher, F. H. (2012). Micellar interpolyelectrolyte complexes. *Chemical Society Reviews*, *41*(21), 6888–6901.

Phan-Xuan, T., Durand, D., Nicolai, T., Donato, L., Schmitt, C., & Bovetto, L. (2011). On the crucial importance of the pH for the formation and self-stabilization of protein microgels and strands. *Langmuir*, *27*(24), 15092–15101.

Philipp, S., Ravi, S., & Jorg, H. (2013). The effect of adding whey protein particles as inert filler on thermophysical properties of fat-reduced semihard cheese type Gouda. *International Journal of Dairy Technology*, *66*(2), 220–230.

Pieters, B. J. G. E., van Eldijk, M. B., Nolte, R. J. M., & Mecinovic, J. (2016). Natural supramolecular protein assemblies. *Chemical Society Reviews*, *45*(1), 24–39.

Pizones Ruiz-Henestrosa, V. M., Martinez, M. J., Patino, J. M. R., & Pilosof, A. M. R. (2012). A dynamic light scattering study on the complex assembly of glycinin soy globulin in aqueous solutions. *Journal of the American Oil Chemists' Society*, *89*(7), 1183–1191.

Qi, J., Yao, P., He, F., Yu, C., & Huang, C. (2010). Nanoparticles with dextran/chitosan shell and BSA/chitosan core—Doxorubicin loading and delivery. *International Journal of Pharmaceutics*, *393*(1), 177–185.

Qin, X.-S., Luo, Z.-G., & Peng, X.-C. (2018). Fabrication and characterization of quinoa protein nanoparticle-stabilized food-grade pickering emulsions with ultrasound treatment: Interfacial adsorption/arrangement properties. *Journal of Agricultural and Food Chemistry*, *66*(17), 4449–4457.

Raviña, M., Cubillo, E., Olmeda, D., Novoa-Carballal, R., Fernandez-Megia, E., Riguera, R., et al. (2010). Hyaluronic acid/chitosan-g-poly(ethylene glycol) nanoparticles for gene therapy: An application for pDNA and siRNA delivery. *Pharmaceutical Research*, *27*(12), 2544–2555.

Riek, R. (1996). Infectious Alzheimer's disease? *Nature*, *444*, 429.

Ron, N., Zimet, P., Bargarum, J., & Livney, Y. D. (2010). Beta-lactoglobulin–polysaccharide complexes as nanovehicles for hydrophobic nutraceuticals in non-fat foods and clear beverages. *International Dairy Journal*, *20*(10), 686–693.

Schmitt, C., Bovay, C., Vuilliomenet, A.-M., Rouvet, M., Bovetto, L., Barbar, R., et al. (2009). Multiscale characterization of individualized β-lactoglobulin microgels formed upon heat treatment under narrow pH range conditions. *Langmuir*, *25*(14), 7899–7909.

Schmitt, C., Moitzi, C., Bovay, C., Rouvet, M., Bovetto, L., Donato, L., et al. (2010). Internal structure and colloidal behaviour of covalent whey protein microgels obtained by heat treatment. *Soft Matter*, *6*(19), 4876–4884.

Schmitt, C., & Turgeon, S. L. (2011). Protein/polysaccharide complexes and coacervates in food systems. *Advances in Colloid and Interface Science*, *167*(1), 63–70.

Schorsch, C., Carrie, H., & Norton, I. T. (2000). Cross-linking casein micelles by a microbial transglutaminase: Influence of cross-links in acid-induced gelation. *International Dairy Journal*, *10*(8), 529–539.

Semo, E., Kesselman, E., Danino, D., & Livney, Y. D. (2007). Casein micelle as a natural nano-capsular vehicle for nutraceuticals. *Food Hydrocolloids*, *21*(5–6), 936–942.

Sperber, B. L. H. M., Stuart, M. A. C., Schols, H. A., Voragen, A. G. J., & Norde, W. (2010). Overall charge and local charge density of pectin determines the enthalpic and entropic contributions to complexation with b-lactoglobulin. *Biomacromolecules*, *11*(12), 3578–3583.

Steffolani, M. E., Villacorta, P., Morales-Soriano, E. R., Repo-Carrasco, R., León, A. E., & Pérez, G. T. (2016). Physicochemical and functional characterization of protein isolated from different quinoa varieties (*Chenopodium quinoa* Willd.). *Cereal Chemistry*, *93*(3), 275–281.

Such, G. K., Johnston, A. P. R., & Caruso, F. (2011). Engineered hydrogen-bonded polymer multilayers: From assembly to biomedical applications. *Chemical Society Reviews*, *40*(1), 19–29.

Sukhishvili, S. A., & Granick, S. (2002). Layered, erasable polymer multilayers formed by hydrogen-bonded sequential self-assembly. *Macromolecules*, *35*(1), 301–310.

Swenson, S. E., & Curatolo, W. J. (1992). (C) Means to enhance penetration: (2) Intestinal permeability enhancement for proteins, peptides and other polar drugs: Mechanisms and potential toxicity. *Advanced Drug Delivery Reviews*, *8*(1), 39–92.

Tan, H., Tu, Z., Jia, H., Gou, X., & Ngai, T. (2018). Hierarchical porous protein scaffold templated from high internal phase emulsion costabilized by gelatin and gelatin nanoparticles. *Langmuir*, *34*(16), 4820–4829.

Taylor, J., Taylor, J. R. N., Belton, P. S., & Minnaar, A. (2009a). Formation of kafirin microparticles by phase separation from an organic acid and their characterisation. *Journal of Cereal Science*, *50*(1), 99–105.

Taylor, J., Taylor, J. R. N., Belton, P. S., & Minnaar, A. (2009b). Kafirin microparticle encapsulation of catechin and sorghum condensed tannins. *Journal of Agricultural and Food Chemistry*, *57*(16), 7523–7528.

Thomar, P., & Nicolai, T. (2015). Dissociation of native casein micelles induced by sodium caseinate. *Food Hydrocolloids*, *49*, 224–231.

Thomson, N. H., Miles, M. J., Popineau, Y., Harries, J., Shewry, P., & Tatham, A. S. (1999). Small angle X-ray scattering of wheat seed-storage proteins: α-, γ- and ω-gliadins and the high molecular weight (HMW) subunits of glutenin. *Biochimica et Biophysica Acta (BBA)—Protein Structure and Molecular Enzymology*, *1430*(2), 359–366.

Uzun, S., Ilavsky, J., & Padua, G. W. (2017). Characterization of zein assemblies by ultra-small-angle X-ray scattering. *Soft Matter*, *13*(16), 3053–3060.

van der Burgh, S., de Keizer, A., & Cohen Stuart, M. A. (2004). Complex coacervation core micelles. Colloidal stability and aggregation mechanism. *Langmuir*, *20*(4), 1073–1084.

Voets, I. K., de Keizer, A., & Cohen Stuart, M. A. (2009). Complex coacervate core micelles. *Advances in Colloid and Interface Science*, *147–148*, 300–318.
Wang, P., Jin, Z., & Xu, X. (2015). Physicochemical alterations of wheat gluten proteins upon dough formation and frozen storage—A review from gluten, glutenin and gliadin perspectives. *Trends in Food Science & Technology*, *46*(2, Pt. A), 189–198.
Wang, F., Yang, Y., Ju, X., Udenigwe, C. C., & He, R. (2018). Polyelectrolyte complex nanoparticles from chitosan and acylated rapeseed cruciferin protein for curcumin delivery. *Journal of Agricultural and Food Chemistry*, *66*(11), 2685–2693.
Wang, J.-S., Zhao, M.-M., & Zhao, Q.-Z. (2007). Correlation of glutenin macropolymer with viscoelastic properties during dough mixing. *Journal of Cereal Science*, *45*(2), 128–133.
Weinbreck, F., de Vries, R., Schrooyen, P., & de Kruif, C. G. (2003). Complex coacervation of whey proteins and gum Arabic. *Biomacromolecules*, *4*(2), 293–303.
Wieser, H., & Kieffer, R. (2001). Correlations of the amount of gluten protein types to the technological properties of wheat flours determined on a micro-scale. *Journal of Cereal Science*, *34*(1), 19–27.
Wu, N.-N., Zhang, J.-B., Tan, B., He, X.-T., Yang, J., Guo, J., et al. (2014). Characterization and interfacial behavior of nanoparticles prepared from amphiphilic hydrolysates of β-conglycinin–dextran conjugates. *Journal of Agricultural and Food Chemistry*, *62*(52), 12678–12685.
Xiao, J., Li, Y., Li, J., Gonzalez, A. P., Xia, Q., & Huang, Q. (2015). Structure, morphology, and assembly behavior of kafirin. *Journal of Agricultural and Food Chemistry*, *63*(1), 216–224.
Xiao, J., Nian, S., & Huang, Q. (2015). Assembly of kafirin/carboxymethyl chitosan nanoparticles to enhance the cellular uptake of curcumin. *Food Hydrocolloids*, *51*, 166–175.
Xiao, J., Wang, X. A., Perez Gonzalez, A. J., & Huang, Q. (2016). Kafirin nanoparticles-stabilized Pickering emulsions: Microstructure and rheological behavior. *Food Hydrocolloids*, *54*, 30–39.
Xu, A. Y., Melton, L. D., Jameson, G. B., Williams, M. A. K., & McGillivray, D. J. (2015). Structural mechanism of complex assemblies: Characterisation of beta-lactoglobulin and pectin interactions. *Soft Matter*, *11*(34), 6790–6799.
Ye, A., Flanagan, J., & Singh, H. (2006). Formation of stable nanoparticles via electrostatic complexation between sodium caseinate and gum arabic. *Biopolymers*, *82*(2), 121–133.
Yuan, D. B., Hu, Y. Q., Zeng, T., Yin, S. W., Tang, C. H., & Yang, X. Q. (2017). Development of stable pickering emulsions/oil powders and pickering HIPEs stabilized by gliadin/chitosan complex particles. *Food & Function*, *8*(6), 2220–2230.
Zhang, G., & Hamaker, B. R. (2003). A three component interaction among starch, protein, and free fatty acids revealed by pasting profiles. *Journal of Agricultural and Food Chemistry*, *51*(9), 2797–2800.
Zhang, G., Maladen, M., Campanella, O. H., & Hamaker, B. R. (2010). Free fatty acids electronically bridge the self-assembly of a three-component nanocomplex consisting of amylose, protein, and free fatty acids. *Journal of Agricultural and Food Chemistry*, *58*(16), 9164–9170.
Zhang, X., Qi, J.-R., Li, K.-K., Yin, S.-W., Wang, J.-M., Zhu, J.-H., et al. (2012). Characterization of soy β-conglycinin–dextran conjugate prepared by Maillard reaction in crowded liquid system. *Food Research International*, *49*(2), 648–654.
Zhen, X., Wang, X., Xie, C., Wu, W., & Jiang, X. (2013). Cellular uptake, antitumor response and tumor penetration of cisplatin-loaded milk protein nanoparticles. *Biomaterials*, *34*(4), 1372–1382.
Zhu, K., Ye, T., Liu, J., Peng, Z., Xu, S., Lei, J., et al. (2013). Nanogels fabricated by lysozyme and sodium carboxymethyl cellulose for 5-fluorouracil controlled release. *International Journal of Pharmaceutics*, *441*(1), 721–727.

Zhu, X.-F., Zheng, J., Liu, F., Qiu, C.-Y., Lin, W.-F., & Tang, C.-H. (2017). The influence of ionic strength on the characteristics of heat-induced soy protein aggregate nanoparticles and the freeze-thaw stability of the resultant pickering emulsions. *Food & Function*, *8*(8), 2974–2981.

Zhou, J. J., Wang, S. Y., & Gunasekaran, S. (2009). Preparation and characterization of whey protein film incorporated with $TiO_2$ nanoparticles. *Journal of Food Science*, *74*, N50–N56.

# Nano-scale polysaccharide materials in food and agricultural applications

**Elessandra da Rosa Zavareze, Dianini Hüttner Kringel, Alvaro Renato Guerra Dias***
Department of Agroindustrial Science and Technology, Federal University of Pelotas, Pelotas, RS, Brazil
*Corresponding author: e-mail address: alvaro.guerradias@gmail.com

## Contents



*Advances in Food and Nutrition Research*, Volume 88
ISSN 1043-4526
https://doi.org/10.1016/bs.afnr.2019.02.013

## Abstract

Potential applications of nanotechnology in food and agriculture include: (1) the encapsulation of functional compounds; (2) production of reinforcing materials; (3) delivery of nutraceuticals in foods; (4) food safety, for detection and control of chemical and microbiological risks; (5) active and intelligent food packaging; (6) incorporation of protective substances of seeds; (7) addition of nutrients in the soil; (8) use of controlled release pesticides. Natural polysaccharides and their derivatives are widely used in the production of nano-scale materials. This chapter examines, the use of polysaccharides, such as starch, cellulose, lignin, pectin, gums, and cyclodextrins for the production of nano-scale materials, including nanocrystals, nanoemulsions, nanocomplexes, nanocapsules, and nanofibers.

## 1. Introduction

Nanotechnology is a multidisciplinary field involving the production, characterization, and manipulation of nanometer-sized substances, by utilizing physics, chemistry, biology, and engineering concepts. In general, nanomaterials are those that have one or more dimensions in the nanoscale - typically referring to length scale of 1 and 100 nm, where properties of a material are likely to change from its bulk counterparts (FAO/WHO, 2010). Depending on the magnitude of the dimensions and mode of assembly, nanomaterials exist in different forms, such as nanoparticles, nanotubes, nanocapsules, nanospheres, nanofibers, nanofilms, nanolayers, nanocomposites, nanocoatings, and nanosensors (Handford et al., 2014). The nanoscale dimension and high surface area of these materials confers them with new and specific properties that can be exploited in various applications (Carrillo-Inungaray, Trejo-Ramirez, Reyes-Munguia, & Carranza-Alvarez, 2018).

In the context of food and nutrition, nanotechnological advancements have provided many opportunities for innovation throughout the entire food chain (Rossi et al., 2014). For example, with regard to food safety, nanotechnology can be used to detect and control chemical and microbiological risks, to deliver antimicrobials in active packaging for shelf life extension, and in intelligent packaging systems where sensors are designed to provide information about the quality or freshness of sealed food products (Dasgupta et al., 2015). Nanotechnology can be used in agricultural applications, for the incorporation of protective substances of seeds, controlled release of nutrients in soils according to the crop's phenological needs, and for improving the efficiency of pesticides. Other promising areas include the use of nanoencapsulated animal feed with enriched supplements, use of smart sensors for the diagnosis of animal diseases, and detection of pathogens in water.

Polysaccharides are composed of monosaccharide units bound together by glycosidic linkages. Polysaccharides, and their derivatives, are versatile polymeric materials for the production of nanosystems. Polysaccharides can be obtained from natural sources and are often by-products of the food industry, mostly through low-cost processing. The main sources of polysaccharides are plants (starch, cellulose, lignin, pectin, β-glucan, and gum guar), animals (chitin and chitosan), algae (alginate, agar, and carrageenan), and microorganisms (dextran, pullulan, xanthan gum, and bacterial cellulose). The specific and diverse properties of these macromolecules are related to their different molecular weights, chemical composition, and presence of reactive groups that allow numerous chemical modifications to meet specific end-use requirements (Thielemans, Belgacem, & Dufresne, 2006). In general, polysaccharides exhibit many desirable properties, such as good stability, semi-crystallinity, non-toxicity, biodegradability, and biocompatibility. These characteristics are important when exploiting polysaccharides as nano-scale materials for food and agricultural applications.

Physical and chemical properties of polysaccharide-based nano-scale materials are dependent on their chemical composition and functional groups, which affect the susceptibility of polysaccharide nanomaterials to chemical and enzymatic reactions (Matalanis, Jones, & McClements, 2011). Electrostatic interactions, as affected by the intrinsic charges on polysaccharide chains, are essential during the formation of nanosystems. Polysaccharides can be classified as neutral (e.g., amylose, amylopectin, cellulose, and gum guar), anionic (e.g., alginates, carrageenans, gellan, gum Arabic, and xanthan gum), and cationic (e.g., chitosan) (Thakur & Thakur, 2016). This chapter overviews the use of vegetal polysaccharides (starch, cellulose, lignin, pectin, gums, and cyclodextrins) for the production of nanoscale materials, including nanocrystals, nanoemulsions, nanocomplexes, nanocapsules, and nanofibers, and their applications in food production and agriculture.

## 2. Native and modified starches

Starch is a naturally-occurring, abundant, renewable, and biodegradable polymer. It is found in various parts of plants, such as roots, stems, fruits, and seeds, as a source of storage energy. Unmodified starch extracted from plants is called "native starch." When it undergoes one or more physical, chemical, or enzymatic modifications, it is "modified starch."

Starch has multi-scale structures consisting of granules (2–100 μm), comprised of growth rings (120–500 nm) that are composed of blocks (20–50 nm). The latter are alternating bands of amorphous and crystalline lamellae (9 nm) that contain amylose and amylopectin chains (0.1–1 nm) (Le Corre, Bras, & Dufresne, 2010). The shape, particle size, and thickness of these layers depend on the botanical origin. Within the crystalline clusters, amylopectin is organized in double helices stacking alternating crystalline and amorphous lamellae, with a regular repeated distance of 9–10 nm. These are embedded in rounded asymmetric structures known as blocklets that have a diameter ranging from 50 to 100 nm.

Native and modified starches are commonly used in the food industry and recently, nano-scale starches are being used as novel materials that present unique physical, chemical, rheological, and functional properties owing to their small size and large surface area. The most relevant examples are nanocrystals obtained by acid hydrolysis of native starches, nanocapsules obtained from nanoemulsions of modified starches, nanofibers obtained by electrospinning, and nanocomposites derived from starches and nano-clay. In this section, selected examples of starch-based nanocrystals, nanocapsules, nanoemulsions, and nanofibers are discussed.

## 2.1 Starch nanocrystals

Several terminologies are used to describe starch nanocrystals: starch crystallite, starch nanocrystal, microcrystalline starch, and hydrolyzed starch—all refer to the crystalline portion of starch obtained by hydrolysis or physical disintegration (Le Corre et al., 2011). On the other hand, starch nanoparticles are obtained from the amorphous regions of the granules. The large surface area and high surface activity provided high grafting per unit mass of particles. Due to these characteristics, starch nanocrystals have been used as reinforcement fillers in nanocomposite systems (Angellier, Molina-Boisseau, Belgacem, & Dufresne, 2005). However, according to Xie, Pollet, Halley, and Averous (2013), phase transitions of the nanocrystals from the ordered granular structure to a disordered state in water-plasticized nanocomposite materials may present limitations.

Starch nanocrystals and nanoparticles can be prepared using different methods (i.e., acid or enzymatic hydrolysis, physical treatments, combined hydrolysis and ultrasonication, complex formation, emulsion-crosslinking, and so on) (Kim, Park, & Lim, 2015). Select examples of the main sources of starches, production conditions, and nanocrystal characteristics are presented in Table 1. Among starch nanocrystal production methods, acid hydrolysis is relatively common since the amorphous regions of starch can be hydrolyzed and removed, leaving only rigid crystalline regions intact (Le Corre et al., 2010). Unlike the rod-shaped cellulose crystals, starch nanocrystals have a platelet-like morphology (Le Corre et al., 2011). The hierarchical organization of starch and its semi-crystalline structure influence the preparation of nanocrystals using controlled acid hydrolysis. For example, high amylose content of starches can hinder acid hydrolysis and consequently lead to reduced nanocrystal yield (Kim et al., 2015). Other authors reported that nanocrystals presented a lower size and higher crystallinity with decreased amylose content (Le Corre, Bras, & Dufresne, 2011). Duan, Sun, Wang, and Yang (2011) studied hydrolysis of waxy maize starch with sulfuric acid, in an aqueous solution, to produce starch nanocrystals and reported that both amorphous and crystalline regions of starch granules were rapidly hydrolyzed during initial stage of hydrolysis. These authors explained the presence of weakly organized crystalline regions in the starch granules were hydrolyzed rapidly in the early stages, while the remaining well-organized nanocrystals showed greater resistance to acid in later stages. They also reported after 10 days of hydrolysis, only 10 wt% insoluble starch nanocrystals were obtained, whereas a 6-day hydrolysis yielded 24 wt% insoluble starch nanocrystals with a size of 40–80 nm and relative crystallinity of 63%. Thus, the increase in the hydrolysis time decreased starch nanocrystal yield and increased crystallinity. Angellier, Choisnard, Molina-Boisseau, Ozil, and Dufresne (2004) also studied waxy maize starch nanocrystals finding nanocrystals 6–8 nm thick in size, 40–60 nm long and 15–30 nm wide.

Hao et al. (2018) produced waxy potato starch nanocrystals with and without glucoamylase enzymatic pretreatment, followed by acid hydrolysis. Enzymatic pretreatment decreases the duration of acid hydrolysis. Kim and Lim (2009) prepared crystalline starch nanoparticles by complex formation with *n*-butanol and successive enzymatic hydrolysis (amylolysis) and complex formation mainly involves amylose. The starch nanoparticles produced presented diameters of 10–20 nm. The yield of nanoparticles was low, owing to the need to purify the amorphous matrices in the amylose-*n*-butanol complex and the significant loss occurring because of hydrolysis (85–90%) of the initially complexed starch. An important factor to be considered is the aggregation of nanocrystals after starch hydrolysis. Generally, after

**Table 1** Starch sources, preparation conditions, and size of starch nanoparticles.

| Starch sources | Preparation conditions | Average size (nm) | References |
|---|---|---|---|
| Potato | Acid hydrolysis (2.2 M HCl, 15 days at 35 °C) | 150 | Dufresne, Cavaille, and Helbert (1996) |
| Waxy potato | Glucoamylase hydrolysis followed by acid hydrolysis (3.16 M $H_2SO_4$ for 5 days at 40 °C) and high-speed homogenization for 5 min at 13,000 rpm | 20–80 | Hao, Chen, Li, and Gao (2018) |
| Waxy maize | Acid hydrolysis (3.16 M $H_2SO_4$, 2 days at 40 °C) and ultrasonication | 70–120 | Kim, Han, Kweon, Park, and Lim (2013) |
| Waxy rice | Enzymatic hydrolysis (α-amylase, 3 h at 37 °C) | 500 | Kim, Park, and Lim (2008) |
| Waxy rice | Enzymatic hydrolysis (α-amylase, 3 h at 37 °C) and ultrasonication | >500 | Kim et al. (2008) |
| Normal maize | Nanoprecipitation of starch paste solution with ethanol | 50–300 | Ma, Chang, and Yu (2008) |
| Normal maize | Extrusion of a mixture of starch, water, and glycerol (55–110 °C) | 160 | Song, Thioc, and Deng (2011) |
| High-amylose maize | Complex formation of starch and n-butanol, and enzymatic hydrolysis | 10–20 | Kim and Lim (2009) |
| High-amylose maize | Physical treatment (microfluidizer under pressure of 207 MPa) | 10–20 | Liu, Wu, Chen, and Chang (2009) |
| Amadumbe | Acid hydrolysis (3.16 M $H_2SO_4$, 5 days at 40 °C) | 50–100 | Mukurumbira, Mariano, Dufresne, Mellem, and Amonsou (2017) |
| Pea | Acid hydrolysis (3.16 M $H_2SO_4$ for 5 days at 40 °C) | 60–150 | Zheng, Ai, Chang, Huang, and Dufresne (2009) |
| Proso millet | Enzymatic hydrolysis (pullulanase) followed by retrogradation | 20–100 | Sun, Li, Dai, Ji, and Xiong (2014) |
| Cassava | Gamma radiation of 20 kGy | 20–30 | Lamanna, Morales, García, and Goyanes (2013) |
| Soluvel starch | High-pressure homogenization followed by cross-linking with sodium trimetaphosphate | 50–250 | Shi, Li, Wang, Li, and Adhikari (2011) |

obtaining nanocrystals, they must undergo mechanical treatment to disintegrate formed aggregates. According to Xu et al. (2010), the aggregation of starch nanocrystals has been attributed to hydrogen bonding that results from hydroxyl group interactions on the surface of nanocrystals.

Although acid hydrolysis is straightforward, the process presents some practical difficulties due to low yield, long treatment period, and use of concentrated acid. Methodologies based on physical disintegration are overcome these challenges, including high pressure homogenization, ultrasonication, gamma irradiation, and extrusion (Bel Haaj, Magnin, Pétrier, & Boufi, 2013; Song et al., 2011). These physical treatments disintegrate starch granules by releasing nanoblocks of different sizes, the extent depends on the starch source, structural organization, granule size, pores on the surface, amylose content, and the amount of lipid-complexed amylose chains in the starches (Le Corre et al., 2011). Song et al. (2011) produced starch nanocrystals using high-pressure homogenization, and reported a large particle size distribution, ranging from 5 to 200 nm. Lamanna et al. (2013) used gamma irradiation to generate active free radicals and promote starch hydrolysis, which forms nanoparticles ranging from 20 to 30 nm. High-intensity ultrasound, another physical method, forms ~40 nm nanoparticles (Bel Haaj et al., 2013). However, the starch nanoparticles obtained using physical disintegration did not have well-defined shapes and cause significant decreases in crystallinity.

In addition to these treatments, starch nanocrystals have been chemically modified, involving reaction with small molecules (Angellier et al., 2005), grafting onto polymer chains with coupling agents (Xu et al., 2010) or polymerization of a monomer (Thielemans et al., 2006), covalent cross-linking (Zhou, Tong, Su, & Ren, 2016) and so on. Modifications enhanced the hydrophobicity and dispersibility, as compared to those of unmodified starch nanocrystals, and decreased the nanocrystal size.

## 2.2 Starch nanocapsules/nanoemulsions

Nanoencapsulation involves the incorporation, absorption, or dispersion of bioactive compounds into small carrier capsules with diameters in the nanometer scale. These structures are intended to protect the bioactives (e.g., vitamins, polyphenols, flavors, lipids, enzymes, and carotenoids) against deleterious environmental factors (e.g., oxygen, light, and free radicals), while capable of delivering these compounds when triggered by external stimuli (e.g., pH, enzyme, relative humidity, and temperature). The main encapsulation methods include: spray drying, co-precipitation, antisolvent precipitation, extrusion, physical trapping, amylose inclusion complexation, electrospinning, and nanoemulsification. Native and modified starches from various botanical sources have been used in nanoencapsulation of a wide range of compounds with good encapsulation efficiency. Chemical characteristics of the starches can be manipulated through chemical modification, such as cross-linking, oxidation, substitution (e.g., octenyl succinic anhydride (OSA), phosphorylation, acetylation), and enzymatic hydrolysis (Zhu, 2017).

Nanoemulsions are kinetically stable and do not phase-separate after prolonged storage. These systems are excellent hydrophobic bioactive carriers. Nanoemulsions are formed by processes such as high-pressure homogenization, microfluidization, ultrasonic homogenization, and high-speed devices (Jin et al., 2016), producing droplets of sizes between 90 and 290 nm. Native starches are water insoluble and have limited emulsifying properties. However, modified starches can act as effective emulsifiers for nanoemulsions. For instance, octenyl succinic anhydride-modified starch (OSA-starch) has hydrophobic groups, conferring it with an amphiphilic character useful for the encapsulation of hydrophilic and hydrophobic food ingredients and agri-chemicals. The modified starch molecules with hydrophobic and hydrophilic groups can disperse in the aqueous phase and be adsorbed at the oil-water interface, which enables the stabilization of oil droplets in aqueous environments (Jin, Li, & Nik, 2018). The OSA-starch contains negatively charged carboxylic acid, with a degree of substitution of 0.01–0.03. FDA allows a maximum of 3% OSA treatment, on a dry-weight basis, for starch to be used as a food additive. According to Jin et al. (2018), a combination of hydrophobic treatment and hydrolysis can improve the water-solubility, thereby enhancing the emulsification properties of OSA-starch. Increasing OSA-starch concentration in the emulsion reduces the droplet diameter and increases emulsion stability (Dokić, Krstonošić, & Nikolić, 2012).

## 2.3 Starch nanofibers

Starch can be converted into nanofibers by electrospinning. It is a versatile technology for the preparation of nanofibers with controllable morphologies. Briefly, the technique is based on the application of a high electric field to generate nanofibers from a charged polymer solution or melt. Detail discussions of this process are presented elsewhere in this volume (see chapter "Electrospinning and electrospraying technologies for food applications" by Lim et al.). The production of electrospun nanofibers relies on the ability of the polymer chains to entangle, which is important to prevent the breaking up of polymer jet into droplets during the electrospinning process. In high-amylose starches, the polysaccharide chains entangled more readily than low-amylose counterparts that have a relatively higher amylopectin content (Kong & Ziegler, 2012). In addition to polymer chain entanglement, establishing electrostatic charges on the polymer jet is essential to induce Coulombic repulsion and electric field forces for stretching the polymer into ultrafine fibers. To facilitate electrospinning, other polymers may be added as a spinning aid, such as poly(ethylene oxide) and poly(vinyl alcohol).

There are few studies regarding the production of starch nanofibers. Kong and Ziegler (2012) evaluated the rheological properties of the high amylose starch solutions (Gelose 80) prepared in dimethylsulfoxide at different starch concentrations. Using an electro-wet-spinning process, starch microfibers were obtained. These authors also evaluated the ability of other starches (i.e., Hylon VII, Hylon V, Melojel, Amioca and mung bean) for producing fibers using. They reported that microfibers could be obtained using Hylon VII starch, but not Hylon V and mung bean starches (15–35% of amylose, respectively). However, Lancuški, Vasilyev, Putaux, and Zussman (2015) were able to produce high amylose starch nanofibers (Hylon VII) using 17% (w/w) aqueous formic acid dispersions. Recently, our group electrospun ultrafine fibers of pure soluble potato starch with a normal amylose content, using aqueous formic acid (75%, w/v) as a solvent (Fonseca et al., 2018). These ultrafine fibers presented a cylindrical and homogeneous morphology of some beads and average diameters in the range of 128–143 nm.

## 2.4 Applications of nano-scale starch in food and agriculture

Starches have been explored in agriculture for a number of applications: (1) controlled release of fertilizer; (2) reduction of fertilizer loss and thus reducing environmental pollution; (3) agricultural superabsorbent; (4) improvement of fertilizer utilization efficiency; and (5) remediation of saline soil. For controlled release of fertilizer, since native starches are hydrophilic, their retention of the fertilizer is limited when exposed to water from rainfall or irrigation (Li & Luo, 2016). Modification with hydrophobic functional groups is needed.

Other nanoparticles such as nanoclay, cellulose nanoparticles, or carbon nanotubes are incorporated into starch matrices to develop nanocomposites with mechanical and barrier properties greater than those of micro-scale composites. Other nano-scale starch applications include nanofibers and nanoemulsions that are used to encapsulate bioactive compounds. Modified starch nanoemulsions containing essential oils are applied to food preservation, such as pathogen control in food products. Other main applications of starch nanoparticles include stabilization of oil-in-water emulsions against coalescence, fat replacement to improve taste and mouthfeel owing to the small particle size, use as a packaging component, use as an absorbent to remove organic pollutants from water, and so on (Kim et al., 2015).

## 3. Cellulose and cellulose derivatives

Cellulose is one of the most abundant and renewable polysaccharides. Its chemical structure consists of a straight carbohydrate polymer chain, composed of (1,4) linked β-D-glucose units with a large number of hydroxyl (OH) groups, which are capable of forming strong hydrogen bonds essential in contributing to the multiscale structures of cellulose, as well as providing opportunities for chemical modifications (Hedjazi & Razavi, 2018; Plackett & Iotti, 2013). Cellulose and its derivatives have attracted considerable interest in encapsulation and different fields (Mendes, Stephansen, & Chronakis, 2017).

"Nanocellulose" or "cellulose nanomaterial" refers to a range of materials derived from cellulose with at least one dimension in the nanometer range. The isolation of crystalline cellulosic regions was first reported in 1947, by acid hydrolysis (Dufresne, 2017). Selected nanometer-sized cellulose and their derivatives are discussed in this section.

### 3.1 Cellulose nanocrystals

Cellulose nanocrystals can be classified into two categories, namely cellulose nanocrystals (CNCs) and cellulose nanofibrils (CNFs) (Tang, Sisler, Grishkewich, & Tam, 2017). CNCs are obtained by acid treatment, while CNFs are produced mainly by mechanical disintegration (Kargarzadeh et al., 2017). The main differences between CNCs and CNFs lie in their dimension and crystallinity. CNFs consist of mixtures of amorphous and crystalline cellulose chains with lengths up to several microns. On the other hand, CNCs are highly crystalline, with lengths that are typically $<500$ nm. The properties of these materials are dependent on the preparation conditions and cellulose sources. More information on cellulose nanocrystals can be found elsewhere in this volume.

## 3.2 Cellulose nanocapsules

In its native form, cellulose has limited solubility in water and other common solvents, which limits its use for encapsulation purposes. To enhance its functionality for encapsulation, cellulose can be modified using physical, chemical, or biochemical processes (Đorđević et al., 2016). Cellulose derivatives that are useful as encapsulating agents and for the formation of nanoemulsion include: (1) carboxymethyl cellulose, a water-soluble derivative of cellulose to enhance the bioavailability and stability of active compounds, such as curcumin; (2) methylcellulose with efficient oxygen and lipid barrier properties; (3) cellulose ethers that are water-soluble typically used to mask colors and undesirable tastes; (4) hydroxypropyl cellulose, which acts as a barrier to oil and fat; (5) cellulose acetate for improving the encapsulation efficiency and targeted release of bioactive compounds; ethyl cellulose, a good substance for encapsulating water-soluble vitamins which reduces the water permeability of the capsules (Abbas et al., 2015; Katouzian & Jafari, 2016; Shishir, Xie, Sun, Zheng, & Chen, 2018).

Research on introducing antimicrobial functionalities into cellulose and their derivatives has received considerable interest over the last few years (Lam, Wong, Boyer, & Qiao, 2018). Fernández, Picouet, and Lloret (2010) studied the antimicrobial activity of cellulose-silver nanoparticles (5–35 nm), during the storage of minimally processed melons and reported that nanoparticles retarded the senescence of cut melons, decreased °Brix values, and increased product shelf-life. Liakos et al. (2016) reported that cellulose acetate-lemongrass essential oil nanocapsules showed promising effects against bacterial growth, and had diameters ranging from 95 to 185 nm.

## 3.3 Cellulose nanofibers

Nanofibers have unique properties different from larger fibers comprised of the same material (Maftoonazad & Ramaswamy, 2018). Currently, nanofibers are mainly produced by melt spinning, solution blowing, and electrospinning. The main application of cellulose nanofibers is for composite material production owing to the high stiffness of cellulose crystals that can be used as a filler material for reinforcement. The first study using cellulosic nanofibers as a reinforcing material in nanocomposites was reported two decades ago. Since then, many studies have focused on cellulose nanofibers. Besides exhibiting excellent mechanical characteristics for the production of cellulosic fibers at the nano-scale, these materials are considered sustainable and environmental friendly (Khalil et al., 2014).

The formation of cellulose nanofibers by electrospinning is mostly dependent on polymer molecular weight and the solvent used for the preparation of the spin dope solution. Typical solvents for the electrospinning of cellulose include *N*-methylmorpholine, *N*-oxide/water, lithium chloride/dimethyl acetamide, ionic liquids, and ethylene diamine/salt (Mendes et al., 2017). The residual solvent in the cellulose nanofiber after electrospinning can be removed by the use of coagulant baths. However, due to toxicity of these solvents, most studies are focused on the use of cellulose derivatives (e.g., cellulose acetate, cellulose triacetate, ethyl cellulose, methylcellulose, and hydroxypropyl cellulose) that are soluble in food-grade solvents (Konwarh, Karak, & Misra, 2013; Rezaei, Nasirpour, & Fathi, 2015; Wen, Zong, Linhardt, Feng, & Wu, 2017). Production methods and applications of selected cellulose nanofibers are summarized in Table 2.

## 3.4 Applications of nano-scale cellulose in food and agriculture

Nano-scale cellulose materials, dispersed in compatible polymer matrices, can result in nanocomposites of enhanced mechanical/barrier properties (Zafar et al., 2016). Due to their nontoxic nature, nano-scale cellulose has great potential in packaging materials to protect the packaged foods (Maftoonazad & Ramaswamy, 2018). Additionally, studies have reported antimicrobial particles grafted onto nanocellulose materials, used for food packaging, increased the shelf life of different food products (Kargarzadeh et al., 2017). For example, cellulose-based nanocomposites containing silver nanoparticles with improved bactericidal properties can be promising for active food packaging and agriculture applications (Moura, Mattoso, & Zucolotto, 2012).

In edible applications, Jafari, Bahrami, Dehnad, and Shahidi (2018) studied the effects of nanocellulose coating on physicochemical properties and storage quality of saffron. The combination of maltodextrin and nanocellulose showed higher crocin (color ingredient that provides red color of saffron) retention during storage, as compared to maltodextrin alone. In another study, Azeredo et al. (2009) incorporated commercially available cellulose nanofibers (Novacel® PH-101) to reinforce mango puree based edible films. In their study, the addition of 10% nanocellulose in the edible film formulation resulted in significant increases in tensile strength and modulus, while a modest improvement in water vapor barrier properties were observed. Nanocellulose has been used in agricultural settings as protective coatings for seeds, plants, and foodstuffs to improve long-term

**Table 2** Methods, average size, and some applications of cellulose nanofibers.

| Aim | Method to produce nanofibers | Average size (nm)[a] | Properties or applications | References |
|---|---|---|---|---|
| To obtain cellulose nanofibers from waste brown algae and investigate their potential as milk thickeners | Oxidation by NaClO in a TEMPO-mediated system | L: 11000<br>w: 4 | Cellulose nanofibers exhibited good thickening behavior in milk as a result of the cellulose absorption of casein micelles through hydrogen bonds, to form a weak gel-like structure | Gao et al. (2018) |
| To carry out deacetylation of cellulose acetate (CA) nanofibers, in order to produce cellulose nanofibers using ultrasonic energy | Electrospinning | CA nanofiber<br>d: 260–390<br>US-deacetylated cellulose nanofibers<br>d: 340–480 | The use of the rapid method along with ultrasonic energy reduced deacetylation time from 30h to merely 1h | Ahmed et al. (2017) |
| To obtain cellulose nanofibers from two different industrial bio-residues: wastes from the juice industry (carrot) and the beer brewing process (BSG) | Mild pretreatment and ultrafine grinding | Carrot nanofibers w: 12<br>BSG nanofibers w: 30 | The yield from the carrot residue was 10% higher than that of BSG | Berglund, Noël, Aitomäki, Öman, and Oksman (2016) |
| To obtain cellulose nanofibers from banana peels using a combination of chemical and mechanical treatments, using a high-pressure homogenizer | Combination of chemical and mechanical treatments with high-pressure homogenizer | d: 335–455 | The most suitable mechanical treatment condition for the preparation of cellulose nanofibers was five passages through the high-pressure homogenizer | Pelissari, Andrade-Mahecha, Sobral, and Menegalli (2017) |

*Continued*

**Table 2** Methods, average size, and some applications of cellulose nanofibers.—cont'd

| Aim | Method to produce nanofibers | Average size (nm)[a] | Properties or applications | References |
|---|---|---|---|---|
| To obtain starch-based composite films reinforced by cellulose nanofibers | Chemo-mechanical process | d: 72 | Cellulose nanofibers increased the tensile properties of the nanocomposite, and decreased the water vapor transmission rate to 86% and oxygen permeability to 94% | Fazeli, Keley, and Biazar (2018) |
| To obtain tree-like cellulose nanofiber membranes via the electrospinning method | Electrospinning | d: 80–200 | Continuous uniform tree-like cellulose nanofiber membranes were obtained via the deacetylation of cellulose acetate nanofibers, and they exhibited hydrophilicity, good solvent resistance, and good mechanical properties | Zhang et al. (2017) |
| To obtain cellulose from *Agave tequilana* Weber var. *azul* and elaboration of cellulose nanofibers by electrospinning | Electrospinning | d: 54.57–171 | Cellulose obtained from agave bagasse produced nanofibers that can be used in delivery systems, such as for the encapsulation of bioactive compounds and reinforcing materials | Robles-García et al. (2018) |
| To achieve the annealing and saponification of electrospun cellulose-acetate (CA) nanofibers | Electrospinning | d: 430 | The increase in the annealing time of CA nanofibers at 50°C from 0 to 12h increased the crystallinity from 37% to 41% | Inukai, Kurokawa, and Hotta (2018) |

[a]L: length; w: width; d: diameter.

stability (Iavicoli, Leso, Beezhold, & Shvedova, 2017). Jung, Deng, Simonsen, Bastías, and Zhao (2016) investigated hydrophobic coatings (Innofresh™) containing cellulose nanofiber for reducing cherry rain-cracking. The addition of plasticizer (glycerol) and surfactants (Tween 80 and Span 80) is needed to provide optimal wettability, elasticity, and water resistance. Field application of cherry fruits with these nanocomposite coatings are effective in preventing cherry cracking without affecting fruit growth and quality.

In other applications, cellulose nanomaterials have been used for water remediation due to their high surface area-to-volume ratio and biocompatibility. Nanostructured systems stabilized with carboxymethyl cellulose have been studied for the decontamination of heavy metals, dyes, or halogenated compounds in agricultural soils (Li, Chen, Zhuang, & Chen, 2016). Quiñones, Mardare, Hassel, and Brüggemann (2017) developed spherical cellulose nanoparticles with sizes ranging from 50 to 300 nm for agrochemical applications. These authors reported a sustained steroid release and good stimulatory agrochemical activity in a radish cotyledon assay.

## 4. Lignin

Lignin is a natural biopolymer, consisting of lignocellulose, cellulose, and hemicellulose. Lignin is highly branched and amorphous; its composition depends on plant source. Lignin is considered a unique biopolymer owing to its characteristics such as the lack of a defined primary structure, heterogeneity, and aromatic functionality (Norgren & Edlund, 2014). The chemical structures of lignin result from the oxidative coupling of three major $C_6$-$C_3$ (phenylpropane) units, namely syringyl alcohol, guaiacyl alcohol, and ρ-hydroxyl alcohol (Sathawong, Sridach, & Techato, 2018). Lignin can be obtained from renewable sources useful for many applications due to high recovery, low cost, low density and high availability (Rangan, Manchiganti, Thilaividankan, Kestur, & Menon, 2017).

Lignin is considered the second most abundant biopolymer in nature. Due to its renewable origin, it has attracted attention as a natural source of aromatic compounds and antioxidant agents. Additionally, lignin is being studied for its other bioactivities, including anti-inflammatory, antibacterial, or anti-carcinogenic activities, and for its properties such as low density, good stiffness, high carbon content, and protection from ultraviolet (UV) radiation (Mishra, Ha, Verma, & Tiwari, 2018).

Traditionally, lignin is extracted from wood, although lignins derived from non-wood biomasses are increasing, including wheat, sugarcane, straw, bamboo, alfalfa, kenaf, and flax fiber have increased (Arni, 2018). Lignin can be processed by the sulfite, kraft, and soda lignin processes. The isolation of lignin from natural sources can result in substantial chemical modification of the three-dimensional network, due to the cleavage of bonds between different lignin monomers, or covalent attachment to polysaccharides (Norgren & Edlund, 2014). The extent of structural changes and molecular weight reduction during the extraction process will determine its physicochemical properties (Mishra et al., 2018).

## 4.1 Lignin nanoparticles

The complex and heterogeneous chemical structure of lignin contains multiple functional groups such as thiol, aliphatic, hydroxyl, and phenolic hydroxyl, which can act as reducing and stabilizing agents. Lignin-based materials at the nano-scale have enhanced material properties, such as solubility, antioxidantive, and UV protection activity (Yin et al., 2018).

By and large, lignin nanoparticles (LNPs) can be produced three ways: (1) polymerization of the monomer to form nanostructures; (2) physical processing treatment of insoluble polymer to form nanoparticles; and (3) cross-linking of the soluble polymer (Beisl, Friedl, & Miltner, 2017). LNPs are mainly used for the production of novel nanocomposites as a reinforcement agent. The reinforcing effect of lignin depends on the particle size and strong interfacial bonding with the polymer matrix (Yang, Kenny, & Puglia, 2015). Shankar and Rhim (2017) exploited lignin as a reducing agent to produce Ag nanoparticles (AgNPs), using $AgNO_3$ as a precursor. Their composite agar films containing the 1 wt% AgNPs exhibited increased mechanical, moisture, and UV-light barrier properties. Moreover, the films exhibited antibacterial activity against food-borne pathogenic bacteria, *Escherichia coli* and *Listeria monocytogenes*. Similarly, Hossain et al. (2015) produced lignin-based silver nanoparticles with a spherical shape, average size of 45–55 nm, with antibacterial potency against *E. coli*. Mattinen et al. (2018) treated LNP (prepared from softwood Kraft lignin) dispersions in water with *Trametes hirsuta* and *Melanocarpus albomyces* laccases to improve the colloidal stability through surface and intraparticle cross-linking. The stable LNP colloidal dispersion may be attractive in food and bionanomaterial applications. Gupta, Mohanty, and Nayak (2014) synthesized lignin nanoparticles using the nanoprecipitation method. The particles produced showed a spherical shape, average particle size of around 80–104 nm, high antioxidant activity, and UV layer protection. Other applications of lignin nanoparticles are summarized in Table 3.

**Table 3** Methods applied, average sizes, and applications of lignin-based nanocapsules and nanofibers.

| Aim | Product | Method | Average size (nm)[a] | Properties or applications | References |
|---|---|---|---|---|---|
| Preparation and characterization of lignin nanoparticles (LNPs) using green solvent and acid technology | Nanoparticle | Acid precipitation | 30–100 | The LNPs showed an absence of particle aggregation, improvement in thermal stability, good biocompatibility, and zero toxicity to the environment | Azimvand, Didehban, and Mirshokrai (2018) |
| To obtain lignin nanoparticle gelatin complex from switchgrass for the capture of *Staphylococcus aureus* and *Escherichia coli* | Nanoparticle | Ultrasonic-assisted alkali | 220.2 | LNPs exhibited efficient flocculation capacity for *S. aureus* and *E. coli*. For both indicator strains, flocculation efficiency (>95%) was achieved within 30 min at pH 4.5 | Yin et al. (2018) |
| Preparation of cross-linked and decolorized lignin nanoparticles | Nanoparticle | Enzymatic and chemical oxidation | 200 | LNPs were enzymatically reinforced using laccase catalyzed cross-linking reactions. LNP dispersions were obtained by mild chemical treatment; they had a faded brown color and did not emit an unpleasant odor | Mattinen et al. (2018) |
| Synthesis and characterization of biodegradable lignin nanoparticles with tunable surface properties | Nanoparticle | pH-induced flash precipitation | 80 | The nanoparticle dispersions showed stable behavior between pH 3.2 and 8.5. Additionally, the stability was evaluated by adding NaCl to the dispersion, which resulted in a stable dispersion up to a concentration of 70 mM | Richter et al. (2016) |

*Continued*

**Table 3** Methods applied, average sizes, and applications of lignin-based nanocapsules and nanofibers.—cont'd

| Aim | Product | Method | Average size (nm)[a] | Properties or applications | References |
|---|---|---|---|---|---|
| Preparation of lignin-rich nanoparticles from lignocellulosic fibers | Nanoparticle | Enzymatic hydrolysis | 20–100 | The crystallinity and cellulosic content in nanoparticles were reduced by enzymatic hydrolysis. Lignin-rich nanoparticles were uniform in size and shape | Rangan et al. (2017) |
| Synthesis of soy protein-lignin nanofibers by solution electrospinning | Nanofiber | Electrospinning | 124–400 | The addition of polyethylene oxide as a coadjutant facilitated the formation of bead-free fibers, whose diameter increased with the lignin concentration | Salas, Ago, Lucia, and Rojas (2014) |
| Developing lignin-based bio-nanofibers by centrifugal spinning technique | Nanofiber | Centrifugal spinning | Below 500 | Optimum conditions for spinnability: A solution with a polymer concentration of 20% by weight, a lignin/TPU (thermoplastic polyurethane) ratio of 1:1, angular velocity of 8500 rpm for nozzles with a diameter of 0.5 mm, and a distance of 30 cm would be required | Stojanovska, Kurtulus, Abdelgawad, Candan, and Kilic (2018) |
| Development of polycaprolactone/lignin nanofibers | Nanofiber | Electrospinning | 214–383 | The addition of lignin enhanced the mechanical and antioxidant properties of PCL nanofibers | Wang et al. (2018) |

[a]Average size (nm) of the lignin-based nanocapsules and nanofibers.

## 4.2 Lignin nanofibers

The conversion of lignin into fibers remains a challenge owing to its low molecular weight, polydispersity, and amorphous structure, which make the conversion of pure lignin into fibers difficult. However, fractionalization of lignin with solvents or blending it with other polymers could overcome this problem to improve polymer chain entanglement and intermolecular interactions essential for producing uniform fibers. Compatibility between the two polymers will determine the feasibility of fiber spinning. For example, Stojanovska et al. (2018) produced lignin-based fibers by blending lignin with thermoplastic urethane (TPU) with dimethylformamide as a solvent. Using a centrifugal spinning technique, spin dope solution of 20 wt% polymer concentration at 1:1 lignin:TPU ratio, were spun into fibers of $<500$ nm with consistent morphologies. Lignin generally has a strong affinity toward polar polymer matrices because of its available hydroxyl groups (Duval & Lawoko, 2014).

## 4.3 Applications of nano-scale lignin in food and agriculture

In food application, Ago et al. (2016) dissolved Kraft and organosolv lignins in dimethylformamide and atomized in an aerosol flow reactor to produce lignin particles ranging from 30 nm to 2 μm. At as low as 0.1 wt% concentration, these particles are effective in the stabilization of oil-in-water Pickering emulsion, without detectable phase speatation for over 2 months. The lignin nanoparticles can be very attractive for forming surfactant-free emulsion with high colloidal stability. The nanolignin particles may be incorporated in polymers (e.g., polyethylene, polypropylene, and polystyrene) for reinforcement. Due to their antioxidant properties, lignin may be incorporated to stabilize polymers against UV degradation or thermo-oxidation (Kun & Pukánszky, 2017). In the agricultural field, lignin is being used for the controlled release of fertilizers and herbicides (Norgren & Edlund, 2014). Du, Li, and Lindström (2014) modified industrial softwood Kraft lignin by an amination process, with and without phenolation pretreatments. The aminated lignins formed stable aqueous colloidal suspensions with particle sizes ranging from 39 to 391 nm. Potential applications of these aminated lignins include surfactant chemicals and polycationic carrier for sustain-release of fertilizers.

# 5. Pectin

Pectin is a polysaccharide extracted from plant cell walls, mainly from citrus fruits, apple pomace, and sugar beet root (Assadpour, Jafari, & Maghsoudlou, 2017). The chemical structure of pectin consists of α-(1–4)-D-galacturonic acid with some methyl ester and L-rhamnose groups (Shishir et al., 2018). The pectin side chains are composed of >20 types of neutral sugars, including L-rhamnose, L-arabinose, and D-galactose that are often attached to the polysaccharide backbone. Pectin polymers can be divided into homogalacturonan, rhamnogalacturonan-I, and rhamnogalacturonan-II, according to the side chains in the pectin molecular structure and composition of the backbone chain (Cao & Li, 2018).

Pectins are usually classified as high methoxyl pectin (HMP, degree of esterification >50) or low methoxyl pectin (LMP, degree of esterification <50), based on their degree of esterification or number of methoxy groups. The degree of esterification directly influences the physico-chemical properties, complexation, and gel formation of pectin (Mungure, Roohinejad, Bekhit, Greiner, & Mallikarjunan, 2018). For example, HMP tends to form gels under acidic conditions in the presence of high sugar concentrations, while LMP forms gel in the presence of divalent cations (Hua, Yang, Din, Chi, & Yang, 2018). Similar as other polysaccharides, pectin can be modified by physical, chemical, and enzymatic treatments to increase water solubility, enhance interfacial properties, and improve biocompatibility and bioactivity (Cao & Li, 2018).

Pectin has been used as a viscosity modifying agent for food. Due to its amphiphilic character, pectin can be an effective surfactant for the preparation of emulsions (Guerra-Rosas, Morales-Castro, Ochoa-Martínez, Salvia-Trujillo, & Martín-Belloso, 2016). Additionally, pectin can be considered as a source of dietary fiber, acting as a prebiotic that has health-promoting effects, including a desirable fermentation profile in the gut (Ferreira-Lazarte, Kachrimanidou, Villamiel, Rastall, & Moreno, 2018; Naqash, Masoodi, Rather, Wani, & Gani, 2017).

## 5.1 Pectin nanocapsules

Pectin forms three-dimensional gels that are resistant to gastrointestinal conditions for the delivery of probiotics (Shishir et al., 2018). Khorasani and Shojaosadati (2017) studied the encapsulation of probiotics by pectin-nanofiber biocomposites under simulated gastrointestinal conditions, and showed that the biocomposites have a great potential not only for the delivery of probiotics but also for enhancing the stability of probiotic in foods during storage.

Pectin nanocapsules have been explored as carriers of bioactive compounds. Being an anionic polysaccharide, pectin can be used to construct the polyelectrolyte capsule wall with another cationic polymer. For example, Ji et al. (2017) developed a novel pectin-based nanocapsule, via a layer-by-layer technique, to form multilayer pectin-chitosan nanocapsules. The nanocapsules produced showed potential as novel anticancer compound carriers because of their pH-sensitivity, good colloidal stability, and anticancer activity. Nanoencapsulation helps not only in improving the bioactive molecule properties, but also in masking the unwanted odor and taste of the compound (Chai et al., 2018).

## 5.2 Pectin nanocomplex

The formation of polysaccharide-protein complex is a promising nanoencapsulation technique (Ghasemi, Jafari, Assadpour, & Khomeiri, 2018). When a protein and a polysaccharide are mixed together in a liquid medium, electrostatic attraction occurs when the protein is positively charged (i.e., $pH < pI$), while the polysaccharide is anionic at $pH > pK_a$. Similarly, complex formation will also occur when the protein is negatively charged ($pH > pI$), in the presence of cationic polysaccharides such as (Ghasemi, Jafari, Assadpour, & Khomeiri, 2017).

Recently, Chang, Wang, Hu, and Luo (2017) studied zein-caseinate-pectin complex nanoparticle as a potential oral delivery system for curcumin with high encapsulation efficiency. The protein-polysaccharide carrier controlled the release of the curcumin and enhanced is antioxidant activity. Moreover, the nanoparticles were re-dispersible after drying. Similarly, Veneranda et al. (2017) developed zein-caseinate-pectin complex nanoparticles for the encapsulation of eugenol by a nano spray drying method. The eugenol-loaded complex nanoparticles had an average diameter of 140 nm, spherical in shape, and uniform in size distribution. These particles have many potential applications in the agricultural and food industry.

## 5.3 Pectin nanoemulsions

An emulsion can be defined as a mixture of two or more immiscible liquid phases, in which one liquid is dispersed as droplets into the other, continuous phase. Nanoemulsions are susceptible to destabilization after prolonged storage. Pectin can form water-soluble protein-polysaccharide complexes, stabilizing aggregates in solution, or acting as a stabilizer of nanoemulsion systems (Oliveira, Amaro, & Pintado, 2018).

Pectin has an amphiphilic character that aids in the reduction of interfacial tension between immiscible phases. The stability of pectin nanoemulsions can be influenced by various factors, including the type of pectin, degree of esterification and concentration of pectin in nanoemulsion production (Mungure et al., 2018). Xu, Liu, Luo, Liu, and McClements (2017) studied the effect of pectin on the physical and oxidative stabilities of emulsions coated by hydrolyzed rice. Guerra-Rosas et al. (2016) reported the long-term stability of nanoemulsions containing essential oils (oregano, thyme, lemongrass, or mandarin). Guerra-Rosas, Morales-Castro, Cubero-Marquez, Salvia-Trujillo, and Martín-Belloso (2017) reported the antimicrobial activity of HMP-essential oil (oregano, thyme, lemongrass, or mandarin) nanoemulsions against *E. coli* and *Listeria innocua*. Wang, Soyama, and Luo (2016) developed a functional drink using sodium caseinate-pectin and peppermint oil as a potential antimicrobial agent. Sharma et al. (2017) prepared clove oil and sodium caseinate/pectin nanoemulsions by high-speed homogenization.

### 5.4 Applications of nano-scale pectin in food and agriculture

Pectin can impart functional properties and improve stability in emulsions through complex interaction with protein. The proteins preferentially arrange at the oil/water interface, while polysaccharides form a thicker stabilizing layer that prevents droplet aggregation (Naqash et al., 2017). Nano-scale pectin materials have diverse potential applications, which include prebiotic delivery vehicle, gelling agent, emulsion stabilizer, fat replacer, and so on. Studies related to pectin-based nano-scale materials are summarized in Table 4.

## 6. Other polysaccharide gums

Polysaccharide gums are high-molecular-weight hydrophilic polysaccharides that are not form a part of the cell wall, but are exudates or slims (Ribeiro et al., 2016). Polysaccharide gums are organized into four categories based on their origin: seed (e.g., gum guar), plant exudates (Arabic gum), microbial exudates (xanthan gum), and sea weed (carrageenans). Some of these hydrocolloids are anionic (e.g., Arabic and xanthan gums), while others are neutral (i.e., guar gum) (Shishir et al., 2018). Gums and their derivatives are widely used in foods due to their unique functionalities,

**Table 4** Some applications of nano-scale pectin in food and agriculture.

| Aim | Type of pectin-based product | Properties or applications | References |
|---|---|---|---|
| Formation and characterization of zein/caseinate/pectin complex nanoparticles | Complex nanoparticles | The complex nanoparticles had a particle size that was lesser than 200 nm, and a narrow distribution, spherical shape, and strong negative charge. The pectin coating resulted in stabilization effects under simulated gastrointestinal conditions | Chang et al. (2017) |
| Formation, characterization, and protection of β-lactoglobulin-triligand-(β-LG) pectin complex particle | Nanocomplex | The particle size was between 342 and 615 nm as the concentration of pectin increased from 0.1 to 1 mg/mL. The formation of particles containing pectin improved the stability of retinol and eliminated the negative effect of the protein on resveratrol | Cheng, Fang, Liu, Gao, and Liang (2018) |
| Evaluation of lipoprotein (LDL)/pectin complex nanogels as potential oral delivery vehicles for curcumin | Complex nanogels | Nanogels had sizes that were lesser than 60 nm. The LDL/pectin nanogels showed excellent stability under simulated gastrointestinal conditions, in the presence of digestive enzymes and enabled the controlled release of curcumin | Zhou, Wang, Hu, and Luo (2016) |
| Investigation of the stability of food-grade nanoemulsions obtained from essential oils containing high levels of methoxyl pectin | Nanoemulsions | The droplet size of nanoemulsions was below 50 nm. Nanoemulsions containing lemongrass or mandarin essential oil presented a high stability during storage, with a constant droplet size of $<100$ nm during 56 days and the absence of creaming | Guerra-Rosas et al. (2016) |

*Continued*

**Table 4** Some applications of nano-scale pectin in food and agriculture.—cont'd

| Aim | Type of pectin-based product | Properties or applications | References |
|---|---|---|---|
| Evaluation of folic acid release from spray dried powder particles of pectin-whey protein (WPC) nanocapsules | Nanocapsules | The pectin-WPC complex had wrinkles and dents at their particle surface, and samples of WPCs alone had smooth surfaces and spherical-shaped particles | Assadpour et al. (2017) |
| Nanoencapsulation of D-limonene within nanocarriers produced by pectin-whey protein complexes | Nanocapsules | The WPC-pectin nanocomplexes were composed of spherical nanoparticles with an average size of 100nm had an encapsulation efficiency of 88%. The most effective ratio between WPC and pectin that resulted in the highest amount of complex formed was 4:1 | Ghasemi et al. (2018) |
| Investigating the influence of $TiO_2$ on the thermo-mechanical stability and antimicrobial efficiency against food pathogens of pectin aerogels | Nanocomposite aerogels | In the presence of $TiO_2$ nanoparticles, the mechanical, thermal, and antimicrobial properties of pectin-based aerogels were improved in comparison to those of the control | Nešić et al. (2018) |
| To study the influence of essential oils and pectin on nanoemulsion formulation | Nanoemulsion | An increase of up to 2% (w/w) in the pectin concentration enlarged the particle sizes of emulsions, suggesting that the emulsification efficiency was decreased | Artiga-Artigas, Guerra-Rosas, Morales-Castro, Salvia-Trujillo, and Martín-Belloso (2018) |

including non-toxicity, biocompatibility, biodegradability, and are safe for human consumption (Postulkova et al., 2017). Natural gums are part of protective mechanism in plants. They entrap large amounts of water between their polysaccharide chains, causing them to swell and form gels. These gums are used as stabilizers, thickeners, gelling agents, sugar/fat substitutes, film/coating formers, sources of dietary fibers, and so on (Jo, Bak, & Yoo, 2018). Since these gums can be degraded by the gut microflora, they are often used as a delivery matrix for bioactive compounds to ensure their release (Mudgil, Barak, Patel, & Shah, 2018; Santos, Zavareze, Dias, & Vanier, 2018).

Among the polysaccharide gums, guar and Arabic gums are the most widely used in the food industry. Guar gum is a natural biopolymer obtained from the seeds of *Cyamopsis tetragonolobus*, by separation of the endosperm portion of the seed from the husk and germ portions. Their molecular structure is composed of a linear backbone chain of β-1,4-linked mannose with a side chain consisting of α-1,6-linked galactose units (Sharma et al., 2018). Gum Arabic is a natural polysaccharide extracted from the bark exudate of *Acacia Senegal* tree. The polysaccharide has a highly branched molecular structure with a molecular weight of about $3 \times 10^5$ g/mol, composed of a complex mixture of glycoproteins and sugars (Wolf, Gasparin, & Paulino, 2018). Its molecular structure consists of: arabinogalactan (80–90%), glycoprotein (2–4%), and arabinogalactan-protein (10–20%). Due to its higher water solubility and lower viscosity as compared to other polysaccharides, gum Arabic is an effective emulsifier (Mehrnia, Jafari, Makhmal-Zadeh, & Maghsoudlou, 2017). Its propensity to stabilize proteins via complex formation can be attributable to the electrostatic attraction caused by opposite charges between proteins and the polysaccharide (Wu, Kong, Zhang, Hua, & Chen, 2018).

## 6.1 Gum nanocapsules, nanoparticles, and nanoemulsion

Gums have a great potential to encapsulate bioactive compounds for controlled release due to their limited digestion in the body (Dag, Kilercioglu, & Oztop, 2017). They are often used in conjunction with other polymers (e.g., starches, chitosan, gelatin, maltodextrin) to modify encapsulation matrices for improving stability and enhancing protection of bioactive compounds (Santos, Zavareze, et al., 2018; Shishir et al., 2018). For example, Lv, Yang, Li, Zhang, and Abbas (2014) combined gum Arabic with gelatin at 1:1 (w/w) ratio to develop heat-resistant nanocapsules for jasmine essential oil with diameters of ~112 nm. Transglutaminase cross-linked nanocapsules endure 80 °C heating for 7 h, although the jasmine essential oil began to deteriorate above 5 h. Ilyasoglu and El (2014) encapsulated fish oil in multi-layered capsules prepared by electrostatic attraction between gum Arabic and sodium caseinate. The gum-protein complex provided around 79% encapsulation efficiency and particle size of ~230 nm. The complexes are useful for nanoencapsulation of hydrophobic compounds such eicosapentaenoic and docosahexaenoic acids in fortified beverage applications. Katouzian and Jafari (2016) investigated gum Arabic for the nanoencapsulation of vitamin C via complex coacervation to produce spherical nanocapsules with a low hygroscopicity that facilitated storage and handling (Katouzian & Jafari, 2016). Recently, several studies focused on exploring new seed mucilage and gum exudates for encapsulation purposes (Dabestani, Kadkhodaee, Phillips, & Abbasi, 2018; Hashtjin & Abbasi, 2015; Salarbashi & Tafaghodi, 2018). Khoshakhlagh, Koocheki, Mohebbi, and Allafchian (2017) studied *Alyssum homolocarpum* seed gum as a wall material for the encapsulation of D-limonene produced via electrospraying oil-in-water emulsions containing 0.5% (w/w) of the gum, 10–30% (w/w of gum weight) of D-limonene, and 0.1% (w/w) Tween 20. At 10 and 20% D-limonene, spherical nanocapsules were obtained, but nanofibers were observed when D-limonene content increased to 30%. Encapsulation efficiency values ranged from 87% to 93%. Other studies related to gum-based nanocapsules are summarized in Table 5.

### 6.2 Applications of nano-scale polysaccharide gums in food and agriculture

Polysaccharide gums have a wide range of industrial food and agriculture applications as stabilizing agents, thickeners, gelling agents, sugar/fat substitutes, prebiotics, encapsulation matrix, and so on (Jo et al., 2018; Mudgil et al., 2018). Guar gum is considered as a potential candidate to produce nanocomposites; according to Tang, Zhang, Zhao, Guo, and Zhang (2018), the addition of nanocrystalline cellulose in chitosan/gum, guar/nanocrystalline cellulose nanocomposites improved their mechanical properties (tensile strength) and reduced the air permeability, rendering them as promising materials for food packaging.

## 7. Dextrin/maltodextrin/cyclodextrin

Dextrins are water-soluble and hydrolyzed starches. They can be classified according to their dextrose equivalent value (DE); dextrins with DE <20 are named as maltodextrins, while those with DE values >20 are

**Table 5** Methods, average sizes, and applications of polysaccharide gum-based nanocapsules and nanoemulsions.

| Aim | Product | Method | Average size (nm)[a] | Properties or applications | References |
|---|---|---|---|---|---|
| To investigate the effect of pH and gum Arabic addition on the stability of wheat gliadin nanoparticles | Nanoparticle | Anti-solvent precipitation | 248.2 (pH 5) 270.4 (pH 7) | Gum Arabic-coated nanoparticles showed a good stability at pH 4.0–7.0 and a low particle size with elevated ionic strengths, besides showing good thermal stability at 80 °C | Wu et al. (2018) |
| Polysaccharide-based nanoparticles as delivery systems to encapsulate, stabilize, and deliver curcumin effectively | Nanoparticle | Polyelectrolyte complexation | 250–290 | Nanoparticles improved the stability and delayed the release of curcumin in a simulated gastrointestinal environment, suggesting that they can be used to deliver hydrophobic bioactive ingredients in functional foods | Tan, Xie, Zhang, Cai, and Xia (2016) |
| Synthesis of gum guar propionate nanoparticles for antimicrobial applications | Nanoparticle | Solvent shifting | 350.8 | The propionate groups were attached to the free hydroxyl groups of purified gum guar. Zeta potential studies confirmed the stability of the suspension, and the nanoparticles were useful to inhibit the growth of the fungus strain | Das, Abdullah, Kundu, and Mukherjee (2018) |
| To obtain redispersible polyelectrolyte complex nanoparticles from gallic acid-chitosan conjugate (GA-CS) and gum Arabic | Nanoparticle | Ionic gelation and nano spray drying | 112.2 | The mass ratio between GA-CS and gum Arabic and the pH of the GA-CS solution played important roles in determining the size of nanoparticles | Hu, Gerhard, et al. (2016) and Hu, Wang, Zhou, Xue, and Luo (2016) |

*Continued*

**Table 5** Methods, average sizes, and applications of polysaccharide gum-based nanocapsules and nanoemulsions.—cont'd

| Aim | Product | Method | Average size (nm)[a] | Properties or applications | References |
|---|---|---|---|---|---|
| Stabilization of pickering emulsion gels by zein/gum Arabic complex colloidal nanoparticles | Nanoparticle | Anti-solvent precipitation | 110 | Nanoparticles were prepared through hydrogen bonding and electrostatic interactions. The gel structure with nanoparticles showed long-term storage stability (30 days) without coalescence | Dai, Sun, Wei, Mao, and Gao (2018) |
| To obtain eugenol oil nanoemulsions using gum Arabic and lecithin as emulsifiers and antimicrobial | Nanoemulsion | Nano spray drying and freeze drying | 103.6 | Spherical, smooth, and ultra-small powders of the nanoemulsion exhibiting excellent re-dispersibility and antimicrobial efficacy were obtained by nano spray drying | Hu, Gerhard, et al., (2016) |
| Nano-emulsions containing crocin prepared with Angum gum (AG), gum Arabic (GA) and whey protein (WP) | Nanoemulsion | Spontaneous method | 695 | The AG biopolymer showed a lower emulsifying capacity than GA and WP, resulting in bigger droplet diameters | Mehrnia et al. (2017) |
| Emulsification properties of persian gum and its soluble (SFPG) and insoluble (IFPG) fractions in water | Nanoemulsion | Sonication | 12.68 | Optimal conditions for producing nanoemulsions include sonication amplitude, sonication time, and process temperature of 94%, 138 s and 37 °C, respectively. The nanoemulsion was completely stable even over a storage period of 3 months | Hashtjin and Abbasi (2015) |

[a]Average size (nm) of the gum-based nanocapsules and nanoemulsions.

denoted as glucose solids or corn syrup solids (Shishir et al., 2018). Maltodextrins with higher DE value are shown to have a lower molecular weight and higher solubility, resulting in a higher moisture content in the final product. To enhance bioactive retention, control of release profile, reduction of oxygen permeability of the wall matrix, and improved emulsifying characteristics, maltodextrins are mixed with other materials, including gums, pectins, alginate, and whey protein (Shishir et al., 2018). Maltodextrins are highly water-soluble, producing solutions of low viscosity. Maltodextrin is widely used for the encapsulation of food components to protect nutrients, colors, antioxidants, and bioactive compounds from thermal and oxidative degradation during storage and drying. Besides, they are low in cost, widely available, and compatible with different materials (Churio & Valenzuela, 2018).

Cyclodextrins (CDs) are cyclic oligosaccharides composed of six (α-CD), seven (β-CD), eight (γ-CD), or more linked glucopyranose subunits that are obtained by the enzymatic modification of starch (Kringel et al., 2017). CDs have hydrophobic inner cavity which forms inclusion complexes with a wide range of guest molecules, while the hydrophilic exterior enhances CD solubility in water and other polar solvents (Yuan, Xu, Cui, & Wang, 2019). The formation of inclusion complex can protect the complexed guest molecules from degradation caused by external conditions, such as heat, pH, or light. CDs are versatile for solubilizing food colors and vitamins, masking of undesired odors or tastes, and enabling the controlled release of bioactive compounds (Ceborska, 2018).

## 7.1 Dextrin/maltodextrin/cyclodextrin nanostructures

Lipophilic vitamins (e.g., A, D, E and K) can be encapsulated in CDs to enhance the solubility of guest molecules in aqueous products, although the stability can be affected by other factors, such as pH (Katouzian & Jafari, 2016). To further increase water solubility, chemical modifications can be performed on CD structure. For example, hydroxypropyl β-cyclodextrin (HP-β-CD) has higher water-solubility and biocompatibility than unmodified CD (Fathi, Martín, & McClements, 2014). Electrospinning techniques have been used by researchers to produce ultrafine CD fibers for active food packaging and functional food applications. Additionally, the electrospinning process is simple, versatile, and non-thermal, which is ideal for encapsulating heat-labile ingredients (Mendes et al., 2017; Mercante, Scagion, Migliorini, Mattoso, & Correa, 2017; Prakash et al., 2018; Wen et al., 2017). Electrospun cyclodextrin nanofibers can be used for enzymatic immobilization owing to their large surface area-to-volume ratio and high porosity, which enables greater enzyme loading, improves access of substrates to immobilized enzymes, promotes controlled release and increases catalytic ability of enzymes (Santos, Flôres, Rios, & Chisté, 2018). Nanocapsules and nanofibers produced with maltodextrins and cyclodextrins, as well as the methods used, the average size, and their applications are summarized in Tables 6 and 7, respectively.

## 7.2 Applications of nano-scale cyclodextrins in food and agriculture

The development of novel techniques capable of monitoring and preventing the spoilage and contamination in food and agricultural products is important (Mercante et al., 2017). To this end, nanotechnology has been exploited to improve seed germination, minimize fertilizer losses, protect/remedy water and soils pollution, maximize cop yield, and so on (Iavicoli et al., 2017; Peters et al., 2016). Cyclodextrins have been used for the development of new agrochemicals, such as herbicides, insecticides, fungicides, and growth regulators (Seglie et al., 2012; Villaverde, Rubio-Bellido, Lara-Moreno, Merchan, & Morillo, 2018). Cyclodextrins could be potentially used for the removal of hydrophobic contaminants from soils (Morillo & Villaverde, 2017). Among the applications of cyclodextrin in food, one important development has been with cyclodextrin-based nanosponges (CD-NS), which are made up of highly cross-linked insoluble 3D networks of cyclodextrins. Their inner cavities act as pores of a sponge that accommodate the guest molecules (Caldera, Tannous, Cavalli, Zanetti, & Trotta, 2017; Sherje, Dravyakar, Kadam, & Jadhav, 2017). Singh et al. (2018) functionalized the surface of β-cyclodextrin nanosponge (β-CD-NS) with cholesterol and concluded that β-CD-NS can be used as a carrier for low water-soluble molecules to improve their solubility and bioavailability. Similarly, Pushpalatha, Selvamuthukumar, and Kilimozhi (2018) utilized CD-NS to deliver curcumin using diphenyl carbonate (DPC) and pyromellitic dianhydride (PMDA) as cross-linkers. They reported that PMDA resulted in CD-NS with higher solubility and curcumin release than those cross-linked by DPC. The former also had higher cytotoxicity than the DPC cross-linker.

# 8. Final considerations

The development of nano-scale materials has increased considerably over the past few years as they have significant advantages when compared

**Table 6** Methods applied, average sizes, and applications of nanocapsules produced using maltodextrin or cyclodextrin.

| Aim | Method | Average size (nm)[a] | Properties or applications | References |
|---|---|---|---|---|
| Production, characterization, and estimation of stability of nanoencapsulates of saffron hydrophilic apocarotenoids in maltodextrin | Spray-drying | Not mentioned | Nanoencapsulation resulted in spherical particles with high product yield and encapsulation efficiency. There was an increase in the stability of crocins and picrocrocin under thermal and gastrointestinal conditions | Kyriakoudi and Tsimidou (2018) |
| Anti-gastrointestinal cancer activity of propolis encapsulated by cyclodextrin (CD) | Inclusion complex (IC) | 170 (γ-CD)<br>700 (IC with propolis) | The complexes resulted in anti-inflammatory effects in vitro with respect to the cytokine TNF-α, and showed strong lipid anti-oxidant activity | Catchpole, Mitchell, Bloor, Davis, and Suddes (2018) |
| To study the antioxidative and antimicrobial activities of liquid smoke nanocapsules using chitosan (CS) and maltodextrin (MD) and its application for fish preservation | Polyelectrolyte complexation (MD)<br>Ionic gelation (CS) | 16.08 (MD + CS)<br>14.87 (MD)<br>13.43 (CS) | The nanocapsule mixture of CS and MD showed a higher total phenolic, total acid and radical scavenging activity. The addition of nanocapsules of CS and MD (> 5.0%) maintained the freshness of fish until 48 h at room temperature | Saloko, Darmadji, Setiaji, and Pranoto (2014) |

*Continued*

**Table 6** Methods applied, average sizes, and applications of nanocapsules produced using maltodextrin or cyclodextrin.—cont'd

| Aim | Method | Average size (nm)[a] | Properties or applications | References |
|---|---|---|---|---|
| In vitro release study of β-CD/ chitosan nanoparticle loaded *Cinnamomum zeylanicum* essential oil (CEO) | Binary system via ionic gelation | 123.3–326.4 | Spherical nanoparticles were obtained with an encapsulation efficiency of 58.03%. In vitro release profiles showed an overall CEO release of 71% and a controlled release for over 120 h | Matshetshe, Parani, Manki, and Oluwafemi (2018) |
| To study inclusion complexes between red bell pepper pigments and 2-hydroxypropyl-β-cyclodextrin (HPβCD) in different mass ratios | Ultrasonic homogenization | 338.6 (1:4)<br>363.6 (1:6)<br>312.4 (1:8)<br>287.8 (1:10) | The inclusion process occurred for all tested ratios and the complex formation with 2-HPβCD increased the aqueous solubility of red bell pepper extracts by up to 660 times | Petito, Dias, Costa, Falcão, and Araujo (2016) |
| Investigation of the nanoparticle formation of an inclusion complex of lycopene and β- cyclodextrin | Supercritical antisolvent precipitation | 40 | Small spherical particles were obtained. At high pressure, high temperature, high $CO_2$ flow rate, and low solution flow rate, the smallest particles with an average size of 40 nm were obtained | Nerome et al. (2013) |

[a]Average size (nm) of the nanocapsules produced using maltodextrin or cyclodextrin.

**Table 7** Methods, average sizes, and some applications of cyclodextrin (CD) nanofibers.

| Aim | Average size (nm)[a] | Properties or applications | References |
|---|---|---|---|
| Study of water-solubility, shelf-life, and photostability of vitamin E/cyclodextrin inclusion complex (IC) electrospun nanofibers | Vitamin E/IC (1:1): 630 Vitamin E/IC (1:2): 735 | Vitamin E/HPβCD-IC showed effective antioxidant activity and provided enhanced photostability for the sensitive Vitamin E. It maintained its antioxidant activity even after exposure to UV light | Celebioglu and Uyar (2017) |
| Investigation of fast-dissolving and antibacterial activity of cyclodextrin/linalool-inclusion complex nanofibers (CD/linalool-IC-NFs) | HPβCD: 700 MβCD: 655 HPγCD: 840 | CD/linalool-IC-NFs were dissolved in water within two seconds. The release of linalool from CD/linalool-IC-NFs inhibited the growth of model gram-negative (*E. coli*) and gram-positive (*S. aureus*) | Aytac, Yildiz, Kayaci-Senirmak, Tekinay, and Uyar (2017) |
| Studying the release behavior and antioxidant activity of the gallic acid/cyclodextrin inclusion complex, which was encapsulated in electrospun polylactic acid nanofibers (PLA/GA/HPβCD-IC-NF) | 235 | PLA/GA/HPβCD-IC-NF showed controlled release in three different mediums and also showed high antioxidant activity | Aytac, Kusku, Durgun, and Uyar (2016) |
| Immobilization of xylanase and the xylanase–β-cyclodextrin complex in polyvinyl alcohol via electrospinning | 200–600 | Electrospun fibers exhibited a smooth surface without beads. Xylanase immobilized in β-cyclodextrin-PVA nanofibers exhibited a higher activity at pHs of 4, 5, 7, and 8, as compared to that of free xylanase | Santos, Flôres, et al. (2018) |

*Continued*

**Table 7** Methods, average sizes, and some applications of cyclodextrin (CD) nanofibers.—cont'd

| Aim | Average size (nm)[a] | Properties or applications | References |
|---|---|---|---|
| Studying plasma-treated poly(ethylene oxide) nanofibers containing tea tree oil/beta-cyclodextrin inclusion complex for antibacterial packaging | 200–400 | Plasma treatment improved the release efficiency of antibacterial agents from nanofibers, and enhanced their antibacterial activity against *Escherichia coli* O157:H7, as shown by tests performed using beef for 7 days at 4°C or 12°C | Cui, Bai, and Lin (2018) |
| Studying the water solubility, high thermal stability, and antioxidant action of the nanofibrous thymol/cyclodextrin inclusion complex | Thymol/HPβCD: 540–1840<br>Thymol/HPγCD: 760–2540<br>Thymol/MβCD: 300–1990 | The inclusion complex formation process improved water solubility, and enhanced the temperature stability and antioxidant property of nanofibrous webs | Celebioglu, Yildiz, and Uyar (2018) |
| Studying the antibacterial poly (ethylene oxide) electrospun nanofibers containing cinnamon essential oil/beta-cyclodextrin (CEO/β-CD) proteoliposomes | 317–364 | The CEO/β-CD proteoliposomes showed antibacterial activity against *Bacillus cereus* and improved stability after nanofiber encapsulation | Lin, Dai, and Cui (2017) |

[a]Average size (nm) of the cyclodextrins nanofibers.

with those for macro-scale materials. Different materials are used in nanotechnology, and a variety of food-grade materials derived from polysaccharides have been used as environment-friendly alternatives that are generally recognized as safe. In this review, trends and applications of polysaccharide-based nano-materials in the fields of food and agriculture are discussed. Advancements in food involve the development of food additives at the nano-scale to enhance product characteristics, increase bioavailability, and improve food stability during processing and storage. In agriculture, nanotechnology is used mainly for production of nano-formulated pesticides and agro-chemicals, to achieve an increased efficacy as compared to that of conventional formulations, thereby resulting in increased productivity.

New processing techniques for production of nano-scale materials are highlighted, such as electrospinning for production of fibers and/or capsules, and the use of cyclodextrins for production of nano-sponges. Highly cross-linked 3D network of cyclodextrins are being developed as carriers for hydrophobic molecules to improve solubility and bioavailability, as well as for controlled delivery of agrochemicals and other bioactive compounds.

Future studies are needed to scale up industrial-scale production of nano-materials, in order to meet commercial demands. Further studies on the application of nano-sensors for controlling the quality of foods would be important. Additionally, it is also necessary to further understand the correlation between size and some physico-biological properties of nano-materials, as well as their potential toxicity.

## References

Abbas, S., Karangwa, E., Bashari, M., Hayat, K., Hong, X., Sharif, H. R., et al. (2015). Fabrication of polymeric nanocapsules from curcumin-loaded nanoemulsion templates by self-assembly. *Ultrasonics Sonochemistry*, *23*, 81–92.

Ago, M., Huan, S., Borghei, M., Raula, J., Kauppinen, E. I., & Rojas, O. J. (2016). High-throughput synthesis of lignin particles ($\sim$30 nm to $\sim$2 μm) via aerosol flow reactor: Size fractionation and utilization in Pickering emulsions. *Applied Materials & Interfaces*, *8*, 23303–23310.

Ahmed, F., Iftikhar, A., Jatoi, A. W., Khatri, M., Memon, N., Khatri, Z., et al. (2017). Ultrasonic-assisted deacetylation of cellulose acetate nanofibers: A rapid method to produce cellulose nanofibers. *Ultrasonics Sonochemistry*, *36*, 319–325.

Angellier, H., Choisnard, L., Molina-Boisseau, S., Ozil, P., & Dufresne, A. (2004). Optimization of the preparation of aqueous suspensions of waxy maize starchnanocrystals using a response surface methodology. *Biomacromolecules*, *5*, 1545–1551.

Angellier, H., Molina-Boisseau, S., Belgacem, M. N., & Dufresne, A. (2005). Surface chemical modification of waxy maize starch nanocrystals. *Langmuir*, *21*, 2425–2433.

Arni, S. A. (2018). Extraction and isolation methods for lignin separation from sugarcane bagasse: A review. *Industrial Crops and Products*, *115*, 330–339.

Artiga-Artigas, M., Guerra-Rosas, M. I., Morales-Castro, J., Salvia-Trujillo, L., & Martín-Belloso, O. (2018). Influence of essential oils and pectin on nanoemulsion formulation: A ternary phase experimental approach. *Food Hydrocolloids*, *81*, 209–219.

Assadpour, E., Jafari, S., & Maghsoudlou, Y. (2017). Evaluation of folic acid release from spray dried powder particles of pectin-whey protein nano-capsules. *International Journal of Biological Macromolecules*, *95*, 238–247.

Aytac, Z., Kusku, S. I., Durgun, E., & Uyar, T. (2016). Encapsulation of gallic acid/cyclodextrin inclusion complex in electrospun polylactic acid nanofibers: Release behavior and antioxidant activity of gallic acid. *Materials Science and Engineering C*, *63*, 231–239.

Aytac, Z., Yildiz, Z. I., Kayaci-Senirmak, F., Tekinay, T., & Uyar, T. (2017). Electrospinning of cyclodextrin/linalool-inclusion complex nanofibers: Fast-dissolving nanofibrous web with prolonged release and antibacterial activity. *Food Chemistry*, *231*, 192–201.

Azeredo, H. M. C., Mattoso, L. H. C., Wood, D., Williams, T. G., Avena-Bustillos, R. J., & Mchugh, T. (2009). Nanocomposite edible films from mango puree reinforced with cellulose nanofiber. *Journal of Food Science*, *74*, 31–35.

Azimvand, J., Didehban, K., & Mirshokrai, S. A. (2018). Preparation and characterization of lignin polymeric nanoparticles using the green solvent ethylene glycol: Acid precipitation technology. *BioResources*, *13*, 2887–2897.

Beisl, S., Friedl, A., & Miltner, A. (2017). Lignin from micro-to nanosize: Production methods. *International Journal of Molecular Sciences*, *18*, 1–31.

Bel Haaj, S., Magnin, A., Pétrier, C., & Boufi, S. (2013). Starch nanoparticles formation via high power ultrasonication. *Carbohydrate Polymers*, *92*, 1625–1632.

Berglund, L., Noël, M., Aitomäki, Y., Öman, T., & Oksman, K. (2016). Production potential of cellulose nanofibers from industrial residues: Efficiency and nanofiber characteristics. *Industrial Crops and Products*, *92*, 84–92.

Caldera, F., Tannous, M., Cavalli, R., Zanetti, M., & Trotta, F. (2017). Evolution of cyclodextrin nanosponges. *International Journal of Pharmaceutics*, *531*, 470–479.

Cao, J., & Li, Q. (2018). *Modification of pectin. In Reference module in food science*. Elsevier https://doi.org/10.1016/B978-0-08-100596-5.22446-3.

Carrillo-Inungaray, M. L., Trejo-Ramirez, J. A., Reyes-Munguia, A., & Carranza-Alvarez, C. (2018). Use of nanoparticles in the food industry: Advances and perspectives. Chapter 15. In A. M. Grumezescu & A. M. Holban (Eds.), *Vol. 12. Impact of nanoscience in the food industry* (pp. 419–444). Elsevier. Ed. By.

Catchpole, O., Mitchell, K., Bloor, S., Davis, P., & Suddes, A. (2018). Anti-gastrointestinal cancer activity of cyclodextrin-encapsulated propolis. *Journal of Functional Foods*, *41*, 1–8.

Ceborska, M. (2018). Structural investigation of solid state host/guest complexes of native cyclodextrins with monoterpenes and their simple derivatives. *Journal of Molecular Structure*, *1165*, 62–70.

Celebioglu, A., & Uyar, T. (2017). Antioxidant vitamin E/cyclodextrin inclusion complex electrospun nanofibers: Enhanced water-solubility, prolonged shelf-life and photostability of vitamin E. *Journal of Agricultural and Food Chemistry*, *65*, 5404–5412.

Celebioglu, A., Yildiz, Z. I., & Uyar, T. (2018). Thymol/cyclodextrin inclusion complex nanofibrous webs: Enhanced water solubility, high thermal stability and antioxidant property of thymol. *Food Research International*, *106*, 280–290.

Chai, J., Jiang, P., Wang, P., Jiang, Y., Li, D., Bao, W., et al. (2018). The intelligent delivery systems for bioactive compounds in foods: Physicochemical and physiological conditions, absorption mechanisms, obstacles and responsive strategies. *Trends in Food Science and Technology*, *78*, 144–154.

Chang, C., Wang, T., Hu, Q., & Luo, Y. (2017). Zein/caseinate/pectin complex nanoparticles: Formation and characterization. *International Journal of Biological Macromolecules*, *104*, 117–124.

Cheng, H., Fang, Z., Liu, T., Gao, Y., & Liang, L. (2018). A study on β-lactoglobulin-triligand-pectin complex particle: Formation, characterization and protection. *Food Hydrocolloids*, *84*, 93–103.

Churio, O., & Valenzuela, C. (2018). Development and characterization of maltodextrin microparticles to encapsulate heme and non-heme iron. *LWT—Food Science and Technology*, *96*, 568–575.

Cui, H., Bai, M., & Lin, L. (2018). Plasma-treated poly (ethylene oxide) nanofibers containing tea tree oil/beta-cyclodextrin inclusion complex for antibacterial packaging. *Carbohydrate Polymers*, *179*, 360–369.

Dabestani, M., Kadkhodaee, R., Phillips, G. O., & Abbasi, S. (2018). Persian gum: A comprehensive review on its physicochemical and functional properties. *Food Hydrocolloids*, *78*, 92–99.

Dag, D., Kilercioglu, M., & Oztop, M. H. (2017). Physical and chemical characteristics of encapsulated goldenberry (*Physalis peruviana* L.) juice powder. *LWT—Food Science and Technology*, *83*, 86–94.

Dai, L., Sun, C., Wei, Y., Mao, L., & Gao, Y. (2018). Characterization of Pickering emulsion gels stabilized by zein/gum Arabic complex colloidal nanoparticle. *Food Hydrocolloids*, *74*, 239–248.

Das, A., Abdullah, M. F., Kundu, S., & Mukherjee, A. (2018). Synthesis of guar gum propionate nanoparticles for antimicrobial applications. *Materials Today: Proceedings*, *5*, 9683–9689.

Dasgupta, N., Ranjan, S., Mundekkad, D., Ramalingam, C., Shanker, R., & Kumar, A. (2015). Nanotechnology in agro-food: From field to plate. *Food Research International*, *69*, 381–400.

Dokić, L., Krstonošić, V., & Nikolić, I. (2012). Physicochemical characteristics and stability of oil-in-water emulsions stabilized by OSA starch. *Food Hydrocolloids*, *29*, 185–192.

Đorđević, V., Paraskevopoulou, A., Mantzouridou, F., Lalou, S., Panti, C. M., Bugarski, B., et al. (2016). Encapsulation technologies for food industry. In V. P. Nedovic, P. Raspor, J. Levic, V. T. Saponjac, & G. V. Barbosa-Canovas (Eds.), *Emerging and traditional technologies for safe, healthy and quality food* (pp. 329–380). Switzerland: Springer International Publishing.

Du, X. Y., Li, L. B., & Lindström, M. E. (2014). Modification of industrial softwood kraft lignin using Mannich reaction with and without phenolation pretreatment. *Industrial Crops and Products*, *52*, 729–735.

Duan, B., Sun, P., Wang, X., & Yang, C. (2011). Preparation and properties of starch nanocrystals/carboxymethyl chitosan nanocomposite films. *Starch-Starke*, *63*, 528–535. 2011.

Dufresne, A. (2017). Cellulose nanomaterial reinforced polymer nanocomposites. *Current Opinion in Colloid & Interface Science*, *29*, 1–8.

Dufresne, A., Cavaille, J. Y., & Helbert, W. (1996). New nanocomposite materials: Microcrystalline starch reinforced thermoplastic. *Macromolecules*, *29*, 7624–7626.

Duval, A., & Lawoko, M. (2014). A review on lignin-based polymeric, micro- and nanostructured materials. *Reactive and Functional Polymers*, *85*, 78–96.

FAO/WHO. (2010). Expert meeting on the application of nanotechnologies in the food and agriculture sectors. In *Potential food safety implications* (p. 129). Rome, Italy: Food and Agriculture Organization of the United Nations and World Health Organization. Meeting report.

Fathi, M., Martín, A., & McClements, D. J. (2014). Nanoencapsulation of food ingredients using carbohydrate based delivery systems. *Trends in Food Science & Technology*, *39*, 18–39.

Fazeli, M., Keley, M., & Biazar, E. (2018). Preparation and characterization of starch-based composite films reinforced by cellulose nanofibers. *International Journal of Biological Macromolecules*, *116*, 272–280.

Fernández, A., Picouet, P., & Lloret, E. (2010). Cellulose-silver nanoparticle hybrid materials to control spoilage-related microflora in absorbent pads located in trays of fresh-cut melon. *International Journal of Food Microbiology*, *142*, 222–228.

Ferreira-Lazarte, A., Kachrimanidou, V., Villamiel, M., Rastall, R. A., & Moreno, F. J. (2018). In vitro fermentation properties of pectins and enzymatic-modified pectins obtained from different renewable bioresources. *Carbohydrate Polymers*, *199*, 482–491.

Fonseca, L. M., Silva, F. T., Antunes, M. D., Halal, S. L. M., Lim, L.-T., & Dias, A. R. G. (2018). Aging time of soluble potato starch solutions for ultrafine fibers formation by electrospinning. *Starch-Starke*, *71* https://doi.org/10.1002/star.201800089.

Gao, H., Duan, B., Lu, A., Deng, H., Du, Y., Shi, X., et al. (2018). Fabrication of cellulose nanofibers from waste brown algae and their potential application as milk thickeners. *Food Hydrocolloids*, *79*, 473–481.

Ghasemi, S., Jafari, S. M., Assadpour, E., & Khomeiri, M. (2017). Production of pectin-whey protein nano-complexes as carriers of orange peel oil. *Carbohydrate Polymers*, *177*, 369–377.

Ghasemi, S., Jafari, S. M., Assadpour, E., & Khomeiri, M. (2018). Nanoencapsulation of D-limonene within nanocarriers produced by pectin-whey protein complexes. *Food Hydrocolloids*, *77*, 152–162.

Guerra-Rosas, M. I., Morales-Castro, J., Cubero-Marquez, M. A., Salvia-Trujillo, L., & Martín-Belloso, O. (2017). Antimicrobial activity of nanoemulsions containing essential oils and high methoxyl pectin during long-term storage. *Food Control*, *77*, 131–138.

Guerra-Rosas, M. I., Morales-Castro, J., Ochoa-Martínez, L. A., Salvia-Trujillo, L., & Martín-Belloso, O. (2016). Long-term stability of food-grade nanoemulsions from high methoxyl pectin containing essential oils. *Food Hydrocolloids*, *52*, 438–446.

Gupta, A. K., Mohanty, S., & Nayak, S. K. (2014). Synthesis, characterization and application of lignin nanoparticles (LNPs). *Materials Focus*, *3*, 444–454.

Handford, C. E., Dean, M., Henchion, M., Spence, M., Elliott, C. T., & Campbell, K. (2014). Implications of nanotechnology for the agri-food industry: Opportunities, benefits and risks. *Trends in Food Science & Technology*, *40*, 226–241.

Hao, Y., Chen, Y., Li, Q., & Gao, Q. (2018). Preparation of starch nanocrystals through enzymatic pretreatment from waxy potato starch. *Carbohydrate Polymers*, *184*, 171–177.

Hashtjin, A. M., & Abbasi, S. (2015). Nano-emulsification of orange peel essential oil using sonication and native gums. *Food Hydrocolloids*, *44*, 40–48.

Hedjazi, S., & Razavi, S. H. (2018). A comparison of Canthaxanthine Pickering emulsions, stabilized with cellulose nanocrystals of different origins. *International Journal of Biological Macromolecules*, *106*, 489–497.

Hossain, M. M., Scott, I. M., McGarvey, B. D., Conn, K., Ferrant, L., Berruti, F., et al. (2015). Insecticidal and anti-microbial activity of bio-oil derived from fast pyrolysis of lignin, cellulose, and hemicellulose. *Journal of Pest Science*, *88*, 171–179.

Hu, Q., Gerhard, H., Upadhyaya, I., Venkitanarayanan, K., & Luo, Y. (2016). Antimicrobial eugenol nanoemulsion prepared by gum Arabic and lecithin and evaluation of drying technologies. *International Journal of Biological Macromolecules*, *87*, 130–140.

Hu, Q., Wang, T., Zhou, M., Xue, J., & Luo, Y. (2016). Formation of redispersible polyelectrolyte complex nanoparticles from gallic acid-chitosan conjugate and gum Arabic. *International Journal of Biological Macromolecules*, *92*, 812–819.

Hua, X., Yang, H., Din, P., Chi, K., & Yang, R. (2018). Rheological properties of deesterified pectin with different methoxylation degree. *Food Bioscience*, *23*, 91–99.

Iavicoli, I., Leso, V., Beezhold, D. H., & Shvedova, A. A. (2017). Nanotechnology in agriculture: Opportunities, toxicological implications, and occupational risks. *Toxicology and Applied Pharmacology*, *329*, 96–111.

Ilyasoglu, H., & El, S. N. (2014). Nanoencapsulation of EPA/DHA with sodium caseinate-gum Arabic complex and its usage in the enrichment of fruit juice. *LWT—Food Science and Technology*, *56*, 461–468.

Inukai, S., Kurokawa, N., & Hotta, A. (2018). Annealing and saponification of electrospun cellulose-acetate nanofibers used as reinforcement materials for composites. *Composites Part A: Applied Science and Manufacturing*, *113*, 158–165.

Jafari, S. M., Bahrami, I., Dehnad, D., & Shahidi, S. A. (2018). The influence of nanocellulose coating on saffron quality during storage. *Carbohydrate Polymers*, *181*, 536–542.

Ji, F., Li, J., Qin, Z., Yang, B., Zhang, E., Dong, D., et al. (2017). Engineering pectin-based hollow nanocapsules for delivery of anticancer drug. *Carbohydrate Polymers*, *177*, 86–96.

Jin, Y., Li, J. Z., & Nik, A. M. (2018). Starch-based microencapsulation. Chapter 17. In *Starch in foods* (2nd ed., pp. 661–690). Elsevier.

Jin, W., Xu, W., Liang, H., Li, Y., Liu, S., & Li, B. (2016). Nanoemulsions for food: Properties, production, characterization, and applications, Chapter 1. In A. M. Grumezescu (Ed.), *Emulsions. Nanotechnology in the Agri-food industry* (p. Vol. 3). Bucharest, Romania: Elsevier. Ed. By.

Jo, W., Bak, J. H., & Yoo, B. (2018). Rheological characterizations of concentrated binary gum mixtures with xanthan gum and galactomannans. *International Journal of Biological Macromolecules*, *114*, 263–269.

Jung, J., Deng, Z., Simonsen, J., Bastías, R. M., & Zhao, Y. (2016). Development and preliminary field validation of water-resistant cellulose nanofiber based coatings with high surface adhesion and elasticity for reducing cherry rain-cracking. *Scientia Horticulturae*, *200*, 161–169.

Kargarzadeh, H., Mariano, M., Huang, J., Lin, N., Ahmad, I., Dufresne, A., et al. (2017). Recent developments on nanocellulose reinforced polymer nanocomposites: A review. *Polymer*, *132*, 368–393.

Katouzian, I., & Jafari, S. M. (2016). Nano-encapsulation as a promising approach for targeted delivery and controlled release of vitamins: A review. *Trends in Food Science & Technology*, *53*, 34–48.

Khalil, H. P. S. A., Davoudpour, Y., Islam, M. N., Mustapha, A., Sudesh, K., Dungani, R., et al. (2014). Production and modification of nanofibrillated cellulose using various mechanical processes: A review. *Carbohydrate Polymers*, *99*, 649–665.

Khorasani, A. C., & Shojaosadati, S. A. (2017). Pectin-non-starch nanofibers biocomposites as novel gastrointestinal-resistant prebiotics. *International Journal of Biological Macromolecules*, *94*, 131–144.

Khoshakhlagh, K., Koocheki, A., Mohebbi, M., & Allafchian, A. (2017). Development and characterization of electrosprayed *Alyssum homolocarpum* seed gum nanoparticles for encapsulation of D-limonene. *Journal of Colloid and Interface Science*, *490*, 562–575.

Kim, H.-Y., Han, J.-A., Kweon, D.-K., Park, J.-D., & Lim, S.-T. (2013). Effect of ultrasonic treatments on nanoparticle preparation of acid-hydrolyzed waxy maize starch. *Carbohydrate Polymers*, *93*, 582–588.

Kim, J.-Y., & Lim, S.-T. (2009). Preparation of nano-sized starch particles by complex formation with n-butanol. *Carbohydrate Polymers*, *76*, 110–116.

Kim, J.-Y., Park, D.-J., & Lim, S.-T. (2008). Fragmentation of waxy rice starchgranules by enzymatic hydrolysis. *Cereal Chemistry*, *85*, 182–187.

Kim, H.-Y., Park, S. S., & Lim, S.-T. (2015). Preparation, characterization and utilization of starch nanoparticles. *Colloids and Surfaces B: Biointerfaces*, *126*, 607–620.

Kong, L., & Ziegler, G. R. (2012). Role of molecular entanglements in starch fiber formation by electrospinning. *Biomacromolecules*, *13*, 2247–2253.

Konwarh, R., Karak, N., & Misra, M. (2013). Electrospun cellulose acetate nanofibers: The present status and gamut of biotechnological applications. *Biotechnology Advances*, *31*, 421–437.

Kringel, D. H., Antunes, M. D., Klein, B., Crizel, R. L., Wagner, R., Oliveira, R. P., et al. (2017). Production, characterization, and stability of orange or eucalyptus essential oil/β-cyclodextrin inclusion complex. *Journal of Food Science*, *82*, 2598–2605.

Kun, D., & Pukánszky, B. (2017). Polymer/lignin blends: Interactions, properties, applications. *European Polymer Journal, 93*, 618–641.
Kyriakoudi, A., & Tsimidou, M. Z. (2018). Properties of encapsulated saffron extracts in maltodextrin using the Büchi B-90 nano spray-dryer. *Food Chemistry, 266*, 458–465.
Lam, S. J., Wong, E. H. H., Boyer, C., & Qiao, G. G. (2018). Antimicrobial polymeric nanoparticles. *Progress in Polymer Science, 76*, 40–64.
Lamanna, M., Morales, N. J., García, N. L., & Goyanes, S. (2013). Development and characterization of starch nanoparticles by gamma radiation: Potential application as starch matrix filler. *Carbohydrate Polymers, 97*, 90–97.
Lancuški, A., Vasilyev, G., Putaux, J. L., & Zussman, E. (2015). Rheological properties and electrospinnability of high-amylose starch in formic acid. *Biomacromolecules, 16*, 2529–2536.
Le Corre, D., Bras, J., & Dufresne, A. (2010). Starch nanoparticles: A review. *Biomacromolecules, 11*, 1139–1153.
Le Corre, D., Bras, J., & Dufresne, A. (2011). Influence of botanic origin and amylose content on the morphology of starch nanocrystals. *Journal of Nanoparticle Research, 13*, 7193–7208.
Li, Q., Chen, X., Zhuang, J., & Chen, X. (2016). Decontaminating soil organic pollutants with manufactured nanoparticles. *Environmental Science and Pollution Research, 23*, 11533–11548.
Li, J., & Luo, X. (2016). Starch-based blends. Chapter 8. In P. M. Visakh & L. Yu (Eds.), *Starch-based blends, composites and nanocomposites* (pp. 263–325). RSC Green Chemistry.
Liakos, I. L., D'autilia, F., Garzoni, A., Bonferoni, C., Scarpellini, A., Brunetti, V., et al. (2016). All natural cellulose acetate—Lemongrass essential oil antimicrobial nanocapsules. *International Journal of Pharmaceutics, 510*, 508–515.
Lin, L., Dai, Y., & Cui, H. (2017). Antibacterial poly (ethylene oxide) electrospun nanofibers containing cinnamon essential oil/beta-cyclodextrin proteoliposomes. *Carbohydrate Polymers, 178*, 131–140.
Liu, D., Wu, Q., Chen, H., & Chang, P. R. (2009). Transitional properties of starchcolloid with particle size reduction from microto nanometer. *Journal of Colloid and Interface Science, 339*, 117–124.
Lv, Y., Yang, F., Li, X., Zhang, X., & Abbas, S. (2014). Formation of heat-resistant nanocapsules ofjasmine essential oil via gelatin/gum Arabic based complex coacervation. *Food Hydrocolloids, 35*, 305–314.
Ma, X., Chang, P. R., & Yu, J. (2008). Properties of biodegradable thermoplastic peastarch/carboxymethyl cellulose and pea starch/microcrystalline cellulose composites. *Carbohydrate Polymers, 72*, 369–375.
Maftoonazad, N., & Ramaswamy, H. (2018). Novel techniques in food processing: Bionanocomposites. *Current Opinion in Food Science, 23*, 49–56.
Matalanis, A., Jones, O. G., & McClements, D. J. (2011). Structured biopolymer-based delivery systems for encapsulation, protection, and release of lipophilic compounds. *Food Hydrocolloids, 25*, 1865–1880.
Matshetshe, K. I., Parani, S., Manki, S. M., & Oluwafemi, O. S. (2018). Preparation, characterization and in vitro release study of β-cyclodextrin/chitosan nanoparticles loaded Cinnamomum zeylanicum essential oil. *International Journal of Biological Macromolecules, 118*, 676–682.
Mattinen, M. L., Valle-Delgado, J. J., Leskinen, T., Anttila, T., Riviere, G., Sipponen, M., et al. (2018). Enzymatically and chemically oxidized lignin nanoparticles for biomaterial applications. *Enzyme and Microbial Technology, 111*, 48–56.
Mehrnia, M. A., Jafari, S.-M., Makhmal-Zadeh, B. S., & Maghsoudlou, Y. (2017). Rheological and release properties of double nano-emulsions containing crocin prepared with Angum gum, Arabic gum and whey protein. *Food Hydrocolloids, 66*, 259–267.

Mendes, A. C., Stephansen, K., & Chronakis, I. S. (2017). Electrospinning of food proteins and polysaccharides. *Food Hydrocolloids*, *68*, 53–68.

Mercante, L. A., Scagion, V. P., Migliorini, F. L., Mattoso, L. H. C., & Correa, D. S. (2017). Electrospinning-based (bio) sensors for food and agricultural applications: A review. *Trends in Analytical Chemistry*, *91*, 91–103.

Mishra, R. K., Ha, S. K., Verma, K., & Tiwari, S. K. (2018). Recent progress in selected bio-nanomaterials and their engineering applications: An overview. *Journal of Science: Advanced Materials and Devices*. https://doi.org/10.1016/j.jsamd.2018.05.003.

Morillo, E., & Villaverde, J. (2017). Advanced technologies for the remediation of pesticide-contaminated soils. *Science of the Total Environment*, *586*, 576–597.

Moura, M. D., Mattoso, L. H. C., & Zucolotto, V. (2012). Development of cellulose based bactericidal nanocomposites containing silver nanoparticles and their use as active food packaging. *Journal of Food Engineering*, *109*, 520–524.

Mudgil, D., Barak, S., Patel, A., & Shah, N. (2018). Partially hydrolyzed guar gum as a potential prebiotic source. *International Journal of Biological Macromolecules*, *112*, 207–210.

Mukurumbira, A., Mariano, M., Dufresne, A., Mellem, J. J., & Amonsou, E. O. (2017). Microstructure, thermal properties and crystallinity of amadumbe starch nanocrystals. *International Journal of Biological Macromolecules*, *102*, 241–247.

Mungure, T. E., Roohinejad, S., Bekhit, A. E. D., Greiner, R., & Mallikarjunan, K. (2018). Potential application of pectin for the stabilization of nanoemulsions. *Current Opinion in Food Science*, *19*, 72–76.

Naqash, F., Masoodi, F. A., Rather, S. A., Wani, S. M., & Gani, A. (2017). Emerging concepts in the nutraceutical and functional properties of pectin—A review. *Carbohydrate Polymers*, *168*, 227–239.

Nerome, H., Machmudah, S., Fukuzato, R., Higashiura, T., Youn, Y. S., Lee, Y. W., et al. (2013). Nanoparticle formation of lycopene/β-cyclodextrin inclusion complex using supercritical antisolvent precipitation. *The Journal of Supercritical Fluids*, *83*, 97–103.

Nešić, A., Gordić, M., Davidović, S., Radovanović, Ž., Nedeljković, J., Smirnova, I., et al. (2018). Pectin-based nanocomposite aerogels for potential insulated food packaging application. *Carbohydrate Polymers*, *195*, 128–135.

Norgren, M., & Edlund, H. (2014). Lignin: Recent advances and emerging applications. *Current Opinion in Colloid and Interface Science*, *19*, 409–416.

Oliveira, A., Amaro, A. L., & Pintado, M. (2018). Impact of food matrix components on nutritional and functional properties of fruit-based products. *Current Opinion in Food Science*, *22*, 153–159.

Pelissari, F. M., Andrade-Mahecha, M. M., Sobral, P. J. A., & Menegalli, F. C. (2017). Nanocomposites based on banana starch reinforced with cellulose nanofibers isolated from banana peels. *Journal of Colloid and Interface Science*, *505*, 154–167.

Peters, R. J. B., Bouwmeester, H., Gottardo, S., Amenta, V., Arena, M., Brandhoff, P., et al. (2016). Nanomaterials for products and application in agriculture, feed and food. *Trends in Food Science & Technology*, *54*, 155–164.

Petito, N. L., Dias, D. S., Costa, V. G., Falcão, D. Q., & Araujo, K. G. L. (2016). Increasing solubility of red bell pepper carotenoids by complexation with 2-hydroxypropyl-β-cyclodextrin. *Food Chemistry*, *208*, 124–131.

Plackett, D., & Iotti, M. (2013). Preparation of nanofibrillated cellulose and cellulose whiskers. In A. Dufresne, S. Thomas, & L. A. Pothen (Eds.), *Biopolymer nanocomposites: Processing, properties, and applications* (1st ed., pp. 309–338). John Wiley & Sons, Inc.

Postulkova, H., Chamradova, I., Pavlinak, D., Humpa, O., Jancar, J., & Vojtova, L. (2017). Study of effects and conditions on the solubility of natural polysaccharide gum karaya. *Food Hydrocolloids*, *67*, 148–156.

Prakash, B., Kujur, A., Yadav, A., Kumar, A., Singh, P. P., & Dubey, N. K. (2018). Nanoencapsulation: An efficient technology to boost the antimicrobial potential of plant essential oils in food system. *Food Control*, *89*, 1–11.

Pushpalatha, R., Selvamuthukumar, S., & Kilimozhi, D. (2018). Cross-linked, cyclodextrin-based nanosponges for curcumin delivery—Physicochemical characterization, drug release, stability and cytotoxicity. *Journal of Drug Delivery Science and Technology*, *45*, 45–53.

Quiñones, J. P., Mardare, C. C., Hassel, A. W., & Brüggemann, O. (2017). Self-assembled cellulose particles for agrochemical applications. *European Polymer Journal*, *93*, 706–716.

Rangan, A., Manchiganti, M. V., Thilaividankan, R. M., Kestur, S. G., & Menon, R. (2017). Novel method for the preparation of lignin-rich nanoparticles from lignocellulosic fibers. *Industrial Crops and Products*, *103*, 152–160.

Rezaei, A., Nasirpour, A., & Fathi, M. (2015). Application of cellulosic nanofibers in food science using electrospinning and its potential risk. *Comprehensive Reviews in Food Science and Food Safety*, *14*, 269–284.

Ribeiro, A. J., Souza, F. R. L., Bezerra, J. M. N. A., Oliveira, C., Nadvorny, D., Soares, M. F. L. R., et al. (2016). Gums' based delivery systems: Review on cashew gum and its derivatives. *Carbohydrate Polymers*, *147*, 188–200.

Richter, A. P., Bharti, B., Armstrong, H. B., Brown, J. S., Plemmons, D., Paunov, V. N., et al. (2016). Synthesis and characterization of biodegradable lignin nanoparticles with tunable surface properties. *Langmuir*, *32*, 6468–6477.

Robles-García, M. A., Del-Toro-Sanchez, C. L., Márquez-Ríos, E., Barrera-Rodríguez, A., Aguilar, J., Aguilar, J. A., et al. (2018). Nanofibers of cellulose bagasse from Agave tequilana Weber var. *azul* by electrospinning: Preparation and characterization. *Carbohydrate Polymers*, *192*, 69–74.

Rossi, M., Cubadda, F., Dini, L., Terranova, M. L., Aureli, F., Sorbo, A., et al. (2014). Scientific basis of nanotechnology, implications for the food sector and future trends. *Trends in Food Science & Technology*, *40*, 127–148.

Salarbashi, D., & Tafaghodi, M. (2018). An update on physicochemical and functional properties of newly seed gums. *International Journal of Biological Macromolecules*, *119*, 1240–1247.

Salas, C., Ago, M., Lucia, L. A., & Rojas, O. J. (2014). Synthesis of soy protein–lignin nanofibers by solution electrospinning. *Reactive and Functional Polymers*, *85*, 221–227.

Saloko, S., Darmadji, P., Setiaji, B., & Pranoto, Y. (2014). Antioxidative and antimicrobial activities of liquid smoke nanocapsules using chitosan and maltodextrin and its application on tuna fish preservation. *Food Bioscience*, *7*, 71–79.

Santos, P. P., Flôres, S. H., Rios, A. O., & Chisté, R. C. (2018). Biodegradable polymers as wall materials to the synthesis of bioactive compound nanocapsules. *Trends in Food Science & Technology*, *53*, 23–33.

Santos, J. P., Zavareze, E. R., Dias, A. R. G., & Vanier, N. L. (2018). Immobilization of xylanase and xylanase–β-cyclodextrin complex in polyvinyl alcohol via electrospinning improves enzyme activity at a wide pH and temperature range. *International Journal of Biological Macromolecules*, *118*, 1676–1684.

Sathawong, S., Sridach, W., & Techato, K. (2018). Lignin: Isolation and preparing the lignin based hydrogel. *Journal of Environmental Chemical Engineering*, *6*, 5879–5888.

Seglie, L., Spadaro, D., Trotta, F., Devecchi, M., Gullino, M. L., & Scariot, V. (2012). Use of 1-methylcylopropene in cyclodextrin-based nanosponges to control grey mould caused by Botrytis cinerea on Dianthus caryophyllus cut flowers. *Postharvest Biology and Technology*, *64*, 55–57.

Shankar, S., & Rhim, J. W. (2017). Preparation and characterization of agar/lignin/silver nanoparticles composite films with ultraviolet light barrier and antibacterial properties. *Food Hydrocolloids*, *71*, 76–84.

Sharma, M., Mann, B., Sharma, R., Bajaj, R., Athira, S., Sarkar, P., et al. (2017). Sodium caseinate stabilized clove oil nanoemulsion: Physicochemical properties. *Journal of Food Engineering*, *212*, 38–46.

Sharma, G., Sharma, S., Kumar, A., Al-Muhtaseb, A. H., Naushad, M., Ghfar, A. A., et al. (2018). Guar gum and its composites as potential materials for diverse applications: A review. *Carbohydrate Polymers*, *199*, 534–545.

Sherje, A. P., Dravyakar, B. R., Kadam, D., & Jadhav, M. (2017). Cyclodextrin-based nanosponges: A critical review. *Carbohydrate Polymers*, *173*, 37–49.

Shi, A.-M., Li, D., Wang, L.-J., Li, B.-Z., & Adhikari, B. (2011). Preparation of starch-based nanoparticles through high-pressure homogenization and miniemulsion cross-linking: Influence of various process parameters on particle size and stability. *Carbohydrate Polymers*, *83*, 1604–1610.

Shishir, M. R. I., Xie, L., Sun, S., Zheng, X., & Chen, X. (2018). Advances in micro and nano-encapsulation of bioactive compounds using biopolymer and lipid-based transporters: A review. *Trends in Food Science & Technology*, *78*, 34–60.

Singh, P., Ren, X., Guo, T., Wu, L., Shakya, S., He, Y., et al. (2018). Biofunctionalization of β-cyclodextrin nanosponges using cholesterol. *Carbohydrate Polymers*, *190*, 23–30.

Song, D., Thioc, Y. S., & Deng, Y. (2011). Starch nanoparticle formation via reactive extrusion and related mechanism study. *Carbohydrate Polymers*, *85*, 208–214.

Stojanovska, E., Kurtulus, M., Abdelgawad, A., Candan, Z., & Kilic, A. (2018). Developing lignin-based bio-nanofibers by centrifugal spinning technique. *International Journal of Biological Macromolecules*, *113*, 98–105.

Sun, Q., Li, G., Dai, L., Ji, N., & Xiong, L. (2014). Green preparation and characterisation of waxy maize starch nanoparticles through enzymolysis and recrystallisation. *Food Chemistry*, *162*, 223–228.

Tan, C., Xie, J., Zhang, X., Cai, J., & Xia, S. (2016). Polysaccharide-based nanoparticles by chitosan and gum Arabic polyelectrolyte complexation as carriers for curcumin. *Food Hydrocolloids*, *57*, 236–245.

Tang, J., Sisler, J., Grishkewich, N., & Tam, K. C. (2017). Functionalization of cellulose nanocrystals for advanced applications. *Journal of Colloid and Interface Science*, *494*, 397–409.

Tang, Y., Zhang, X., Zhao, R., Guo, D., & Zhang, J. (2018). Preparation and properties of chitosan/guar gum/nanocrystalline cellulose nanocomposite films. *Carbohydrate Polymers*, *197*, 128–136.

Thakur, V. K., & Thakur, M. K. (2016). *Handbook of sustainable Polymers*. CRC Press Web.

Thielemans, W., Belgacem, M. N., & Dufresne, A. (2006). Starch nanocrystals with large chain surface modifications. *Langmuir*, *22*, 4804–4810.

Veneranda, M., Hu, Q., Wang, T., Luo, Y., Castro, K., & Madariaga, J. M. (2017). Formation and characterization of zein-caseinate-pectin complex nanoparticles for encapsulation of eugenol. *LWT—Food Science and Technology*, *89*, 596–603.

Villaverde, J., Rubio-Bellido, M., Lara-Moreno, A., Merchan, F., & Morillo, E. (2018). Combined use of microbial consortia isolated from different agricultural soils and cyclodextrin as a bioremediation technique for herbicide contaminated soils. *Chemosphere*, *193*, 118–125.

Wang, T., Soyama, S., & Luo, Y. (2016). Development of a novel functional drink from all natural ingredients using nanotechnology. *LWT—Food Science and Technology*, *73*, 458–466.

Wang, J., Tian, L., Luo, B., Ramakrishna, S., Kai, D., Loh, X. J., et al. (2018). Engineering PCL/lignin nanofibers as an antioxidant scaffold for the growth of neuron and Schwann cell. *Colloids and Surfaces B: Biointerfaces*, *169*, 356–365.

Wen, P., Zong, M.-H., Linhardt, R. J., Feng, K., & Wu, H. (2017). Electrospinning: A novel nano-encapsulation approach for bioactive compounds. *Trends in Food Science & Technology*, *70*, 56–68.

Wolf, M., Gasparin, B. C., & Paulino, A. T. (2018). Hydrolysis of lactose using β-D-galactosidase immobilized in a modified Arabic gum-based hydrogel for the production of lactose-free/lowlactose milk. *International Journal of Biological Macromolecules*, *115*, 157–164.
Wu, W., Kong, X., Zhang, C., Hua, Y., & Chen, Y. (2018). Improving the stability of wheat gliadin nanoparticles—Effect of gum Arabic addition. *Food Hydrocolloids*, *80*, 78–87.
Xie, F., Pollet, E., Halley, P. J., & Averous, L. (2013). Starch-based nano-biocomposites. *Progress in Polymer Science*, *38*, 1590–1628.
Xu, Y., Ding, W., Liu, J., Li, Y., Kennedy, J. F., Gu, Q., et al. (2010). Preparation and characterization of organic-soluble acetylated starch nanocrystals. *Carbohydrate Polymers*, *80*, 1078–1084.
Xu, X., Liu, W., Luo, L., Liu, C., & McClements, D. J. (2017). Influence of anionic polysaccharides on the physical and oxidative stability of hydrolyzed rice glutelin emulsions: Impact of polysaccharide type and pH. *Food Hydrocolloids*, *72*, 185–194.
Yang, W., Kenny, J. M., & Puglia, D. (2015). Structure and properties of biodegradable wheat gluten bionanocomposites containing lignin nanoparticles. *Industrial Crops and Products*, *74*, 348–356.
Yin, H., Liu, L., Wang, X., Wang, T., Zhou, Y., Liu, B., et al. (2018). A novel flocculant prepared by lignin nanoparticles-gelatin complex from switchgrass for the capture of *Staphylococcus aureus* and *Escherichia coli*. *Colloids and Surfaces A: Physicochemical and Engineering Aspects*, *545*, 51–59.
Yuan, C., Xu, D., Cui, B., & Wang, Y. (2019). Gelation of κ-carrageenan/Konjac glucommanan compound gel: Effect of cyclodextrins. *Food Hydrocolloids*, *87*, 158–164.
Zafar, R., Zia, K. M., Tabasum, S., Jabeen, F., Noreen, A., & Zuber, M. (2016). Polysaccharide based bionanocomposites, properties and applications: A review. *International Journal of Biological Macromolecules*, *92*, 1012–1024.
Zhang, K., Li, Z., Kang, W., Deng, N., Yan, J., Ju, J., et al. (2017). Preparation and characterization of tree-like cellulose nanofiber membranes via the electrospinning method. *Carbohydrate Polymers*, *183*, 62–69.
Zheng, H., Ai, F., Chang, P. R., Huang, J., & Dufresne, A. (2009). Structure and properties of starch nanocrystal-reinforced soy protein plastics. *Polymer Composites*, *30*, 474–480.
Zhou, J., Tong, J., Su, X., & Ren, L. (2016). Hydrophobic starch nanocrystals preparations through crosslinking modification using citric acid. *International Journal of Biological Macromolecules*, *91*, 1186–1193.
Zhou, M., Wang, T., Hu, Q., & Luo, Y. (2016). Low density lipoprotein/pectin complex nanogels as potential oral delivery vehicles for curcumin. *Food Hydrocolloids*, *57*, 20–29.
Zhu, F. (2017). Encapsulation and delivery of food ingredients using starch based systems. *Food Chemistry*, *229*, 542–552.

## Further reading

LeCorre, D., Vahanian, E., Dufresne, A., & Bras, J. (2012). Enzymatic pretreatment for preparing starch nanocrystals. *Biomacromolecules*, *13*, 132–137.
Sharma, S., Kaur, J., Sharma, G., Thakur, K. K., Chauhan, G. S., & Chauhan, K. (2013). Preparation and characterization of pH-responsive guar gum microspheres. *International Journal of Biological Macromolecules*, *62*, 636–641.

CHAPTER FOUR

# Nanoencapsulation of functional food ingredients

**Jieyu Zhu, Qingrong Huang***
Department of Food Science, Rutgers University, New Brunswick, NJ, United States
*Corresponding author: e-mail address: qhuang@sebs.rutgers.edu

## Contents



## Abstract

Many functional food ingredients are poorly soluble in water, susceptible to chemical degradation, and incompatible with surrounding food matrix. Other issues are related to limited oral bioavailability, unpleasant sensory properties, and poor release profiles. Nanoencapsulation of functional food ingredients can help increase their water solubility/dispersibility in foods and beverages, improve their bioavailability by exhibiting good dose-dependent functionalities, mask undesired flavors/tastes to reduce the adverse effect on mouth-feel, enhance shelf-life and compatibility during production, storage, transportation and utilization of food products, and control release rate or specific delivery environment for better performance on their functionalities. This chapter provides an overview of different delivery systems for different functional food ingredients, the types of materials suitable for wall materials or building blocks of nanocapsules, the fabrication methods to assemble different delivery systems and release these active ingredients under different physiological conditions.

*Advances in Food and Nutrition Research*, Volume 88
ISSN 1043-4526
https://doi.org/10.1016/bs.afnr.2019.03.005

## 1. Introduction

With an increasing awareness of the importance of eating a healthy diet, consumers have tried to incorporate functional food ingredients into foods and beverages. For example, fruit juices are fortified with Vitamin C and calcium; margarines are enriched with phytosterol esters, probiotics and probiotics (Siró, Kápolna, Kápolna, & Lugasi, 2008); green tea extracts were used to promote the oxidative stability and inhibit microbial growth in hamburger patties (Özvural, Huang, & Chikindas, 2016). β-Glucans are incorporated into baked and dairy-based food systems to manipulate food texture (Brennan & Cleary, 2005), as well as to promote health benefits (prevent constipation, reduce the risk of cancer (Mantovani et al., 2008), lower blood cholesterol (Kim, Kim, Choi, & Lee, 2005), promote the growth of health beneficial gut microbiome (Arena et al., 2014 and so on).

The "functional or active food ingredients" are referring to the food ingredients or nutraceuticals encapsulated within a delivery system to provide certain functional attributes, such as inhibiting undesired microbial growth on food products, enhancing flavors or tastes, serving as natural food colorants, promoting the growth of probiotics in gut microbiome, and providing health-beneficial effects by preventing certain diseases (Garti & McClements, 2012). The term "nutraceutical" was firstly coined from "nutrition" and "pharmaceutical" in 1989 by Stephen DeFelice (Brower, 1998). It was defined as "a food (or part of a food) that provides medical or health benefits, including the prevention and/or treatment of a disease" (Brower, 1998). Therefore, when a food product demonstrates beneficial effects relevant to the health, well-being, or the prevention and/or treatment of disease(s) and/or disorder(s) other than anemia, it can also be called as a "nutraceutical" (Kalra, 2003).

Many "functional food ingredients" are incompatible with the food matrix, and cannot be incorporated into food systems through simple mixing in their original forms. Often, they have to be firstly encapsulated using certain delivery systems before they can be successfully incorporated into food systems (Garti & McClements, 2012). This chapter will begin with an introduction of the terminology and classification for different delivery systems. Various functional food ingredients that have been encapsulated will be introduced. Case studies related the efficacies of the nanoencapsulated functional food ingredients are then provided.

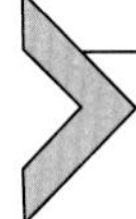

# 2. Food grade delivery systems

## 2.1 Terminology and classification

For the convenience of readers, this section will begin with an introduction of the terminology and classification of various food grade delivery systems (Fig. 1).

*Liposomes*: Liposomes are vesicles formed by amphipathic lipid molecules, usually by phospholipids, which give a spherical bilayer structure with one (unilamellar liposomes) or multiple (multi-lamellar liposomes) aqueous cores.

*Niosomes*: Niosomes are a class of vesicles that have a similar bilayer structure as liposomes, but are stabilized by nonionic surfactants, such as amphiphiles bearing sugar, polyoxyethylene, polyglycerol, crown ether, and amino acid hydrophilic head groups (Uchegbu & Vyas, 1998).

*Colloidosomes*: Colloidosomes are formed by clusters of colloidal particles with a typical hollow shell structure.

*Emulsions*: Emulsions, also known as *conventional emulsions* or *macro-emulsions* (McClements, 2010), are fine dispersion systems comprised of two immiscible phases, in which a dispersed phase is homogeneously dispersed into a continuous phase, with the emulsion droplets being stabilized by amphiphilic surface-active surfactants. A conventional emulsion typically appears to be turbid/opaque with an average droplets radius between 100 nm and 100 μm (McClements, 2011). The term "macro-emulsions" was developed in order to distinguish them from microemulsions (Wennerström, Söderman, Olsson, & Lindman, 1997).

*Nanoemulsions*: Nanoemulsions are a type of conventional emulsions that have an average droplets size between 10 and 100 nm (McClements, 2011).

*Microemulsions*: Microemulsions are thermodynamically stable systems that typically contain particles with radii somewhere in the range of 2 to 50 nm (McClements, 2011).

*Pickering emulsions*: Pickering emulsions refer to emulsions that are stabilized by insoluble amphiphilic materials, which can be stabilized by either spherical particles or rod-like fibers. Due to the larger size of surfactants as compared with conventional emulsions, the average droplet sizes of Pickering emulsions are commonly in the micrometer scales.

*Solid lipid nanoparticles*: Solid lipid nanoparticles are considered as a submicron emulsions, because they are formed in the same way as emulsions by homogenizing two immiscible phases together with the presence of

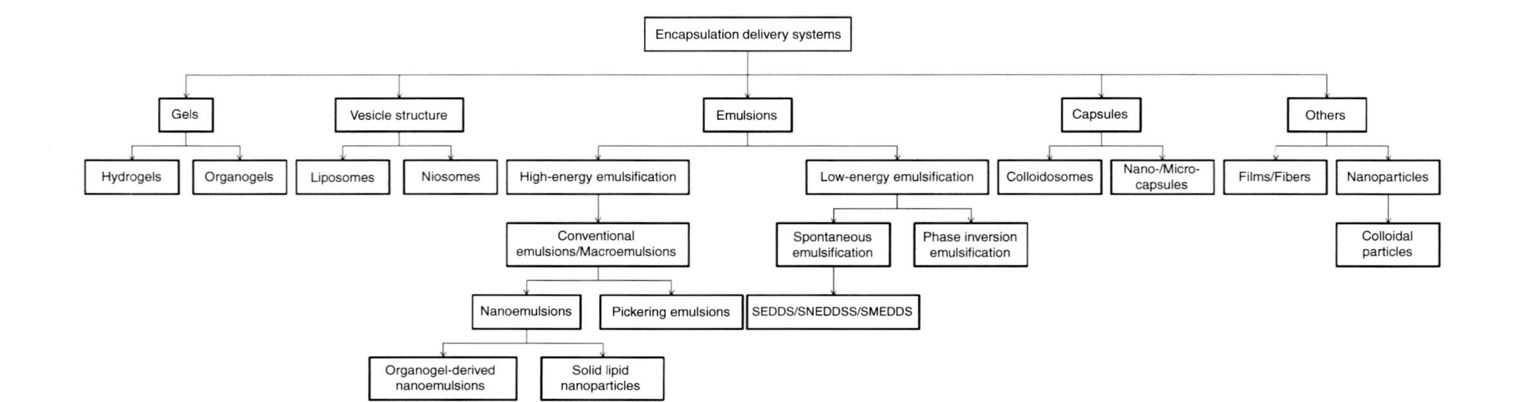


**Fig. 1** Summary of various encapsulation delivery systems

amphiphilic surface-active surfactants. But the melting point of the lipid phase is higher than that of conventional emulsions, and the emulsification should be conducted at an elevated temperature to maintain the lipid phase in a liquid form during the process.

*SEDDS/SMEDDS/SNEDDS*: SEDDS/SMEDDS/SNEDDS stands for self-(micro/nano)-emulsifying drug delivery system. In this system, stable emulsions can be formed simply by mild mixing.

*Nanoparticles*: Nanoparticles refer to particles that have at least one dimension between 1 and 100 nm.

*Colloidal particles*: Colloidal particles are microscopic solid particles that suspended in a fluid, which are small enough to ensure an equilibration with the suspending fluid, while large enough that their positions and motions can be measured by optical methods (Lu & Weitz, 2013).

*Gel*: A gel is a solid-like network with crosslinked polymers swollen in a liquid medium. For gels with solid-like characteristics, a storage modulus, G', which exhibits a pronounced plateau extending to times at least of the order of seconds, and the loss modulus, G", is considerably smaller than the storage modulus in the plateau region (Almdal, Dyre, Hvidt, & Kramer, 1993).

*Hydrogels*: Hydrogels are defined as two- or multi-component systems consisting of a three-dimensional polymer network and the water that fills the space between macromolecules (Ahmed, 2015).

*Organogels*: Organogels are defined as self-standing, thermo-reversible, anhydrous, and viscoelastic materials with a three-dimensional supramolecular network of self-assembled small molecules dissolved in an organic liquid at concentrations no greater than their percolation threshold, usually $\leq 2\%$ (w/w) (Rogers, Wright, & Marangoni, 2009). The surfactant-like molecules may include a variety of compounds, such as hydroxylated fatty acids, fatty acids, fatty alcohols, fatty acid–fatty alcohol mixtures, phytosterol–oryzanol mixtures, sorbitan monostearate, waxes, and lecithin–sorbitan tristearate mixtures (Rogers et al., 2009).

*Capsules*: Capsules have a typical core–shell structure where they have a solid shell and the core can take all aggregate states (Neubauer, Poehlmann, & Fery, 2014), such as liquid state, solid state, or gas/vacant state. Their sizes range from several hundreds of nanometers to tens (in few cases to hundreds) of micrometers (Neubauer et al., 2014).

*Nano-/micro-capsules*: Capsules are formed in the nano- or micro-range.

*Films*: Films refer to thin layers or coatings similar to membranes.

## 2.2 Emulsions

As discussed above, there are many different delivery systems for the encapsulation of functional food ingredients. This chapter will focus on detail discussion of emulsion delivery systems.

### *2.2.1 High-energy emulsification*

In high-energy methods, intense disruptive forces are applied to break the dispersed phase into small droplets using devices such as high speed/shear homogenizer (HSH), high-pressure homogenizers (HPH), microfluidizers (MF), ultrasonic devices (US), and membrane filtration method. The disruptive forces applied must exceed the restorative forces so that large droplets can be broken down into small droplets (Schubert & Engel, 2004). To achieve smaller and more even-distributed droplets, a two-step operation is usually applied, by first using the HSH technique to form primary coarse emulsion, followed by processing the coarse emulsion in HPH or MF systems to form nanoemulsions.

In HPH, the coarse emulsion is pumped into a chamber on its backstroke and then forced out through a narrow valve at the end of the chamber. Larger droplets are broken down into smaller ones under intense disruptive forces caused by turbulence, shear and cavitation, as the emulsion passes through the valve. The droplet size decreases with increases in homogenization pressure, the number of passes, and adsorption rate of emulsifiers (Tadros, Izquierdo, Esquena, & Solans, 2004). A reduction in particle size is desirable for oral formulations containing hydrophobic bioactives to provide higher oral bioavailability (Ting, Li, Wang, Ho, & Huang, 2015). Besides reducing the droplet size, HPH can also influence the morphology of the encapsulated compounds. For example, Ting et al. evaluated the effect of the polymethoxyflavones (PMFs) loading concentration (0.3–2.0%), high pressure levels (500, 1000, or 1500 bar), and processing temperatures (55, 65, and 75 °C) on the resulting crystals in the emulsion stabilized by 1.5% rapeseed lecithin (Ting, Li, Wang, et al., 2015). They found out that the higher the processing pressure, the larger the crystals size (Ting, Li, Wang, et al., 2015), which may be due to the shift of melting temperature toward a higher temperature, eventually resulting in the increase of the degree of undercooling and of the nucleation rates (Ito, Tsutsumi, Minagawa, Takimoto, & Koyama, 1995).

In MF, similar to HPH, a pump is used to force a coarse emulsion through a narrow orifice at high pressures into an inlet channel (Mahdi Jafari, He, & Bhandari, 2006). This channel is designed to split the coarse emulsion into two streams, collide with each other at high speed, and form fine emulsions (McClements, 2011). On the other hand, in US, an ultrasonic probe containing a piezoelectric crystal generates high-intensity ultrasonic waves (frequency >20 kHz) to turn a pre-existing coarse emulsion into a fine emulsion with very small droplet sizes (Kentish et al., 2008). The driving forces for the disruption of droplets include cavitation, turbulence and interfacial waves (Kentish et al., 2008). Finally, in membrane filtration method, pressure is applied to drive the emulsion through a semipermeable membrane with a specific pore size, forming droplets ranging from micro (microfiltration) to nano or micro (ultrafiltration) scale. Membrane filtration is ideal for processing emulsions that contain heat-sensitive materials, because the process does not generate heat.

#### 2.2.1.1 Organogel-derived nanoemulsions

Organogel-based delivery systems are relatively new to the food industry. Organogels are formed by liquid oils trapped by the extensive crystalline networks of organogelators, such as monoglycerides, fatty acids and fatty alcohols (Pernetti, van Malssen, Flöter, & Bot, 2007).

Organogel-based nanoemulsion systems were first developed by Huang et al., involving a thermoreversible semisolid organogel to immobilize highly solubilized curcumin (Yu & Huang, 2012). Other nutraceuticals have also been encapsulated in the organogel-derived nanoemulsions, such as capsaicin (8-methyl-*N*-vanillyl-6-nonenamide) (Lu, Cao, Ho, & Huang, 2016) extracted from capsicum plants (Hayman & Kam, 2008), and D-limonene (4-isopropenyl-1-methylcyclohexene) extracted from the citrus essential oils (Zahi, Wan, Liang, & Yuan, 2014).

The emulsification of the organogel-derived nanoemulsions can be achieved through ultrasonication or high pressure homogenization (Lu et al., 2016). Organogel systems have the advantages of high stability and high encapsulation efficiency for water-insoluble nutraceuticals (Lu et al., 2016). Organogel-derived nanoemulsion systems also have the capacity to improve the dispersibility, water solubility, and oral bioavailability of hydrophobic compounds (Lu et al., 2016).

#### 2.2.1.2 Solid-lipid nanoparticles

Solid lipid nanoparticles (SLNs) are fabricated by homogenizing an oil phase and an aqueous phase together at temperatures above the melting point of the lipid phase, together with one or more surfactants. The elevated temperatures should be maintained during the entire emulsification procedure to prevent the formation of solidified lipid crystals that may block or damage the homogenizer (McClements, 2010). When the emulsion is cooled under controlled conditions, the lipid phase begins to crystallize. The spatial orientation of the lipid crystals and the stability of the lipid nanoparticles/droplets depend on the amount and type of lipids used, the property of the surfactant(s), the initial droplet size, concentration, and the cooling conditions (McClements, 2010).

Due to the high melting temperature of the core material, SLNs contain a solid or semi-solid lipid core, which slows down the release rate of the nutraceuticals from the crystalline lipid. Moreover, the solid core of SLNs makes them more resistant to breakdown by gastric enzyme activities when compared to emulsions with liquid disperse phase. Therefore, SLNs are excellent controlled-release systems to prevent burst release and prolong the retention time of bioactive compounds during digestion (Ting, Jiang, Ho, & Huang, 2014). However, this delivery system is not suitable to encapsulate heat-sensitive compound.

#### 2.2.1.3 Pickering emulsions

Pickering emulsions are based on the emulsification capacities of amphiphilic colloidal particles. In order to stabilize Pickering emulsions, there are several requirements for the particle stabilizers, including their partial wettability at the two immiscible phases and optimal particle sizes of at least one order of magnitude smaller than the targeted emulsion droplet (Xiao, Li, & Huang, 2016).

There are two major strategies to fabricate colloidal particles. The top-down strategy uses either chemical (e.g., enzyme-catalyzed hydrolysis and acidic hydrolysis) or mechanical (e.g., wet-ball milling machine, cryogenic milling machine, and high-pressure homogenizer) methods to break down materials to achieve enhanced swelling properties and emulsification capacities. Pickering emulsions stabilized by starch granules or cellulose crystals are usually prepared by this method. Another strategy is the bottom-up assembling method, which forms colloidal particles through aggregates or complexes through anti-solvent precipitation, crosslinking, and electrostatic interaction.

In the anti-solvent precipitation method, the polymers are firstly dissolved in a good solvent to prepare a stock solution, which is then trickled into a poor solvent under continuous stirring, leading to aggregation of the polymers. The sizes of the polymer aggregates can be precisely controlled by tuning the concentration of the polymer stock solution, the stirring speed, the volume ratio of dispersed stock solution to the bulk poor solvent, and the sequence of addition (i.e., adding the stock solution into the bulk poor solvent or vice versa) (Joye & McClements, 2013). This method can be applied to fabricate prolamin-based colloidal particles, such as zein protein extracted from maize (de Folter, van Ruijven, & Velikov, 2012), kafirin extracted from sorghum (Xiao, Wang, Perez Gonzalez, & Huang, 2016), hordein extracted from barley, gliadin extracted from wheat, and coixin extracted from coix.

In the crosslinking method, crosslinkers are added into the polymer solution to form crosslinked nanoparticles either through covalent crosslinking effect or ionic crosslinking effect (Hu & Huang, 2013). The crosslinking density is determined by the molar ratio between the crosslinkers and the polymer repeating units.

In the electrostatic interaction method, the oppositely charged polyelectrolytes interact with each other through direct electrostatic interaction, leading to the formation of polyelectrolyte complexes (Hu & Huang, 2013). By tuning the pH and the ionic strength of the protein and polysaccharide solutions, protein–polysaccharide complexed nanoparticles are formed for stabilizing Pickering emulsions. Besides the electrostatic attraction, hydrogen bonding and hydrophobic interaction are the major driving forces for formation of protein–polysaccharide complexes (Xiao, Nian, & Huang, 2015). The size and the surface charge of the colloidal particles can be tuned by modifying the pH values and the concentration of protein and polysaccharide solutions, and the mass ratio of protein to polysaccharide.

### 2.2.2 Low-energy emulsification

In the low-energy emulsification method, stable emulsions can be produced through mild mixing or even by hand-shaking. Self-nanoemulsifying drug delivery system (SNEDDS) is a commonly used low-energy method to prepare nanoemulsions or microemulsions. These emulsions are formed by dissolving the water-miscible solvent and/or surfactant in the organic phases, followed by mixing of oil and water phases together, during which the surfactant molecules spontaneously move from the organic phase into the aqueous phase to form stable emulsions through a budding process (McClements, 2011). The driving force for the self-emulsifying process is the movement of a water-miscible component (either solvent or surfactant) from the organic phase into the aqueous phase (Anton & Vandamme, 2009). In order to achieve the optimum composition profile in SNEDDS, typically a series of formulas are being prepared to construction of pseudo-ternary phase diagrams with varying weight percentages of oil, surfactant, and co-surfactant. Then the mixtures are gently diluted (1:100 or 1:1000) with distilled water and inspected for the tendency of self-emulsification (e.g., transparency of the emulsion, time taken to form emulsions). The SNEDDS is useful for improving the aqueous solubility and the oral bioavailability of the encapsulated compounds. For instance, according to an in vivo pharmacokinetic study, the bioavailability of boswellic acids, a potent anti-inflammatory agent, was improved by around two times by using SNEDDS formulation, as compared with the bulk oil suspension (Ting et al., 2018).

The phase inversion temperature (PIT) method is the other low-energy emulsification method, which generates emulsions through changing the optimum curvature (molecular geometry) or solubility of non-ionic surfactants (Anton & Vandamme, 2009). Emulsions formed by phase inversion method undergo a transition from O/W emulsion to W/O emulsion or vice versa (Ting, Li, Ho, & Huang, 2013). Phase inversion can be attributed to changes in the physicochemical properties of the surfactant with temperature (McClements, 2011). For instance, upon increasing the processing temperature, the non-polar lipophilic head of surfactants may expand to much larger size as compared to polar hydrophilic head, which eventually changes the curvature of droplets and turns the emulsion from O/W to W/O. Besides changing the temperature, the phase inversion can be induced by altering the emulsion composition (e.g., types of surfactant and lipid, fraction of aqueous phase) (Ting, Li, et al., 2013). Phase inversion method is capable of generating emulsions with small droplet size in nanometer range using an energy-efficient manufacturing processes (Ting, Li, et al., 2013).

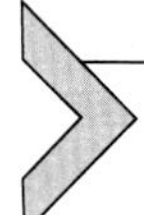

## 3. Case studies for bioactivities of encapsulated functional food ingredients

In this section, we will introduce selected functional food ingredients and their functional/physicochemical properties. The rationales and methods of encapsulation of these ingredients in various delivery systems are discussed.

### 3.1 Health-beneficial phytochemicals

Many phytochemicals exhibit antioxidant, anti-inflammation, anti-diabetic, anti-obesity, and anti-cancer properties. However, their applications for preventing and treating diseases have been limited due to their low levels of water solubility, stability, bioavailability and target specificity in the body (Li, Jiang, Xu, & Gu, 2015; Wang et al., 2014). To illustrate these phenomena, this section will review several phytochemicals including (−)-epigallocatechin gallate (EGCG) in green tea, resveratrol in grapes, polymethoxyflavones in citrus peels, and curcumin in turmeric.

### 3.1.1 (−)-Epigallocatechin gallate

Originating from China, tea, made from leaf and bud of the plant *Camellia sinensis*, has been consumed all over the world for the past 2000 years (Cabrera, Artacho, & Giménez, 2006). There are four major catechins that present in green tea, including (−)-epigallocatechin-3-gallate (EGCG), (−)-epigallocatechin (EGC), (−)-epicatechin-3-gallate (ECG), and (−)-epicatechin (EC). EGCG, as the most abundant and also the most bioactive catechin in tea products, which accounts for 25–55% of total catechins (Wang et al., 2014). According to a database published by the United States Department of Agriculture (USDA), 68.2–70.2 mg of EGCG was detected in 100 g of 1% freshly brewed tea infusion (1 g tea leaves/100 mL boiling water) (Bhagwat, Haytowitz, & Holden, 2014).

Many reviews have systematically summarized the beneficial health effects of green tea, such as antioxidant activity, anti-obesity and anti-diabetic, anti-mutagenic, anti-carcinogenic, anti-hypertensive effects, as well as the reduction on the risks of cardiovascular disease, prevention on dental caries, protection against ultraviolet damage, etc. (Cabrera et al., 2006; McKay & Blumberg, 2002). However, these potential beneficial effects depend not only on the amount of catechins consumed, but also on their bioavailability in vivo (Cabrera et al., 2006).

EGCG has a lower bioavailability than other green tea catechins in humans (Yang et al., 1998). After taking 1.5 g of decaffeinated green tea solids (dissolved in 500 mL of water), the maximum plasma concentration of EGCG, EGC, and EC was 326, 550, and 190 ng/mL, respectively (Yang et al., 1998). After the oral administration of 3 g of decaffeinated green tea solids (in water) to humans, the maximum plasma concentration of EGCG, EGC, and EC (reached between 1.4 and 2.4 h) were 0.57, 1.60, and 0.6 μM respectively (Lu et al., 2003). In another recent human trial, after a single dose of 1.5 mmol of pure EGCG, EGC, and ECG, the average peak plasma concentrations were 1.3 μmol/L for EGCG, 5.0 μmol/L for EGC, and 3.1 μmol/L for ECG (Van Amelsvoort et al., 2001). Similar results were also observed in rats and mice when they were given 0.6% green tea polyphenol solution as the drinking fluid. The plasma concentrations of EGCG were much lower than those of EGC or EC (Kim et al., 2000).

After oral administration of green tea to rats, about 14% of EGC, 31% of EC, and <1% of EGCG appeared in the blood (Lu et al., 2003).

The poor bioavailability of EGCG was due to low bio-absorption and extensive metabolism during digestion. Tea catechins undergo Phase II metabolism through extensive O-methylation, glucuronidation, and sulfation reactions. These reactions, together with the efflux pumps and other transporters, may play an important role in reducing the bioavailability of tea catechins (Lu et al., 2003). As for the tissue distribution of catechins after green tea consumption, it was observed that EGCG levels were higher in the esophagus and large intestine, but lower in other organs in an animal study, when rats were given 0.6% green tea polyphenol in their drinking water over a period of 28 days. The observation was probably due to the poor systemic absorption of EGCG (Kim et al., 2000). Both animal and human studies have confirmed that EGC and EC were excreted through urine, while most of the EGCG ingested was excreted in the bile, and then excreted through feces (Chen, Lee, Li, & Yang, 1997; Kim et al., 2000; Yang et al., 1998).

In order to improve the bioabsorption (e.g., cellular uptake) and prolong the release profile of EGCG, researchers have developed chitosan-tripolyphosphate (CS-TPP) nanoparticles with EGCG encapsulation efficiencies of 25.8–47.4%, wherein 45–60% of EGCG was released from the nanoparticles in 12 h (Hu et al., 2008). They further improved the formulation by using the bioactive caseinophosphopeptide (CPP) to form self-assembling nanoparticles with CS (Hu, Ting, Zeng, & Huang, 2013). The CS-CPP nanoparticles had lower cellular toxicity, as revealed by MTT assay on Caco-2 cells (Hu, Ting, Zeng, & Huang, 2012) with the IC50 value of 0.95 mg/mL compared with 0.35 mg/mL for CS-TPP nanoparticles (Hu, Ting, Yang, et al., 2012). Moreover, the encapsulation efficiency of EGCG in CS-CPP nanoparticles increased to 70.5–81.7%, while a more sustainable EGCG release profile was achieved with CS-CPP nanoparticles, at around 23–40% in 12 h (Hu et al., 2013). The reduced release rate was due to the reaction between CPP and EGCG that trapped EGCG in CS-CPP nanoparticles, as confirmed by the QCMD results (Hu, Ting, Yang, et al., 2012).

### *3.1.2 Resveratrol*

Resveratrol (RES), a naturally occurring polyphenol, has been widely studied for its health benefits, such as anti-oxidation, anti-inflammation, anti-obesity, anti-tumor activities, etc. (Amri, Chaumeil, Sfar, & Charrueau, 2012; Davidov-Pardo & McClements, 2014; Szkudelska & Szkudelski, 2010).

RES exhibited good anti-obesity and anti-diabetic effects on mice fed with high-fat diet by increasing insulin sensitivity, reducing the levels of insulin-like growth factor-1, increasing AMP-activated protein kinase, and increasing the amount of mitochondrial number, etc. (Baur et al., 2006). The anti-obesity effect of RES is considered to be related to gut microbiome; it regulates the profile of gut microbes which in turn affects the fat metabolism and changes fat storage in adipose tissues (Qiao et al., 2014). It has been reported that obese microbiome, with lower *Bacteroidetes/Firmicutes* ratio, has a higher capacity to absorb energy from the diet (Turnbaugh et al., 2006). For its functionalities in the lower gastro-intestinal (GI) tract, RES also exhibits therapeutic effects on colitis (Cui et al., 2010) and colon cancer (Schneider et al., 2000; Vanamala, Reddivari, Radhakrishnan, & Tarver, 2010).

However, RES has a very limited bioavailability due to its poor solubility, limited stability, high metabolism rate, and fast excretion after oral administration. RES is soluble in ethanol (~50 mg/mL) and DMSO (~16 mg/mL) (NCBI, 2014), but poorly soluble in water (~0.03 mg/mL) (Amri et al., 2012) and triacylglycerol oils (~0.18 mg/mL) (Hung, Chen, Liao, Lo, & Fang, 2006). Moreover, RES is very susceptible to chemical degradation when exposed to high temperatures (Liazid, Palma, Brigui, & Barroso, 2007; Zupančič, Lavrič, & Kristl, 2015), alkaline conditions (Trela & Waterhouse, 1996; Zupančič et al., 2015), ultraviolet light (Trela & Waterhouse, 1996), and certain enzymes (Pinto, García-Barrado, & Macías, 2003). After oral administration, RES is being absorbed (46–80%) through the small intestine, and then undergone rapid metabolism (sulfation and glucuronidation reactions) in the liver cells and enterocytes (Amri et al., 2012). The peak serum level has been recorded at around 30 min at 22–40 nM levels (Goldberg, Yan, & Soleas, 2003; Walle, Hsieh, DeLegge, Oatis, & Walle, 2004). The majority of RES (71–98%) was rapidly excreted from kidney 7–15 h post-administration (Amri et al., 2012). Therefore, the use of suitable delivery systems is essential for delivering RES into the colon to elicit its anti-colitis and anti-colon cancer activities.

Examples of emulsion-based systems for the delivery of RES include oil-in-water emulsions (Sessa et al., 2014; Wan, Wang, Wang, Yang, & Yuan, 2013), water-in-oil emulsions (Spigno et al., 2013), and water-in-oil-in-water double emulsions (Hemar, Cheng, Oliver, Sanguansri, & Augustin, 2010; Matos, Gutiérrez, Coca, & Pazos, 2014; Matos, Gutiérrez, Iglesias, Coca, & Pazos, 2015). Many of these delivery systems are aimed at improving the encapsulation efficiency (Matos et al., 2015), increasing the bio-absorption in the small intestine (Sessa et al., 2014), improving the oxidative stability of emulsions (Spigno et al., 2013), or improving the chemical stability of RES (Sessa, Tsao, Liu, Ferrari, & Donsì, 2011). On the other hand, SLNs provide a sustained release profile for RES (Teskač & Kristl, 2010). The inclusion of RES in liposomes improved its chemical stability against UV-B-induced oxidative damage (Caddeo, Teskač, Sinico, & Kristl, 2008). RES has also been encapsulated in biodegradable nanoparticles for its controlled-release into the colon (Das, Lin, Ho, & Ng, 2008; Das & Ng, 2010a, 2010b).

### 3.1.3 Polymethoxyflavones

Citrus flavonoids are widely present in the peels of tangerine (*Citrus tangerina*), sweet orange (*Citrus sinensis*) and bitter orange (*Citrus aurantium*) (Li, Wang, Guo, Zhao, & Ho, 2014). They can be either glycosides (e.g., hesperidin and naringin) or O-methylated aglycones of flavones (e.g., nobiletin and tangeretin), which are also known as polymethoxyflavones (PMFs) (Li et al., 2014). PMFs are abundant in citrus (*Citrus reticulata* Blanco) peels, especially in aged citrus peels, which are commonly used as a traditional Chinese medicine for many years to relieve stomach upset, cough, skin inflammation, muscle pain, and ringworm infections, as well as for lowering blood pressure (Guo et al., 2016; Li et al., 2009). In terms of health-beneficial effects, various studies have shown that PMFs have anti-inflammatory (Ho & Kuo, 2014; Murakami et al., 2000; Tominari, Hirata, Matsumoto, Inada, & Miyaura, 2012), anti-cancer (Lai et al., 2007; Manthey & Guthrie, 2002), and anti-atherogenic (Kurowska & Manthey, 2004; Miyata et al., 2011; Saito, Abe, & Sekiya, 2007) properties. The proposed mechanisms for these functional properties have been systematically summarized in a review paper by Li et al. (2009). The anti-obesity effect of *Chenpi* extract (31.28% nobiletin, 22.82% tangeretin, and 9.22% 5-OH nobiletin) was firstly reported in a mice study conducted by Guo et al. (2016). After oral administration of 0.25% and 0.5% chenpi extract for over 15 weeks, they reported that the high fat diet-induced obesity, hepatic steatosis, and diabetic symptoms were significantly reduced, which may be due to the activation of 5′-adenosine monophosphate-activated protein kinase (AMPK) in adipose tissue (Guo et al., 2016).

PMFs are hydrophobic molecules with poor solubility in water, as well as oil at ambient and body temperatures (Li, Zheng, Xiao, & McClements, 2012). PMFs tend to be rapidly metabolized during the digestion in the

GI tract. One study in rats showed that after oral administration of nobiletin or tangeretin (50 mg/kg of body weight), the peak serum level was 9.3 μg/mL in 0.5 h and 0.49 μg/mL in 1 h, respectively (Manthey, Cesar, Jackson, & Mertens-Talcott, 2010), indicating that the bio-absorption of tangeretin was much lower than that of nobiletin. The highest concentrations of the two main metabolites of tangeretin, i.e., aglycone metabolites and glucuronic acid conjugates, were 0.38 μg/mL at 4–6 h and 1.45 μg/mL at 3–6 h, respectively (Manthey et al., 2010). As for nobiletin, eight metabolites were detected, including mono-demethylnobiletin glucuronides (peak level of 5.4 μg/mL at 8 h), mono-demethylnobiletin aglycones (peak level of 0.50 μg/mL at 8 h) (Manthey et al., 2010), and others due to the demethylation of PMFs at the C-4′ position (Yasuda et al., 2003).

Bioavailability, defined as the fraction that eventually reaches the systemic circulation after oral administration, is related to the following three components, i.e., the amount of bioactive compounds released into the small intestine and becoming bioaccessible for the subsequent bio-absorption, the amount of bioactive compounds absorbed through the gut wall, and the amount of bioactive compound remained intact by the first pass metabolism (Ting et al., 2014). The poor solubility of PMFs limits their bioaccessibility in the human gut system, while high metabolism rate affects the amount of PMFs entering the human circulating system, which eventually leads to a poor bioavailability and reduces the therapeutic dosages they can achieve in the target organs.

Chemical modification is a useful approach to improve the bioavailability of PMFs. For example, Tung et al. employed chemical modification on 5-demethylnobiletin (5-OH-Nob) to produce a acetyl derivative of 5-OH-Nob, that is, 5-acetyloxy-6,7,8,3,4-pentamethoxyflavone (5-Ac-Nob). They reported that 5-Ac-Nob reduced the triacylglycerol in 3T3-L1 preadipocytes to an extent which was even >5-OH-Nob (Tung et al., 2016). However, the synthesized compound may not naturally exist in the food system. For the safety concern and from an acceptance standpoint, the use of nanoencapsulation technique to improve the bioavailability of PMFs is likely more desirable than the chemical modification approach.

Previous studies have shown that when dissolving lipophilic bioactives into oil carriers, such as medium and long chain triglycerides, the bioavailability can be improved. This effect has been attributed to the surface active products produced during lipid digestion, such as monoacylglycerols and free fatty acids, that facilitated the adhesion and absorption through the gut cell walls (Porter, Trevaskis, & Charman, 2007). Researchers have attempted to use emulsions to improve the bioavailability of PMFs. For instance, high amounts of tangeretin (2.3–2.5%) can be loaded into a visco-elastic emulsion system stabilized by rapeseed PC 75 lecithin (Ting, Jiang, Lan, et al., 2015; Ting, Xia, Li, Ho, & Huang, 2013). According to the results obtained from the TNO's gastrointestinal model (TIM-1; a simulating apparatus for the entire human GI tract, including stomach, duodenum, jejunum, and ileum), the bioaccessibility of tangeretin in the viscoelastic emulsion was increased 2.6-fold compared with the one of tangeretin oil suspension (Ting, Jiang, Lan, et al., 2015). In vivo mice study further confirmed that the oral bioavailability of tangeretin was enhanced 2.3-fold in the viscoelastic emulsion as compared with the unformulated oil suspension (Ting, Jiang, Lan, et al., 2015). Other studies also showed that the bio-efficacy of tangeretin against colorectal cancer was also significantly improved in the lecithin stabilized emulsions at the dosages of 12.5 and 25 μM (Ting, Chiou, Pan, Ho, & Huang, 2015).

Since lecithin increased the dosing efficacy and the oral bioavailability of tangeretin, the toxicity effect that each dosage had originally posed for the organisms may change (Ting, Chiou, Jiang, et al., 2015). In a sub-acute study (feeding mice with 50 mg/kg bw or 100 mg/kg bw of tangeretin for over 28 days either as MCT suspension or lecithin emulsion), the emulsion system did not generate significant adverse changes in physiological status, nor induce a significant increase of toxicity response (Ting, Chiou, Jiang, et al., 2015).

### 3.1.4 Curcumin

Curcumin is the main curcuminoid compound found in the rhizome of plant turmeric (*Curcuma longa*), together with demethoxycurcumin (D-Cur) and bisdemethoxycurcumin (BD-Cur) (Yu & Huang, 2012). Curcumin has been widely reported for its various health promoting properties, such as anti-cancer (Aggarwal, Kumar, & Bharti, 2003; Bimonte et al., 2016), anti-inflammation (Aggarwal & Harikumar, 2009; Jurenka, 2009; Menon & Sudheer, 2007), antioxidant (Menon & Sudheer, 2007), antimicrobial activities (Basniwal, Buttar, Jain, & Jain, 2011; Vimala et al., 2011), anti-obesity and anti-diabetic effect (Aggarwal, 2010; Alappat & Awad, 2010), etc. Curcumin is relatively safe even at an extremely high dosage. For instance, curcumin showed no toxicity in human at 36–180 mg per day for 4 months, and dose-limiting toxicity was not observed (Sharma et al., 2001). Even though curcumin has poor bioavailability, Phase I clinical trials showed that it is safe even at a dosage of as high as 12 g/day in humans (Anand, Kunnumakkara, Newman, & Aggarwal, 2007).

Curcumin has a low oral bioavailability due to its poor solubility in water, weak absorption, rapid metabolism, and quick systemic elimination (Anand et al., 2007). In one study, after oral administration of a high dose of curcumin extract (440 and 2200 mg/day containing 36–180 mg of curcumin) on 15 patients with advanced colorectal cancer, neither curcumin nor its metabolites were detected in blood or urine, but it was recovered from feces (Sharma et al., 2001). In another pharmacokinetic study involving 24 subjects, curcumin was not detected in serum after oral administration of 500, 1000, 2000, 4000, 6000 or 8000 mg (Lao et al., 2006). But when the dose was increased to 11,000 or 12,000 mg, curcumin was detected in only two subjects with the highest plasma concentrations of 0.051 and 0.057 μg/mL, respectively, whereas no plasma level of curcumin was detected in the remaining subjects at the same dose levels (Lao et al., 2006). In another study where rats were fed with curcumin at a dose of 1 g/kg or 2 g/kg, a maximum plasma concentration of 0.22 μg/mL or $1.35 \pm 0.23$ μg/mL was observed, respectively, while 2 g of curcumin orally administrated in humans resulted in either undetectable or extremely low plasma levels ($0.006 \pm 0.005$ μg/mL at 1 h) (Shoba, Joy, Joseph, Rajendran, & Srinivas, 1998).

As mentioned before, the oral bioavailability of curcumin is influenced by its solubility, permeability, and pre-/post-absorption metabolism. The water solubility of curcumin is $3 \times 10^{-5}$ μmol/mL (Kaminaga et al., 2003). The solubilized curcumin permeates through cells by passive diffusion with a fairly high rate (Yu & Huang, 2011). During and after the permeation process, curcumin is being metabolized rapidly, by biotransformation to dihydrocurcumin and tetrahydrocurcumin (THC), followed by subsequent conversion to monoglucuronide conjugates (Pan, Huang, & Lin, 1999). Curcumin sulfates have also been identified in the feces of a human subject in a clinical study (Sharma et al., 2001). These results suggested that curcumin-glucuronoside, dihydrocurcumin-glucuronoside, THC-glucuronoside, and THC are the main metabolites of curcumin in vivo (Pan et al., 1999).

Up to now, various formulations, such as micelles (Yu, Li, Shi, & Huang, 2011), conventional O/W emulsions (Wang et al., 2008), Pickering emulsions (Xiao, Li, & Huang, 2015), self-emulsifying system (Cui et al., 2009), nanoparticles (Yallapu, Gupta, Jaggi, & Chauhan, 2010), organogel (Yu, Shi, Liu, & Huang, 2012) and organogel derived emulsions (Yu & Huang, 2012), have been developed to encapsulate curcumin. Curcuminoids encapsulated by octenyl succinic anhydride modified ε-polylysine (M-EPL) micelles protected curcumin against the rapid hydrolysis at weak basic conditions, which facilitated the movement of curcuminoids into/onto the HepG2 cells, and improved the cellular antioxidant activity due to the increased soluble concentration in micelle solution (Yu et al., 2011). In a study involving a mouse ear inflammation model, enhanced anti-inflammation activity was achieved by using 10% of Tween 20 to stabilize 10% of curcumin in medium chain triacylglycerols (MCT) oil solutions under high-speed and high-pressure homogenization, with an average droplet size ranging from 618.6 nm to 79.5 nm (Wang et al., 2008). Kafirin nanoparticle-stabilized Pickering emulsions (KPEs) provided a better protection effect on curcumin against UV radiation and the better oxidative stability against lipid oxidation, as compared with bulk oil (BE) or Tween 80 stabilized nanoemulsions (TEs) (Xiao, Li, et al., 2015). The bioaccessibility of curcumin in TEs was higher than that in KPE and BE, since KPEs were broken down by pepsin during the gastric digestion (Xiao, Li, et al., 2015). Cui et al. reported that the solubility of curcumin was significantly increased to 21 mg/g in a self-microemulsifying drug delivery system (SMEDDS), consisting of 57.5% surfactant (emulsifier OP:Cremorphor EL = 1:1), 30.0% co-surfactant (PEG 400) and 12.5% oil (ethyl oleate) (Cui et al., 2009).

Curcumin nanoparticles can be applied to improve its anti-cancer potentials. Cell proliferation and clonogenic assays have demonstrated that curcumin nanoparticles increased the cellular uptake by two and six times as compared to free curcumin, in cisplatin resistant A2780CP ovarian and metastatic MDA-MB-231 breast cancer cells, respectively (Yallapu et al., 2010). Xiao et al. fabricated kafirin/carboxymethyl chitosan nanoparticles (Kaf/CMC) to enhance the cellular uptake of curcumin with an encapsulation efficiency and loading efficiency of 86.1 ± 2.1% and 6.1 ± 0.2%, respectively (Xiao, Nian, et al., 2015). After 180 min of incubation, the intracellular uptake of curcumin in cells treated with curcumin-Kaf/CMC nanoparticles was 42% higher than those treated with curcumin-Kaf nanoparticles (Xiao, Nian, et al., 2015). The surface carboxyl groups from CMC play important roles in internalization of nanoparticles (Dausend et al., 2008) and mitigates the surface hydrophobicity of kafirin nanoparticles, thereby facilitating the cellular uptake of curcumin (Xiao, Nian, et al., 2015).

The bioaccessibility of curcumin was improved by encapsulating it in a food-grade organogel system, consisting of MCT, Span 20 as the surfactant, and monostearin as an organogelator (Yu et al., 2012). Compared with unformulated curcuminoids, MCT-Span 20 oil and organogel increased bioaccessibility by 107-fold and 100-fold in the fasted state, respectively. While in the fed state, the bioaccessibility increased by 25-fold and 21-fold, respectively (Yu et al., 2012). What's more, organogels can also be applied in nanoemulsions. For example, novel organogel-based nanoemulsion, stabilized by Tween 20, improved the oral bioavailability of curcumin by 9-fold as compared with unformulated curcumin according to an in vivo pharmacokinetics analysis on mice (Yu & Huang, 2012). To further improve the antimicrobial efficacy of curcumin, chitosan-poly(vinyl alcohol) (PVA) silver nanocomposite films have been developed, which displayed enormous growth inhibition of *Escherichia coli* compared to unformulated curcumin and the blank film (Vimala et al., 2011).

## 3.2 Flavors

Flavors play an important role in the food industry by directly influencing the quality and acceptability of food products. Natural flavors are more acceptable among consumers as compared with artificial/synthetic counterparts. However, the price of natural flavors can be costly. For instance, the price of pure natural vanillin flavor is 4000 USD/kg, while the one of synthetic equivalent is just around 12 USD/kg (Ramachandra Rao & Ravishankar, 2000). Preservation of flavors in food matrices is challenging due to evaporation, degradation, reaction, or migration phenomena that occur during production, transportation, and storage of the food products. There are many techniques that can be applied to encapsulate flavors, such as spray drying, spray chilling/cooling, extrusion, freeze drying, coacervation and molecular inclusion, which have been systematically reviewed in a paper (Madene, Jacquot, Scher, & Desobry, 2006). There are a plethora of flavor compounds being investigated by researchers. For brevity, this chapter will use citral as an example to illustrate a general idea on how to encapsulate flavors using emulsions.

Citral (3,7-dimethyl-2,6-octadienal), a mixture of two geometrical isomers neral and geranial, is an important lemon flavor component extracted from citrus essential oils (Yang, Tian, Ho, & Huang, 2011a). Due to its monoterpene structure with an aldehyde group, it is vulnerable to acid-catalyzed cyclization and oxidation (Kimura, Iwata, & Nishimura, 1982). In order to improve the stability of citral, Yang et al. encapsulated citral in conventional nanoemulsions (Yang, Tian, Ho, & Huang, 2011b) and multilayer nanoemulsions (Yang et al., 2011a) using Alcolec PC 75 soy lecithin as the emulsifier. The multilayer emulsions were developed via layer-by-layer deposition of positively charged biopolymers (chitosan or ε-poly-lysine) onto negatively charged emulsion droplets (Yang et al., 2011a). The amount of remaining citral after 28-day storage at 25 °C was >2-fold higher in lecithin–CS-stabilized nanoemulsion (Yang et al., 2011a). These observations show that multilayer nanoemulsions exhibited better protection effect on citral than single-layer nanoemulsions. The combination of nanoencapsulation and additional antioxidants can significantly improve the stability of citral. For instance, 0.10 wt% of ubiquinol-10 (Q10H2) in citral-loaded oil-in-water (O/W) nanoemulsions (Q10H2/citral ratio 1:1) effectively protected citral from chemical degradation and oxidation, where the major citral oxidation off-flavor compounds (*p*-cresol, α,*p*-dimethylstyrene, *p*-methylacetophenone) and some lipid degradation products were significantly reduced to lower levels (Zhao, Ho, & Huang, 2013).

## 3.3 Antimicrobial agents

The microbial safety of food products has always been a major concern to consumers, regulatory agencies, and food industries all over the world (Negi, 2012). Based on the concept of hurdle technology, various food preservation methods have been used in combination to control food contamination and microbial-induced food spoilage by controlling the temperature, water activity, pH values, salt/sugar concentration, and processing pressure, etc. For the safety concern and also the exploration of more healthy and natural products, natural antimicrobial agents are more acceptable to consumers as compared with synthetic antimicrobials, such as potassium sorbate, sodium benzoate, sodium nitrite, and sulfur dioxide, etc.

The natural anti-microbial agents can be directly added into food and beverages to control microbial spoilage, or indirectly added into food packaging material to prolong shelf-life of food product (Donsì, Annunziata, Sessa, & Ferrari, 2011; Negi, 2012).

### 3.3.1 Essential oils

Essential oils (EOs) are well known for their anti-bacterial, anti-fungal, anti-oxidant, and insecticidal effects, etc. (Burt, 2004; Tajkarimi, Ibrahim, & Cliver, 2010). EOs can be extracted from a variety of natural sources like flowers, fruits, leaves, bark, and buds, through steam distillation, solvent extraction, fermentation, or other approaches (Burt, 2004). In order to improve the antimicrobial efficacy of EOs, various techniques are used, such as nanoemulsions (Donsì, Annunziata, Vincensi, & Ferrari, 2012; Ghosh, Mukherjee, & Chandrasekaran, 2014; Salvia-Trujillo, Rojas-Graü, Soliva-Fortuny, & Martín-Belloso, 2015; Topuz et al., 2016), organogel-derived nanoemulsions (Zahi, Liang, & Yuan, 2015), micro-emulsions (Ma & Zhong, 2015; Zhang, Critzer, Davidson, & Zhong, 2014), and films (Ma et al., 2016). For instance, the anise (*Pimpinella anisum*) oil-incorporated nanoemulsions, stabilized by soy lecithin using high pressure homogenization, showed better bactericidal effect against common foodborne pathogens, *Listeria monocytogenes* and *E. coli* O157: H7, in comparison with the coarse emulsion achieved by high speed homogenization (Topuz et al., 2016). The increased antimicrobial activity of nanoemulsions may be due to their effects on facilitating the accessibility of the essential oils to the bacterial cells (Salvia-Trujillo et al., 2015). For instance, in the case of Gram-negative bacteria, such as *E. coli*, proteins (porins) existing in their outer membrane form channels (pores) which allow the influx of hydrophilic particles, while inhibit the permeation of lipophilic compounds, such as the lipophilic components in EOs. On the other hand, nanoemulsion could increase the solubility/dispersity of lipophilic EOs, facilitate the binding and the interaction between the porins and EOs, thereby resulting in higher permeability of EOs (Salvia-Trujillo et al., 2015). Similarly, nanoscale delivery systems can improve the bioactivity of the encapsulated compounds through the activation of passive mechanisms of cell absorption (Donsì et al., 2012). Various nanoencapsulation formulations of essential oils were summarized in Table 1.

**Table 1** Nanoencapsulation formulations of selected essential oils.

| Essential oils | EOs percentage | Surfactant | Target microbes | Reference |
|---|---|---|---|---|
| Anise oil | 0.25–1% (v/v) | Alcolec soy lecithin PC75 (5%, w/w) | *Listeria monocytogenes* and *Escherichia coli* O157:H7 | Topuz et al. (2016) |
| Peppermint oil | 2–10% (v/v) | Purity Gum 2000 s (a succinylated waxy maize starch, 12%, w/w) | *Listeria monocytogenes* Scott A and *Staphylococcus aureus* | Liang et al. (2012) |
| *Thymus daenensis* | 2% (w/w) | Tween 80 (2%, w/w) and lecithin (0.001%, w/w) | *Escherichia coli* | Moghimi, Ghaderi, Rafati, Aliahmadi, and McClements (2016) |
| Thyme oil | 5% (w/w) (mixed with corn oil in different ratio) | Tween 80 (neutral, 0.5%, w/w), or lauric arginate (cationic, 0.8%, w/w), or sodium dodecyl sulfate (anionic, 0.32%, w/w) | *Zygosaccharomyces bailii*, *Saccharomyces cerevisiae*, *Brettanomyces bruxellensis*, and *Brettanomyces naardenensis* | Ziani, Chang, McLandsborough, and McClements (2011) |
| Thyme oil | 1% (w/v) | Sodium caseinate (4% or 2%, w/v), and soy lecithin (0.5% or 0.25%, w/v) | *Escherichia coli* O157:H7, *Listeria monocytogenes*, and *Salmonella Enteritidis* | Xue, Davidson, and Zhong (2015) |
| Eucalyptus oil | 16.66% (v/v) | Tween 80 (16.66%, v/v) | *Staphylococcus aureus* | Sugumar, Ghosh, Nirmala, Mukherjee, and Chandrasekaran (2014) |
| Lemongrass oil | 1% (v/v) | Tween 80 (1%, v/v), and sodium alginate (1%, w/v) in water | *Escherichia coli* | Salvia-Trujillo et al. (2015) |
| D-limonene | 4% (w/w) | Tween 80 (10%, or 6%, w/w) | *Escherichia coli*, *Staphylococcus aureus*, *Bacillus subtilis*, *Saccharomyces cerevisiae* | Zahi et al. (2015); Zhang, Vriesekoop, Yuan, and Liang (2014) |

### 3.3.2 Antimicrobial peptides

Some functional peptides extracted from food products have antimicrobial properties. For example, kefir is a traditional milk beverage fermented by lactic acid bacteria, acetic acid bacteria, and yeasts (Abraham & de Antoni, 1999). Miao et al. found that a cell-penetrating peptide extracted from kefir displayed anti-microbial activity against a broad spectrum of microorganisms including both Gram-positive and Gram-negative bacteria, as well as several fungi (Miao et al., 2016). Nisin is another well-known antimicrobial peptide produced by certain strains of *Lactococcus lactis*, which have been commercially used as food preservatives in processed cheese, various pasteurized dairy products, and canned vegetables for several decades (Delves-Broughton, Blackburn, Evans, & Hugenholtz, 1996).

In order to improve the stability and anti-microbial efficacy of nisin against *L. monocytogenes*, Were et al. encapsulated nisin in phospholipid liposomes. The encapsulated nisin in 100% phosphatidylcholine (PC) and PC-cholesterol liposomes inhibited bacterial growth by >2 log CFU/mL compared with free nisin (Were, Bruce, Davidson, & Weiss, 2004). Chopra et al. synthesized by the ionotropic complexation method, nisin incorporated in chitosan/carageenan nanocapsules showed stronger antibacterial effect in vitro and in tomato juice for prolonged periods as compared to the unformulated nisin (Chopra, Kaur, Bernela, & Thakur, 2014). They showed that even though the initial inhibition zone of free nisin was smaller for the free nisin, its antibacterial effect only lasted for 3 days, while the antimicrobial effect of the encapsulated nisin sustained for at least up to 20 days (Chopra et al., 2014). These observations show that the nanoencapsulated nisin is able to provide a sustained antimicrobial effect during the shelf-life of food products. A similar higher and longer-lasting inhibitory effect of nisin was reported by Kopermsub et al. when the antimicrobial peptide was encapsulated in niosomes prepared by Span 80 with sodium stearoyl lactate (SSL) and PEG400 (Kopermsub, Mayen, & Warin, 2011).

## 3.4 Natural colorants

The color of food plays a vital role in food industry, because it is directly related to the appearance, safety, and nutritional value of the food products. Hence, color has a great influence on the product acceptability among consumers. Due to the increasing trend on the demand of "green" food and "clean" label, consumers tend to prefer naturally extracted food colorants over the synthetic ones, even though the synthetic colorants exhibited higher stability, better hues, and lower cost (Sigurdson, Tang, & Giusti, 2017). The application of natural colorants in the food and beverages is quite limited, mainly due to their low stability, weak tinctorial strength, interactions with food ingredients, and limited availability to match the desired hues (Sigurdson et al., 2017).

There are many natural colorants available in the nature, such as those extracted from flowers, fruits, herbs, and vegetables, such as anthocyanins, carotenoids, betalains, and chlorophylls, etc. In this chapter, anthocyanins and β-carotenes are used as model compounds to illustrate how natural colorants can be encapsulated, protected, and delivered into food systems.

### 3.4.1 Anthocyanins

Anthocyanins are natural colorants and nutraceuticals widely existing in the red and blue parts of many plants (Cheynier, 2012). During the recent decades, anthocyanins have been widely applied in the food industry to improve the color of food products such as jams, fruit juices, soft drinks, canned food, dairy products, and confectioneries (Montilla, Arzaba, Hillebrand, & Winterhalter, 2011). However, anthocyanins are susceptible to chemical degradation. Moreover, the rate of color fading and the loss of functionalities are highly dependent on pH, light, temperature, oxygen, enzymes, and the interactions with surrounding food ingredients (Bordenave, Hamaker, & Ferruzzi, 2014; West & Mauer, 2013). In order to apply anthocyanin into beverage products as a natural food colorant, Chung et al. investigated the use of biopolymers (1%), including native whey protein, denatured whey protein, citrus pectin, and beet pectin, to inhibit its degradation (Chung, Rojanasasithara, Mutilangi, & McClements, 2015). Ersus Bilek et al. incorporated anthocyanins in whey protein hydrogel and applied into yoghurt to improve their chemical stability and antioxidant activity (Ersus Bilek, Yılmaz, & Özkan, 2017). Also for yoghurt coloring application, Robert et al. used maltodextrin microcapsule to stabilize anthocyanin by a spray drying method (Robert et al., 2010).

### 3.4.2 *β*-Carotene

Carotenoid represent a large group of tetraterpenoids that provide yellow, orange, and red colors (Liang, Shoemaker, Yang, Zhong, & Huang, 2013). Carotenoids can be extracted from fruits and vegetables such as carrots, sweet potatoes, winter squash, cantaloupe, and mango (Van Duyn &

Pivonka, 2000), as well as animal origins such as eggs, shrimp, lobster, salmon, bacteria and fungi (Yonekura & Nagao, 2007). β-Carotene, a common carotenoid, is a well-known active phytochemical with many health-promoting properties, including antioxidant, enhancement of immune system, prevention of Alzheimer's disease, precursor to vitamin A, cell differentiation, synthesis of glycoprotein, mucus secretions from epithelial tissues, reproduction, development of bones, and so on (Dutta, Chaudhuri, & Chakraborty, 2005).

However, β-carotene has poor aqueous solubility and only slightly soluble in oil at room temperature, which limits its application in the food industry (Liang, Shoemaker, et al., 2013). In addition, the high extent of unsaturation in the structure of β-carotene makes it sensitive to pH, oxygen, light, and temperature. It is susceptible to oxidation, isomerization, and photosensitization during production and storage (Jia, Kim, & Min, 2007; López-Rubio & Lagaron, 2011; Zeb & Murkovic, 2010). Moreover, the intestinal absorption of carotenoids has been thought to occur by passive diffusion and receptor-mediated transport in enterocytes (Yonekura & Nagao, 2007). In vitro studies involving the modeling of the gastrointestinal environment demonstrated that carotenoids have to be emulsified and micellized before being absorbed by the intestinal cells (Yonekura & Nagao, 2007). Therefore, nanoencapsulation is essential to facilitate the bioabsorption and health-promoting benefits of carotenoids.

In an effort to improve the stability and bioaccessibility of β-carotene, oil-in-water nanoemulsions stabilized by food-grade biopolymer emulsifiers (OSA-modified starches) were fabricated by Liang et al. using high-pressure homogenization (Liang, Shoemaker, et al., 2013). They showed that the bioaccessibility of β-carotene increased from 3.1 to 35.6% through the nanoencapsulation (Liang, Shoemaker, et al., 2013). To overcome the limitations of liquid emulsion system, Liang et al. spray-dried β-carotene nanoemulsions, stabilized by modified starch, into powders. This process resulted in lower film oxygen permeability and higher retention of β-carotene during storage (Liang, Huang, Ma, Shoemaker, & Zhong, 2013). Besides nanoemulsions, β-carotene have been encapsulated in nanotubes (Yanagi, Miyata, & Kataura, 2006), solid-lipid nanoparticles (Helgason et al., 2009), electrospun fibers (Fernandez, Torres-Giner, & Lagaron, 2009), etc. Examples of nanoencapsulation delivery systems of different functional food ingredients were summarized in Table 2.

**Table 2** Examples of nanoencapsulation of functional food ingredients.

| Categories | Examples | Functional attributes | Reasons for encapsulation | Examples of delivery systems |
|---|---|---|---|---|
| Health beneficial phytochemicals | EGCG | Anti-oxidant, anti-inflammation, anti-obesity, anti-hypertensive effect, anti-cancer, and UV-protection, etc. | Active efflux pumps, rapid Phase II metabolism, rapid elimination from feces, and low bioavailability | Chitosan-tripolyphosphate nanoparticles (Hu et al., 2008), chitosan-caseinophosphopeptide nanoparticles (Hu et al., 2013) |
| | Resveratrol | Anti-oxidation, anti-inflammation, anti-obesity, anti-colitis, and anti-cancer, etc. | Poor solubility, limited stability, high metabolism rate, high absorption in small intestine, rapid elimination from urine, and low bioavailability | Oil-in-water emulsions (Sessa et al., 2014; Wan et al., 2013), water-in-oil emulsions (Spigno et al., 2013), and water-in-oil-in-water double emulsions (Hemar et al., 2010; Matos et al., 2014; Matos et al., 2015), solid lipid nanoparticles (Teskač & Kristl, 2010), liposomes (Caddeo et al., 2008), and biodegradable nanoparticles (Das et al., 2008; Das & Ng, 2010a, 2010b) |
| | Polymethoxylated flavones | Anti-inflammation, anti-cancer, anti-atherogenic, and anti-obesity and anti-diabetic effect, etc. | Poor solubility and low bioavailability | Nanoemulsions (Ting, Jiang, Lan, et al., 2015; Ting, Xia, et al., 2013) |
| | Curcumin | Anti-cancer, anti-inflammation, antioxidant, antimicrobial, and anti-obesity and anti-diabetic effect, etc. | Poor absorption, rapid metabolism, and rapid systemic elimination | Micelles (Yu et al., 2011), conventional O/W emulsions (Wang et al., 2008), Pickering emulsions (Xiao, Li, et al., 2015), self-emulsifying system (Cui et al., 2009), nanoparticles (Yallapu et al., 2010), organogels (Yu et al., 2012) and organogel derived emulsions (Yu & Huang, 2012) |

| | | | | |
|---|---|---|---|---|
| Flavors | Citral | Lemon aroma | Susceptible to acid-catalyzed cyclization and oxygen-induced oxidation | Nanoemulsions (Yang et al., 2011b; Zhao et al., 2013) and multilayer nanoemulsions (Yang et al., 2011a) |
| Anti-microbial agents | Essential oils | Anti-bacterial, anti-fungal, antioxidant, and insecticidal effects | Improve anti-microbial efficacy, prevent un-desired evaporation during food production, and provide sustained release during shelf-life | Nanoemulsions (Donsì et al., 2012; Ghosh et al., 2014; Salvia-Trujillo et al., 2015; Topuz et al., 2016), organogel-derived nanoemulsions (Zahi et al., 2015), microemulsions (Ma & Zhong, 2015; Zhang, Critzer, et al., 2014), and films (Ma et al., 2016) |
| | Anti-microbial peptides | Anti-bacterial and anti-fungal properties | Improve the anti-microbial efficacy and stability, provide sustained release during shelf-life | Liposomes (Were et al., 2004), nanocapsules (Chopra et al., 2014), niosomes (Kopermsub et al., 2011) |
| Natural colorants | Anthocyanins | Colorants and antioxidant | Susceptible to chemical degradation | Polymer complexes (Chung et al., 2015), hydrogels (Ersus Bilek et al., 2017), microcapsules (Robert et al., 2010) |
| | β-Carotenes | Colorants, antioxidant, enhance immune system, prevent Alzheimer's disease, and precursor to vitamin A | Poor solubility, susceptible to chemical degradation, improve bio-absorption and bio-efficacy | Nanoemulsions (Liang, Shoemaker, et al., 2013), spray-dried powders (Liang, Huang, et al., 2013), nanotubes (Yanagi et al., 2006), solid-lipid nanoparticles (Helgason et al., 2009), electrospun fibers (Fernandez et al., 2009) |

## 4. Conclusion

With the development of nanotechnology, more novel delivery systems are being developed than ever for the encapsulation of various functional components to improve the flavors and colors of food products. The goals are to provide safer and "clean-label" food products, to increase the acceptability of consumers on the food products, to enhance the bioabsorption and bioavailability of nutraceuticals, and to extend product shelf-life. Optimal encapsulation formulations are selected based on the chemical and physical property of the encapsulants. For example, lipophilic bioactive compounds can be incorporated in oil-in-water emulsions, organogels, the hydrophobic bilayers in liposomes, and electrospun fibers, etc., while hydrophilic compounds can be encapsulated in water-in-oil emulsions, oil-in-water-in-oil double emulsions, the hydrophilic compartments in liposomes, hydrogels, and nanoparticles, and so on. Even though there are a plethora of delivery systems available nowadays, studies are needed to further elucidate the digestion profiles, the pharmacokinetic properties, and the safety issues of various delivery systems.

## References

Abraham, A. G., & de Antoni, G. L. (1999). Characterization of kefir grains grown in cows' milk and in soya milk. *Journal of Dairy Research*, *66*, 327–333.

Aggarwal, B. B. (2010). Targeting inflammation-induced obesity and metabolic diseases by curcumin and other nutraceuticals. *Annual Review of Nutrition*, *30*, 173–199.

Aggarwal, B. B., & Harikumar, K. B. (2009). Potential therapeutic effects of curcumin, the anti-inflammatory agent, against neurodegenerative, cardiovascular, pulmonary, metabolic, autoimmune and neoplastic diseases. *The International Journal of Biochemistry & Cell Biology*, *41*, 40–59.

Aggarwal, B. B., Kumar, A., & Bharti, A. C. (2003). Anticancer potential of curcumin: Preclinical and clinical studies. *Anticancer Research*, *23*, 363–398.

Ahmed, E. M. (2015). Hydrogel: Preparation, characterization, and applications: A review. *Journal of Advanced Research*, *6*, 105–121.

Alappat, L., & Awad, A. B. (2010). Curcumin and obesity: Evidence and mechanisms. *Nutrition Reviews*, *68*, 729–738.

Almdal, K., Dyre, J., Hvidt, S., & Kramer, O. (1993). Towards a phenomenological definition of the term 'gel'. *Polymer Gels and Networks*, *1*, 5–17.

Amri, A., Chaumeil, J., Sfar, S., & Charrueau, C. (2012). Administration of resveratrol: What formulation solutions to bioavailability limitations? *Journal of Controlled Release*, *158*, 182–193.

Anand, P., Kunnumakkara, A. B., Newman, R. A., & Aggarwal, B. B. (2007). Bioavailability of curcumin: Problems and promises. *Molecular Pharmaceutics*, *4*, 807–818.

Anton, N., & Vandamme, T. F. (2009). The universality of low-energy nano-emulsification. *International Journal of Pharmaceutics*, *377*, 142–147.

Arena, M. P., Caggianiello, G., Fiocco, D., Russo, P., Torelli, M., Spano, G., et al. (2014). Barley β-glucans-containing food enhances probiotic performances of beneficial bacteria. *International Journal of Molecular Sciences*, *15*, 3025–3039.

Basniwal, R. K., Buttar, H. S., Jain, V., & Jain, N. (2011). Curcumin nanoparticles: Preparation, characterization, and antimicrobial study. *Journal of Agricultural and Food Chemistry*, *59*, 2056–2061.

Baur, J. A., Pearson, K. J., Price, N. L., Jamieson, H. A., Lerin, C., Kalra, A., et al. (2006). Resveratrol improves health and survival of mice on a high-calorie diet. *Nature*, *444*, 337.

Bhagwat, S., Haytowitz, D. B., & Holden, J. M. (2014). *USDA database for the flavonoid content of selected foods, Release 3.1.* Beltsville, MD, USA: US Department of Agriculture.

Bimonte, S., Barbieri, A., Leongito, M., Piccirillo, M., Giudice, A., Pivonello, C., et al. (2016). Curcumin anticancer studies in pancreatic cancer. *Nutrients*, *8*, 433.

Bordenave, N., Hamaker, B. R., & Ferruzzi, M. G. (2014). Nature and consequences of non-covalent interactions between flavonoids and macronutrients in foods. *Food & Function*, *5*, 18–34.

Brennan, C. S., & Cleary, L. J. (2005). The potential use of cereal (1→3,1→4)-β-d-glucans as functional food ingredients. *Journal of Cereal Science*, *42*, 1–13.

Brower, V. (1998). Nutraceuticals: Poised for a healthy slice of the healthcare market? *Nature Biotechnology*, *16*, 728.

Burt, S. (2004). Essential oils: Their antibacterial properties and potential applications in foods—A review. *International Journal of Food Microbiology*, *94*, 223–253.

Cabrera, C., Artacho, R., & Giménez, R. (2006). Beneficial effects of green tea—A review. *Journal of the American College of Nutrition*, *25*, 79–99.

Caddeo, C., Teskač, K., Sinico, C., & Kristl, J. (2008). Effect of resveratrol incorporated in liposomes on proliferation and UV-B protection of cells. *International Journal of Pharmaceutics*, *363*, 183–191.

Chen, L., Lee, M.-J., Li, H., & Yang, C. S. (1997). Absorption, distribution, and elimination of tea polyphenols in rats. *Drug Metabolism and Disposition*, *25*, 1045–1050.

Cheynier, V. (2012). Phenolic compounds: From plants to foods. *Phytochemistry Reviews*, *11*, 153–177.

Chopra, M., Kaur, P., Bernela, M., & Thakur, R. (2014). Surfactant assisted nisin loaded chitosan-carageenan nanocapsule synthesis for controlling food pathogens. *Food Control*, *37*, 158–164.

Chung, C., Rojanasasithara, T., Mutilangi, W., & McClements, D. J. (2015). Enhanced stability of anthocyanin-based color in model beverage systems through whey protein isolate complexation. *Food Research International*, *76*, 761–768.

Cui, X., Jin, Y., Hofseth, A. B., Pena, E., Habiger, J., Chumanevich, A., et al. (2010). Resveratrol suppresses colitis and colon cancer associated with colitis. *Cancer Prevention Research*, *3*, 549–559. https://doi.org/10.1158/1940-6207.CAPR-09-0117.

Cui, J., Yu, B., Zhao, Y., Zhu, W., Li, H., Lou, H., et al. (2009). Enhancement of oral absorption of curcumin by self-microemulsifying drug delivery systems. *International Journal of Pharmaceutics*, *371*, 148–155.

Das, S., Lin, H.-S., Ho, P. C., & Ng, K.-Y. (2008). The impact of aqueous solubility and dose on the pharmacokinetic profiles of resveratrol. *Pharmaceutical Research*, *25*, 2593–2600.

Das, S., & Ng, K.-Y. (2010a). Resveratrol-loaded calcium-pectinate beads: Effects of formulation parameters on drug release and bead characteristics. *Journal of Pharmaceutical Sciences*, *99*, 840–860.

Das, S., & Ng, K.-Y. (2010b). Colon-specific delivery of resveratrol: Optimization of multi-particulate calcium-pectinate carrier. *International Journal of Pharmaceutics*, *385*, 20–28.

Dausend, J., Musyanovych, A., Dass, M., Walther, P., Schrezenmeier, H., Landfester, K., et al. (2008). Uptake mechanism of oppositely charged fluorescent nanoparticles in HeLa cells. *Macromolecular Bioscience*, *8*, 1135–1143.

Davidov-Pardo, G., & McClements, D. J. (2014). Resveratrol encapsulation: Designing delivery systems to overcome solubility, stability and bioavailability issues. *Trends in Food Science & Technology*, *38*, 88–103.
de Folter, J. W., van Ruijven, M. W., & Velikov, K. P. (2012). Oil-in-water Pickering emulsions stabilized by colloidal particles from the water-insoluble protein zein. *Soft Matter*, *8*, 6807–6815.
Delves-Broughton, J., Blackburn, P., Evans, R., & Hugenholtz, J. (1996). Applications of the bacteriocin, nisin. *Antonie Van Leeuwenhoek*, *69*, 193–202.
Donsì, F., Annunziata, M., Sessa, M., & Ferrari, G. (2011). Nanoencapsulation of essential oils to enhance their antimicrobial activity in foods. *LWT—Food Science and Technology*, *44*, 1908–1914.
Donsì, F., Annunziata, M., Vincensi, M., & Ferrari, G. (2012). Design of nanoemulsion-based delivery systems of natural antimicrobials: Effect of the emulsifier. *Journal of Biotechnology*, *159*, 342–350.
Dutta, D., Chaudhuri, U. R., & Chakraborty, R. (2005). Structure, health benefits, antioxidant property and processing and storage of carotenoids. *African Journal of Biotechnology*, *4*, 1510–1520.
Ersus Bilek, S., Yılmaz, F. M., & Özkan, G. (2017). The effects of industrial production on black carrot concentrate quality and encapsulation of anthocyanins in whey protein hydrogels. *Food and Bioproducts Processing*, *102*, 72–80.
Fernandez, A., Torres-Giner, S., & Lagaron, J. M. (2009). Novel route to stabilization of bioactive antioxidants by encapsulation in electrospun fibers of zein prolamine. *Food Hydrocolloids*, *23*, 1427–1432.
Garti, N., & McClements, D. J. (2012). *Encapsulation technologies and delivery systems for food ingredients and nutraceuticals*. Elsevier.
Ghosh, V., Mukherjee, A., & Chandrasekaran, N. (2014). Eugenol-loaded antimicrobial nanoemulsion preserves fruit juice against, microbial spoilage. *Colloids and Surfaces B: Biointerfaces*, *114*, 392–397.
Goldberg, D. M., Yan, J., & Soleas, G. J. (2003). Absorption of three wine-related polyphenols in three different matrices by healthy subjects. *Clinical Biochemistry*, *36*, 79–87.
Guo, J., Tao, H., Cao, Y., Ho, C.-T., Jin, S., & Huang, Q. (2016). Prevention of obesity and type 2 diabetes with aged citrus peel (chenpi) extract. *Journal of Agricultural and Food Chemistry*, *64*, 2053–2061.
Hayman, M., & Kam, P. C. (2008). Capsaicin: A review of its pharmacology and clinical applications. *Current Anaesthesia and Critical Care*, *19*, 338–343.
Helgason, T., Awad, T. S., Kristbergsson, K., Decker, E. A., McClements, D. J., & Weiss, J. (2009). Impact of surfactant properties on oxidative stability of β-carotene encapsulated within solid lipid nanoparticles. *Journal of Agricultural and Food Chemistry*, *57*, 8033–8040.
Hemar, Y., Cheng, L. J., Oliver, C. M., Sanguansri, L., & Augustin, M. (2010). Encapsulation of resveratrol using water-in-oil-in-water double emulsions. *Food Biophysics*, *5*, 120–127.
Ho, S.-C., & Kuo, C.-T. (2014). Hesperidin, nobiletin, and tangeretin are collectively responsible for the anti-neuroinflammatory capacity of tangerine peel (Citri reticulatae pericarpium). *Food and Chemical Toxicology*, *71*, 176–182.
Hu, B., & Huang, Q.-r. (2013). Biopolymer based nano-delivery systems for enhancing bioavailability of nutraceuticals. *Chinese Journal of Polymer Science*, *31*, 1190–1203.
Hu, B., Pan, C., Sun, Y., Hou, Z., Ye, H., & Zeng, X. (2008). Optimization of fabrication parameters to produce chitosan-tripolyphosphate nanoparticles for delivery of tea catechins. *Journal of Agricultural and Food Chemistry*, *56*, 7451–7458.
Hu, B., Ting, Y., Yang, X., Tang, W., Zeng, X., & Huang, Q. (2012). Nanochemoprevention by encapsulation of (−)-epigallocatechin-3-gallate with bioactive peptides/chitosan nanoparticles for enhancement of its bioavailability. *Chemical Communications*, *48*, 2421–2423.

Hu, B., Ting, Y., Zeng, X., & Huang, Q. (2012). Cellular uptake and cytotoxicity of chitosan–caseinophosphopeptides nanocomplexes loaded with epigallocatechin gallate. *Carbohydrate Polymers, 89*, 362–370.
Hu, B., Ting, Y., Zeng, X., & Huang, Q. (2013). Bioactive peptides/chitosan nanoparticles enhance cellular antioxidant activity of (−)-epigallocatechin-3-gallate. *Journal of Agricultural and Food Chemistry, 61*, 875–881.
Hung, C.-F., Chen, J.-K., Liao, M.-H., Lo, H.-M., & Fang, J.-Y. (2006). Development and evaluation of emulsion-liposome blends for resveratrol delivery. *Journal of Nanoscience and Nanotechnology, 6*, 2950–2958.
Ito, H., Tsutsumi, Y., Minagawa, K., Takimoto, J.-i., & Koyama, K. (1995). Simulations of polymer crystallization under high pressure. *Colloid and Polymer Science, 273*, 811–815.
Jia, M., Kim, H. J., & Min, D. B. (2007). Effects of soybean oil and oxidized soybean oil on the stability of β-carotene. *Food Chemistry, 103*, 695–700.
Joye, I. J., & McClements, D. J. (2013). Production of nanoparticles by anti-solvent precipitation for use in food systems. *Trends in Food Science & Technology, 34*, 109–123.
Jurenka, J. S. (2009). Anti-inflammatory properties of curcumin, a major constituent of *Curcuma longa*: A review of preclinical and clinical research. *Alternative Medicine Review, 14*, 141–153.
Kalra, E. K. (2003). Nutraceutical—Definition and introduction. *AAPS PharmSci, 5*, 27–28.
Kaminaga, Y., Nagatsu, A., Akiyama, T., Sugimoto, N., Yamazaki, T., Maitani, T., et al. (2003). Production of unnatural glucosides of curcumin with drastically enhanced water solubility by cell suspension cultures of *Catharanthus roseus*. *FEBS Letters, 555*, 311–316.
Kentish, S., Wooster, T., Ashokkumar, M., Balachandran, S., Mawson, R., & Simons, L. (2008). The use of ultrasonics for nanoemulsion preparation. *Innovative Food Science & Emerging Technologies, 9*, 170–175.
Kim, Y.-W., Kim, K.-H., Choi, H.-J., & Lee, D.-S. (2005). Anti-diabetic activity of β-glucans and their enzymatically hydrolyzed oligosaccharides from Agaricus blazei. *Biotechnology Letters, 27*, 483–487.
Kim, S., Lee, M.-J., Hong, J., Li, C., Smith, T. J., Yang, G.-Y., et al. (2000). Plasma and tissue levels of tea catechins in rats and mice during chronic consumption of green tea polyphenols. *Nutrition and Cancer, 37*, 41–48.
Kimura, K., Iwata, I., & Nishimura, H. (1982). Relationship between acid-catalyzed cyclization of citral and deterioration of lemon flavor. *Agricultural and Biological Chemistry, 46*, 1387–1389.
Kopermsub, P., Mayen, V., & Warin, C. (2011). Potential use of niosomes for encapsulation of nisin and EDTA and their antibacterial activity enhancement. *Food Research International, 44*, 605–612.
Kurowska, E. M., & Manthey, J. A. (2004). Hypolipidemic effects and absorption of citrus polymethoxylated flavones in hamsters with diet-induced hypercholesterolemia. *Journal of Agricultural and Food Chemistry, 52*, 2879–2886.
Lai, C.-S., Li, S., Chai, C.-Y., Lo, C.-Y., Ho, C.-T., Wang, Y.-J., et al. (2007). Inhibitory effect of citrus 5-hydroxy-3,6,7,8,3′,4′-hexamethoxyflavone on 12-O-tetradecanoylphorbol 13-acetate-induced skin inflammation and tumor promotion in mice. *Carcinogenesis, 28*, 2581–2588.
Lao, C. D., Ruffin, M. T., Normolle, D., Heath, D. D., Murray, S. I., Bailey, J. M., et al. (2006). Dose escalation of a curcuminoid formulation. *BMC Complementary and Alternative Medicine, 6*, 10.
Li, Z., Jiang, H., Xu, C., & Gu, L. (2015). A review: Using nanoparticles to enhance absorption and bioavailability of phenolic phytochemicals. *Food Hydrocolloids, 43*, 153–164.
Li, S., Pan, M.-H., Lo, C.-Y., Tan, D., Wang, Y., Shahidi, F., et al. (2009). Chemistry and health effects of polymethoxyflavones and hydroxylated polymethoxyflavones. *Journal of Functional Foods, 1*, 2–12.

Li, S., Wang, H., Guo, L., Zhao, H., & Ho, C.-T. (2014). Chemistry and bioactivity of nobiletin and its metabolites. *Journal of Functional Foods*, *6*, 2–10.
Li, Y., Zheng, J., Xiao, H., & McClements, D. J. (2012). Nanoemulsion-based delivery systems for poorly water-soluble bioactive compounds: Influence of formulation parameters on polymethoxyflavone crystallization. *Food Hydrocolloids*, *27*, 517–528.
Liang, R., Huang, Q., Ma, J., Shoemaker, C. F., & Zhong, F. (2013). Effect of relative humidity on the store stability of spray-dried beta-carotene nanoemulsions. *Food Hydrocolloids*, *33*, 225–233.
Liang, R., Shoemaker, C. F., Yang, X., Zhong, F., & Huang, Q. (2013). Stability and bioaccessibility of β-carotene in nanoemulsions stabilized by modified starches. *Journal of Agricultural and Food Chemistry*, *61*, 1249–1257.
Liang, R., Xu, S., Shoemaker, C. F., Li, Y., Zhong, F., & Huang, Q. (2012). Physical and antimicrobial properties of peppermint oil nanoemulsions. *Journal of Agricultural and Food Chemistry*, *60*, 7548–7555.
Liazid, A., Palma, M., Brigui, J., & Barroso, C. G. (2007). Investigation on phenolic compounds stability during microwave-assisted extraction. *Journal of Chromatography A*, *1140*, 29–34.
López-Rubio, A., & Lagaron, J. M. (2011). Improved incorporation and stabilisation of β-carotene in hydrocolloids using glycerol. *Food Chemistry*, *125*, 997–1004.
Lu, M., Cao, Y., Ho, C.-T., & Huang, Q. (2016). Development of organogel-derived capsaicin nanoemulsion with improved bioaccessibility and reduced gastric mucosa irritation. *Journal of Agricultural and Food Chemistry*, *64*, 4735–4741.
Lu, H., Meng, X. F., Lee, M.-J., Li, C., Maliakal, P., & Yang, C. S. (2003). Bioavailability and biological activity of tea polyphenols. In F. Shahidi, C.-T. Ho, S. Watanabe, & T. Osawa (Eds.), *ACS symposium series*. *Vol. 851*. *Food factors in health promotion and disease prevention*. Washington, DC: American Chemical Society. (chapter 2).
Lu, P. J., & Weitz, D. A. (2013). Colloidal particles: Crystals, glasses, and gels. *Annual Review of Condensed Matter Physics*, *4*, 217–233.
Ma, Q., Zhang, Y., Critzer, F., Davidson, P. M., Zivanovic, S., & Zhong, Q. (2016). Physical, mechanical, and antimicrobial properties of chitosan films with microemulsions of cinnamon bark oil and soybean oil. *Food Hydrocolloids*, *52*, 533–542.
Ma, Q., & Zhong, Q. (2015). Incorporation of soybean oil improves the dilutability of essential oil microemulsions. *Food Research International*, *71*, 118–125.
Madene, A., Jacquot, M., Scher, J., & Desobry, S. (2006). Flavour encapsulation and controlled release–a review. *International Journal of Food Science & Technology*, *41*, 1–21.
Mahdi Jafari, S., He, Y., & Bhandari, B. (2006). Nano-emulsion production by sonication and microfluidization—A comparison. *International Journal of Food Properties*, *9*, 475–485.
Manthey, J. A., Cesar, T. B., Jackson, E., & Mertens-Talcott, S. (2010). Pharmacokinetic study of nobiletin and tangeretin in rat serum by high-performance liquid chromatography–electrospray ionization–mass spectrometry. *Journal of Agricultural and Food Chemistry*, *59*, 145–151.
Manthey, J. A., & Guthrie, N. (2002). Antiproliferative activities of citrus flavonoids against six human cancer cell lines. *Journal of Agricultural and Food Chemistry*, *50*, 5837–5843.
Mantovani, M. S., Bellini, M. F., Angeli, J. P. F., Oliveira, R. J., Silva, A. F., & Ribeiro, L. R. (2008). β-Glucans in promoting health: Prevention against mutation and cancer. *Mutation Research: Reviews in Mutation Research*, *658*, 154–161.
Matos, M., Gutiérrez, G., Coca, J., & Pazos, C. (2014). Preparation of water-in-oil-in-water (W1/O/W2) double emulsions containing trans-resveratrol. *Colloids and Surfaces A: Physicochemical and Engineering Aspects*, *442*, 69–79.
Matos, M., Gutiérrez, G., Iglesias, O., Coca, J., & Pazos, C. (2015). Enhancing encapsulation efficiency of food-grade double emulsions containing resveratrol or vitamin B12 by membrane emulsification. *Journal of Food Engineering*, *166*, 212–220.

McClements, D. J. (2010). Emulsion design to improve the delivery of functional lipophilic components. *Annual Review of Food Science and Technology, 1*, 241–269.

McClements, D. J. (2011). Edible nanoemulsions: Fabrication, properties, and functional performance. *Soft Matter, 7*, 2297–2316.

McKay, D. L., & Blumberg, J. B. (2002). The role of tea in human health: An update. *Journal of the American College of Nutrition, 21*, 1–13.

Menon, V. P., & Sudheer, A. R. (2007). Antioxidant and anti-inflammatory properties of curcumin. In B. B. Aggarwal, Y. J. Surh, & S. Shishodia (Eds.), *The molecular targets and therapeutic uses of curcumin in health and disease* (pp. 105–125). Springer.

Miao, J., Guo, H., Chen, F., Zhao, L., He, L., Ou, Y., et al. (2016). Antibacterial effects of a cell-penetrating peptide isolated from kefir. *Journal of Agricultural and Food Chemistry, 64*, 3234–3242.

Miyata, Y., Tanaka, H., Shimada, A., Sato, T., Ito, A., Yamanouchi, T., et al. (2011). Regulation of adipocytokine secretion and adipocyte hypertrophy by polymethoxyflavonoids, nobiletin and tangeretin. *Life Sciences, 88*, 613–618.

Moghimi, R., Ghaderi, L., Rafati, H., Aliahmadi, A., & McClements, D. J. (2016). Superior antibacterial activity of nanoemulsion of Thymus daenensis essential oil against *E. coli*. *Food Chemistry, 194*, 410–415.

Montilla, E. C., Arzaba, M. R., Hillebrand, S., & Winterhalter, P. (2011). Anthocyanin composition of black carrot (*Daucus carota* ssp. sativus var. atrorubens alef.) cultivars antonina, beta sweet, deep purple, and purple haze. *Journal of Agricultural and Food Chemistry, 59*, 3385–3390.

Murakami, A., Nakamura, Y., Ohto, Y., Yano, M., Koshiba, T., Koshimizu, K., et al. (2000). Suppressive effects of citrus fruits on free radical generation and nobiletin, an anti-inflammatory polymethoxyflavonoid. *BioFactors, 12*, 187–192.

NCBI. (2014). *PubChem substance database*. SID 6374U.S. National Library of Medicine, National Center for Biotechnology Information (NCBI). https://pubchem.ncbi.nlm.nih.gov/compound/445154#section=Top.

Negi, P. S. (2012). Plant extracts for the control of bacterial growth: Efficacy, stability and safety issues for food application. *International Journal of Food Microbiology, 156*, 7–17.

Neubauer, M. P., Poehlmann, M., & Fery, A. (2014). Microcapsule mechanics: From stability to function. *Advances in Colloid and Interface Science, 207*, 65–80.

Özvural, E. B., Huang, Q., & Chikindas, M. L. (2016). The comparison of quality and microbiological characteristic of hamburger patties enriched with green tea extract using three techniques: Direct addition, edible coating and encapsulation. *LWT—Food Science and Technology, 68*, 385–390.

Pan, M.-H., Huang, T.-M., & Lin, J.-K. (1999). Biotransformation of curcumin through reduction and glucuronidation in mice. *Drug Metabolism and Disposition, 27*, 486–494.

Pernetti, M., van Malssen, K. F., Flöter, E., & Bot, A. (2007). Structuring of edible oils by alternatives to crystalline fat. *Current Opinion in Colloid & Interface Science, 12*, 221–231.

Pinto, M. d. C., García-Barrado, J. A., & Macías, P. (2003). Oxidation of resveratrol catalyzed by soybean lipoxygenase. *Journal of Agricultural and Food Chemistry, 51*, 1653–1657.

Porter, C. J., Trevaskis, N. L., & Charman, W. N. (2007). Lipids and lipid-based formulations: Optimizing the oral delivery of lipophilic drugs. *Nature Reviews Drug Discovery, 6*, 231.

Qiao, Y., Sun, J., Xia, S., Tang, X., Shi, Y., & Le, G. (2014). Effects of resveratrol on gut microbiota and fat storage in a mouse model with high-fat-induced obesity. *Food & Function, 5*, 1241–1249.

Ramachandra Rao, S., & Ravishankar, G. A. (2000). Vanilla flavour: Production by conventional and biotechnological routes. *Journal of the Science of Food and Agriculture, 80*, 289–304.

Robert, P., Gorena, T., Romero, N., Sepulveda, E., Chavez, J., & Saenz, C. (2010). Encapsulation of polyphenols and anthocyanins from pomegranate (*Punica granatum*) by spray drying. *International Journal of Food Science & Technology*, *45*, 1386–1394.
Rogers, M. A., Wright, A. J., & Marangoni, A. G. (2009). Oil organogels: The fat of the future? *Soft Matter*, *5*, 1594–1596.
Saito, T., Abe, D., & Sekiya, K. (2007). Nobiletin enhances differentiation and lipolysis of 3T3-L1 adipocytes. *Biochemical and Biophysical Research Communications*, *357*, 371–376.
Salvia-Trujillo, L., Rojas-Graü, A., Soliva-Fortuny, R., & Martín-Belloso, O. (2015). Physicochemical characterization and antimicrobial activity of food-grade emulsions and nanoemulsions incorporating essential oils. *Food Hydrocolloids*, *43*, 547–556.
Schneider, Y., Vincent, F., Duranton, B. t., Badolo, L., Gossé, F., Bergmann, C., et al. (2000). Anti-proliferative effect of resveratrol, a natural component of grapes and wine, on human colonic cancer cells. *Cancer Letters*, *158*, 85–91.
Schubert, H., & Engel, R. (2004). Product and formulation engineering of emulsions. *Chemical Engineering Research and Design*, *82*, 1137–1143.
Sessa, M., Balestrieri, M. L., Ferrari, G., Servillo, L., Castaldo, D., D'Onofrio, N., et al. (2014). Bioavailability of encapsulated resveratrol into nanoemulsion-based delivery systems. *Food Chemistry*, *147*, 42–50.
Sessa, M., Tsao, R., Liu, R., Ferrari, G., & Donsì, F. (2011). Evaluation of the stability and antioxidant activity of nanoencapsulated resveratrol during in vitro digestion. *Journal of Agricultural and Food Chemistry*, *59*, 12352–12360.
Sharma, R. A., McLelland, H. R., Hill, K. A., Ireson, C. R., Euden, S. A., Manson, M. M., et al. (2001). Pharmacodynamic and pharmacokinetic study of oral curcuma extract in patients with colorectal cancer. *Clinical Cancer Research*, 7, 1894–1900.
Shoba, G., Joy, D., Joseph, T., Rajendran, M. M. R., & Srinivas, P. (1998). Influence of piperine on the pharmacokinetics of curcumin in animals and human volunteers. *Planta Medica*, *64*, 353–356.
Sigurdson, G. T., Tang, P., & Giusti, M. M. (2017). Natural colorants: Food colorants from natural sources. *Annual Review of Food Science and Technology*, *8*, 261–280.
Siró, I., Kápolna, E., Kápolna, B., & Lugasi, A. (2008). Functional food. Product development, marketing and consumer acceptance—A review. *Appetite*, *51*, 456–467.
Spigno, G., Donsì, F., Amendola, D., Sessa, M., Ferrari, G., & De Faveri, D. M. (2013). Nanoencapsulation systems to improve solubility and antioxidant efficiency of a grape marc extract into hazelnut paste. *Journal of Food Engineering*, *114*, 207–214.
Sugumar, S., Ghosh, V., Nirmala, M. J., Mukherjee, A., & Chandrasekaran, N. (2014). Ultrasonic emulsification of eucalyptus oil nanoemulsion: Antibacterial activity against Staphylococcus aureus and wound healing activity in Wistar rats. *Ultrasonics Sonochemistry*, *21*, 1044–1049.
Szkudelska, K., & Szkudelski, T. (2010). Resveratrol, obesity and diabetes. *European Journal of Pharmacology*, *635*, 1–8.
Tadros, T., Izquierdo, P., Esquena, J., & Solans, C. (2004). Formation and stability of nano-emulsions. *Advances in Colloid and Interface Science*, *108*, 303–318.
Tajkarimi, M., Ibrahim, S. A., & Cliver, D. (2010). Antimicrobial herb and spice compounds in food. *Food Control*, *21*, 1199–1218.
Teskač, K., & Kristl, J. (2010). The evidence for solid lipid nanoparticles mediated cell uptake of resveratrol. *International Journal of Pharmaceutics*, *390*, 61–69.
Ting, Y., Chiou, Y.-S., Jiang, Y., Pan, M.-H., Lin, Z., & Huang, Q. (2015). Safety evaluation of tangeretin and the effect of using emulsion-based delivery system: Oral acute and 28-day sub-acute toxicity study using mice. *Food Research International*, *74*, 140–150.
Ting, Y., Chiou, Y.-S., Pan, M.-H., Ho, C.-T., & Huang, Q. (2015). In vitro and in vivo anti-cancer activity of tangeretin against colorectal cancer was enhanced by emulsion-based delivery system. *Journal of Functional Foods*, *15*, 264–273.

Ting, Y., Jiang, Y., Ho, C.-T., & Huang, Q. (2014). Common delivery systems for enhancing in vivo bioavailability and biological efficacy of nutraceuticals. *Journal of Functional Foods*, 7, 112–128.

Ting, Y., Jiang, Y., Lan, Y., Xia, C., Lin, Z., Rogers, M. A., et al. (2015). Viscoelastic emulsion improved the bioaccessibility and oral bioavailability of crystalline compound: A mechanistic study using in vitro and in vivo models. *Molecular Pharmaceutics*, *12*, 2229–2236.

Ting, Y., Jiang, Y., Zhao, S., Li, C. C., Nibber, T., & Huang, Q. (2018). Self-nanoemulsifying system (SNES) enhanced oral bioavailability of boswellic acids. *Journal of Functional Foods*, *40*, 520–526.

Ting, Y., Li, S., Ho, C.-T., & Huang, Q. (2013). Emulsion in oral delivery of bioactive lipophilic phytochemicals. In C.-T. Ho, C. Mussinan, F. Shahidi, & E. T. Contis (Eds.), *Nutrition, functional and sensory properties of foods* (pp. 205–223). Cambridge, UK: Royal Society of Chemistry.

Ting, Y., Li, C. C., Wang, Y., Ho, C.-T., & Huang, Q. (2015). Influence of processing parameters on morphology of polymethoxyflavone in emulsions. *Journal of Agricultural and Food Chemistry*, *63*, 652–659.

Ting, Y., Xia, Q., Li, S., Ho, C.-T., & Huang, Q. (2013). Design of high-loading and high-stability viscoelastic emulsions for polymethoxyflavones. *Food Research International*, *54*, 633–640.

Tominari, T., Hirata, M., Matsumoto, C., Inada, M., & Miyaura, C. (2012). Polymethoxy flavonoids, nobiletin and tangeretin, prevent lipopolysaccharide-induced inflammatory bone loss in an experimental model for periodontitis. *Journal of Pharmacological Sciences*, *119*, 390–394.

Topuz, O. K., Özvural, E. B., Zhao, Q., Huang, Q., Chikindas, M., & Gölükçü, M. (2016). Physical and antimicrobial properties of anise oil loaded nanoemulsions on the survival of foodborne pathogens. *Food Chemistry*, *203*, 117–123.

Trela, B. C., & Waterhouse, A. L. (1996). Resveratrol: Isomeric molar absorptivities and stability. *Journal of Agricultural and Food Chemistry*, *44*, 1253–1257.

Tung, Y.-C., Li, S., Huang, Q., Hung, W.-L., Ho, C.-T., Wei, G.-J., et al. (2016). 5-Demethylnobiletin and 5-acetoxy-6, 7, 8, 3′, 4′-pentamethoxyflavone suppress lipid accumulation by activating the LKB1-AMPK pathway in 3T3-L1 preadipocytes and high fat diet-fed C57BL/6 mice. *Journal of Agricultural and Food Chemistry*, *64*, 3196–3205.

Turnbaugh, P. J., Ley, R. E., Mahowald, M. A., Magrini, V., Mardis, E. R., & Gordon, J. I. (2006). An obesity-associated gut microbiome with increased capacity for energy harvest. *Nature*, *444*, 1027.

Uchegbu, I. F., & Vyas, S. P. (1998). Non-ionic surfactant based vesicles (niosomes) in drug delivery. *International Journal of Pharmaceutics*, *172*, 33–70.

Van Amelsvoort, J. M. M., Van Het Hof, K. H., Mathot, J. N. J. J., Mulder, T. P. J., Wiersma, A., & Tijburg, L. B. M. (2001). Plasma concentrations of individual tea catechins after a single oral dose in humans. *Xenobiotica*, *31*, 891–901.

Van Duyn, M. A. S., & Pivonka, E. (2000). Overview of the health benefits of fruit and vegetable consumption for the dietetics professional: Selected literature. *Journal of the American Dietetic Association*, *100*, 1511–1521.

Vanamala, J., Reddivari, L., Radhakrishnan, S., & Tarver, C. (2010). Resveratrol suppresses IGF-1 induced human colon cancer cell proliferation and elevates apoptosis via suppression of IGF-1R/Wnt and activation of p53 signaling pathways. *BMC Cancer*, *10*, 238.

Vimala, K., Yallapu, M. M., Varaprasad, K., Reddy, N. N., Ravindra, S., Naidu, N. S., et al. (2011). Fabrication of curcumin encapsulated chitosan-PVA silver nanocomposite films for improved antimicrobial activity. *Journal of Biomaterials and Nanobiotechnology*, *2*, 55.

Walle, T., Hsieh, F., DeLegge, M. H., Oatis, J. E., & Walle, U. K. (2004). High absorption but very low bioavailability of oral resveratrol in humans. *Drug Metabolism and Disposition, 32*, 1377–1382.

Wan, Z.-L., Wang, J.-M., Wang, L.-Y., Yang, X.-Q., & Yuan, Y. (2013). Enhanced physical and oxidative stabilities of soy protein-based emulsions by incorporation of a water-soluble stevioside-resveratrol complex. *Journal of Agricultural and Food Chemistry, 61*, 4433–4440.

Wang, X., Jiang, Y., Wang, Y.-W., Huang, M.-T., Ho, C.-T., & Huang, Q. (2008). Enhancing anti-inflammation activity of curcumin through O/W nanoemulsions. *Food Chemistry, 108*, 419–424.

Wang, S., Su, R., Nie, S., Sun, M., Zhang, J., Wu, D., et al. (2014). Application of nanotechnology in improving bioavailability and bioactivity of diet-derived phytochemicals. *The Journal of Nutritional Biochemistry, 25*, 363–376.

Wennerström, H., Söderman, O., Olsson, U., & Lindman, B. (1997). Macroemulsions versus microemulsions. *Colloids and Surfaces A: Physicochemical and Engineering Aspects, 123–124*, 13–26.

Were, L. M., Bruce, B., Davidson, P. M., & Weiss, J. (2004). Encapsulation of nisin and lysozyme in liposomes enhances efficacy against Listeria monocytogenes. *Journal of Food Protection, 67*, 922–927.

West, M. E., & Mauer, L. J. (2013). Color and chemical stability of a variety of anthocyanins and ascorbic acid in solution and powder forms. *Journal of Agricultural and Food Chemistry, 61*, 4169–4179.

Xiao, J., Li, C., & Huang, Q. (2015). Kafirin nanoparticle-stabilized Pickering emulsions as oral delivery vehicles: Physicochemical stability and in vitro digestion profile. *Journal of Agricultural and Food Chemistry, 63*, 10263–10270.

Xiao, J., Li, Y., & Huang, Q. (2016). Recent advances on food-grade particles stabilized Pickering emulsions: Fabrication, characterization and research trends. *Trends in Food Science & Technology, 55*, 48–60.

Xiao, J., Nian, S., & Huang, Q. (2015). Assembly of kafirin/carboxymethyl chitosan nanoparticles to enhance the cellular uptake of curcumin. *Food Hydrocolloids, 51*, 166–175.

Xiao, J., Wang, X. a., Perez Gonzalez, A. J., & Huang, Q. (2016). Kafirin nanoparticles-stabilized Pickering emulsions: Microstructure and rheological behavior. *Food Hydrocolloids, 54*, 30–39.

Xue, J., Davidson, P. M., & Zhong, Q. (2015). Antimicrobial activity of thyme oil co-nanoemulsified with sodium caseinate and lecithin. *International Journal of Food Microbiology, 210*, 1–8.

Yallapu, M. M., Gupta, B. K., Jaggi, M., & Chauhan, S. C. (2010). Fabrication of curcumin encapsulated PLGA nanoparticles for improved therapeutic effects in metastatic cancer cells. *Journal of Colloid and Interface Science, 351*, 19–29.

Yanagi, K., Miyata, Y., & Kataura, H. (2006). Highly stabilized β-carotene in carbon nanotubes. *Advanced Materials, 18*, 437–441.

Yang, C. S., Chen, L., Lee, M.-J., Balentine, D., Kuo, M. C., & Schantz, S. P. (1998). Blood and urine levels of tea catechins after ingestion of different amounts of green tea by human volunteers. *Cancer Epidemiology, Biomarkers and Prevention, 7*, 351–354.

Yang, X., Tian, H., Ho, C.-T., & Huang, Q. (2011a). Stability of citral in emulsions coated with cationic biopolymer layers. *Journal of Agricultural and Food Chemistry, 60*, 402–409.

Yang, X., Tian, H., Ho, C.-T., & Huang, Q. (2011b). Inhibition of citral degradation by oil-in-water nanoemulsions combined with antioxidants. *Journal of Agricultural and Food Chemistry, 59*, 6113–6119.

Yasuda, T., Yoshimura, Y., Yabuki, H., Nakazawa, T., Ohsawa, K., Mimaki, Y., et al. (2003). Urinary metabolites of nobiletin orally administered to rats. *Chemical and Pharmaceutical Bulletin, 51*, 1426–1428.

Yonekura, L., & Nagao, A. (2007). Intestinal absorption of dietary carotenoids. *Molecular Nutrition & Food Research, 51*, 107–115.

Yu, H., & Huang, Q. (2011). Investigation of the absorption mechanism of solubilized curcumin using Caco-2 cell monolayers. *Journal of Agricultural and Food Chemistry, 59*, 9120–9126.

Yu, H., & Huang, Q. (2012). Improving the oral bioavailability of curcumin using novel organogel-based nanoemulsions. *Journal of Agricultural and Food Chemistry, 60*, 5373–5379.

Yu, H., Li, J., Shi, K., & Huang, Q. (2011). Structure of modified ε-polylysine micelles and their application in improving cellular antioxidant activity of curcuminoids. *Food & Function, 2*, 373–380.

Yu, H., Shi, K., Liu, D., & Huang, Q. (2012). Development of a food-grade organogel with high bioaccessibility and loading of curcuminoids. *Food Chemistry, 131*, 48–54.

Zahi, M. R., Liang, H., & Yuan, Q. (2015). Improving the antimicrobial activity of d-limonene using a novel organogel-based nanoemulsion. *Food Control, 50*, 554–559.

Zahi, M. R., Wan, P., Liang, H., & Yuan, Q. (2014). Formation and stability of d-limonene organogel-based nanoemulsion prepared by a high-pressure homogenizer. *Journal of Agricultural and Food Chemistry, 62*, 12563–12569.

Zeb, A., & Murkovic, M. (2010). High-performance thin-layer chromatographic method for monitoring the thermal degradation of β-carotene in sunflower oil. *Journal of Planar Chromatography—Modern TLC, 23*, 35–39.

Zhang, L., Critzer, F., Davidson, P. M., & Zhong, Q. (2014). Formulating essential oil microemulsions as washing solutions for organic fresh produce production. *Food Chemistry, 165*, 113–118.

Zhang, Z., Vriesekoop, F., Yuan, Q., & Liang, H. (2014). Effects of nisin on the antimicrobial activity of D-limonene and its nanoemulsion. *Food Chemistry, 150*, 307–312.

Zhao, Q., Ho, C.-T., & Huang, Q. (2013). Effect of ubiquinol-10 on citral stability and off-flavor formation in oil-in-water (O/W) nanoemulsions. *Journal of Agricultural and Food Chemistry, 61*, 7462–7469.

Ziani, K., Chang, Y., McLandsborough, L., & McClements, D. J. (2011). Influence of surfactant charge on antimicrobial efficacy of surfactant-stabilized thyme oil nanoemulsions. *Journal of Agricultural and Food Chemistry, 59*, 6247–6255.

Zupančič, Š., Lavrič, Z., & Kristl, J. (2015). Stability and solubility of trans-resveratrol are strongly influenced by pH and temperature. *European Journal of Pharmaceutics and Biopharmaceutics, 93*, 196–204.

# Electrospinning and electrospraying technologies for food applications

**Loong-Tak Lim[a,*], Ana C. Mendes[b], Ioannis S. Chronakis[b]**
[a]Department of Food Science, University of Guelph, Guelph, ON, Canada
[b]Nano-BioScience Research Group, DTU-Food, Technical University of Denmark, Lyngby, Denmark
*Corresponding author: e-mail address: llim@uoguelph.ca

## Contents



*Advances in Food and Nutrition Research*, Volume 88
ISSN 1043-4526
https://doi.org/10.1016/bs.afnr.2019.02.005

## Abstract

Electrospinning and electrospraying are versatile techniques for the production of nano- to micro-scale fibers and particles. Over the past 2 decades, significant progresses have been made to advance the fundamental understandings of these electrohydrodynamic processes. Researchers have investigated different polymeric and non-polymeric substrates for producing submicron electrospun/electrosprayed materials of unique morphologies and physicochemical properties. This chapter provides an overview on the basic principles of electrospinning and electrospraying, highlighting the effects of key processing and solution parameters. Electrohydrodynamic phenomena of edible substrates, including polysaccharides (xanthan, alginate, starch, cyclodextrin, pullulan, dextran, modified celluloses, and chitosan), proteins (zein, what gluten, whey protein, soy protein, gelatin, etc.), and phospholipids are reviewed. Selected examples are presented on how ultrafine fibers and particles derived from these substrates are being exploited for food and nutraceutical applications. Finally, the challenges and opportunities of the electrostatic methods are discussed.

## 1. Introduction

Electrospinning, also known electrostatic spinning, is a fiber forming technique that exploits electrostatic force to draw polymers into continuous fibers. The technology was developed around 1930 by Formhals and Gladding (Baumgarten, 1971), although it was not being broadly applied until about 2 decades ago. The hallmark of electrospun fibers is their ultrafine thicknesses, typically in the order of hundreds of nanometers, which are in between the dimensions of carbon nanotubes and textile microfibers (Fig. 1). The process results in continuous fibers being laid down as nonwovens with extremely large specific surface areas, endowing them with unique properties, such as enhanced surface activity, increased capacity for surface attachment of functional groups, high porosity, etc. Electrospinning, along with another related process known as electrospraying, continues to captivate industrial and academia interests for various applications, including filtration, textiles, tissue engineering, drug delivery, controlled release, encapsulation, nanocomposite, catalyst/enzyme immobilization, sensors, and so on (Braghirolli, Steffens, & Pranke, 2014; Greiner & Wendorff, 2008; Ko, 2004; Li & Xia, 2004; Teo & Ramakrishna, 2006; Zhang & Yu, 2014).

Over the past decade, electrospinning and electrospraying, collectively known as electrohydrodynamic processing, have started to gain inroads in food- and nutrition-related applications. The nano- and micro-scale fibers

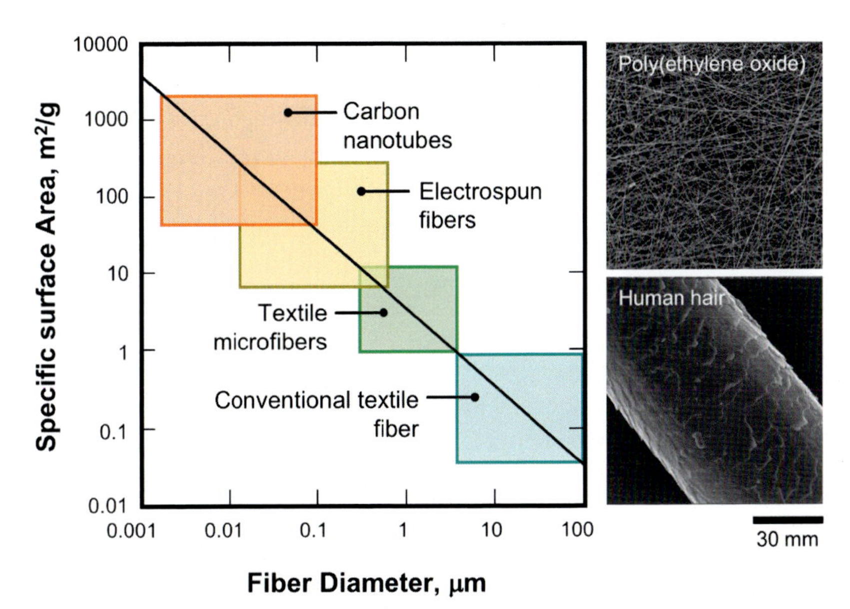


**Fig. 1** Typical diameters of various categories of fibers, showing dramatic increase in material specific surface area as the diameter decreases. The scanning electron micrographs compare the size of electrospun poly(ethylene oxide) fibers and a strand of human hair (author's unpublished data). *Adapted from Gibson, P., Schreuder-Gibson, H., & Rivin, D. (2001). Transport properties of porous membranes based on electrospun nanofibers.* Colloids and Surfaces A: Physicochemical and Engineering Aspects, 187–188, *469–481; Ko, F. K. (2004). Formation of nanofibers and nanotubes production. In: Guceri, S., Gogotsi, Y. G., & V. Kuznetsov (Eds.),* Nanoengineered nanofibrous materials. *NATO Science Series, Vol. 169. Springer, Dordrecht.*

and particles derived are versatile for the encapsulation of bioactives (e.g., micronutrients, nutraceuticals, probiotics). By and large, the objective of encapsulation is to protect these agents from the deleterious environmental factors (e.g., light, oxygen, moisture, acid), unwanted interactions with other components in food matrices, and/or to mask the undesirable sensory characteristics of the bioactives. The selection of encapsulation matrices, which are typically comprised of one or more polymers, is important to ensure the bioactive remains available to the body (i.e., release in the gut). In active packaging applications, it may be desirable to release the bioactives (e.g., antimicrobial, antioxidant) under specific end-use conditions for food preservation. Unlike other particle and fiber forming techniques (e.g., spray drying, melt spinning) that involve substantial heating, the electrohydrodynamic processes are non-thermal, which is beneficial for the encapsulation of nutraceuticals and ingredients that are thermal-labile.

Moreover, the electrostatic charge repulsion prevalent in electrospraying prevents the agglomeration of sprayed droplets, resulting in capsules that are at least two orders of magnitude smaller than the typical spraying process. These submicron particles may be beneficial to increase bioavailability of the bioactives and improve sensory properties of fortified food products.

Electrospinning and electrospraying can be considered a "bottom-up" approach to assemble complex fibrous and particulate materials starting at the molecular level. To meet the end-use objectives, the assembly of these materials relies on the interplay between the polymer, solvent, and other the essential components (e.g., bioactive, surfactant, spinning aid, precursors) in the spin/spray dope solutions. Through manipulating the solution properties (e.g., concentration, viscosity, surface tension, electrical conductivity) and optimization of electrohydrodynamic processing conditions, fibers and particles with diameters ranging from hundreds of nanometers to tens of microns are achievable.

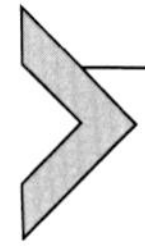

## 2. Basic principles of electrospinning and electrospraying

### 2.1 Electrohydrodynamic processes

Electrospinning of a polymer requires first solubilizing it in a compatible solvent, along with additives, to form a homogeneous spin dope solution. Alternatively, the polymer may be melted using various heating methods (e.g., heated enclosure, laser, heated extruder) and electrospun in a molten state (Ogata et al., 2007; Zhou, Green, & Joo, 2006). The latter approach is less common due to more elaborate experimental setup for temperature control as well as potential thermal degradation of the bioactive agents. For the solution approache, a typical setup involves pumping the spin dope to a spinneret attached to a positive or negative electrode of a direct current power supply (Fig. 2A). An electrically grounded collector, positioned at a distance ranging from several to tens of centimeters away from the spinneret, establishes an electric field (typically 1–5 kV/cm), causing the spin dope at the spinneret tip to elongate toward the collector. The conical geometry of the pendant droplet is known as "Taylor cone," named after Geoffrey Taylor who studied the electrostatic phenomenon of fluids in great detail during the 60s (Taylor, 1966, 1969; Yarin, Koombhongse, & Reneker, 2001). The geometry of the Taylor cone depends on the applied voltage, spin dope flow rate, and electrode configuration. Liquid properties, especially electrical conductivity, viscosity, and surface tension are the

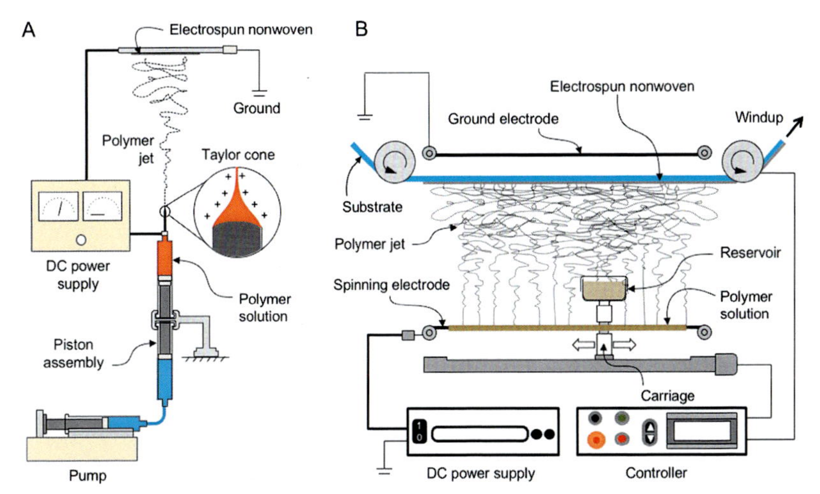


**Fig. 2** Schematic representations of typical single-spinneret (A) and free-surface wire (B) setups for electrospinning of polymers.

determinants that dictate the electrospinnability of the spin dope solution (Rutledge & Fridrikh, 2007). Above a critical applied voltage, the build up of electrostatic charge repulsion overcomes the surface tension of the solution, forming an electrified slender jet that emits from the apex of the Taylor cone and accelerates (up to ~600 m/s$^2$) toward the grounded collector. Due to the bending instability of the polymer jet and concomitant rapid increase in surface area that promotes solvent evaporation, the polymer is being stretched into an ultrafine fiber being laid down on the collector to form a nonwoven (Hou & Reneker, 2004; Reneker, Yarin, Fong, & Koombhongse, 2000).

The spinneret approach typically provides a production throughput of several milliliter per hour of polymer solution. To increase the productivity, various multiple-spinneret and free-surface electrospinning methodologies have been developed (Cengiz, Krucińska, Gliścinsk, Chrzanowski, & Göktepe, 2009; Xiao & Lim, 2018; Zhou, Gong, & Porat, 2009). In one variant of the free-surface approach, the spin dope solution was deposited onto a positively charged stationary wire electrode (~30–50 kV direct current), by a carriage that shutters back and forth along the wire (Fig. 2B). A grounded electrode wire is positioned above the wire electrode to maintain a constant electric field. As the spin dope is electrically charged, numerous jets are formed and ejected toward a collector substrate (e.g., spunbound nonwovens, papers), which is positioned between the two electrodes. Spin dope formulations developed for spinneret electrospinning (Fig. 2A) can be scaled up to free-surface electrospinning (Fig. 2B), although optimization of spin dope formulation may be needed. For example, the enlarged surface area on spin dope on the free surface may result in premature solidification of the polymer solution, especially when a volatile solvent is used. In aqueous-based spin dope systems, humidity and temperature controls are important, since excessive humidity buildup can dissipate electrostatic charge, hindering the jetting process. Moreover, a reduction in water evaporation rate in humid air may result in wet and fused materials on the collector.

## 2.2 Physical factors governing electrospinning and electrospraying

The formation of a polymer jet during electrospinning is mainly governed by Coulombic ($F_C$), electric field ($F_E$), viscoelastic ($F_V$), and surface tension ($F_S$) forces. Other forces, such as air drag ($F_A$) and gravitational ($F_G$) forces are less prominent due to the ultrafine nature of the fibers (Feng, 2003; Ramakrishna, Fujihara, Teo, Lim, & Ma, 2005; Reneker et al., 2000). The effects of these forces on spin jet trajectory can be conceptualized in Fig. 3. Reneker et al. (2000) modeled the jetting phenomena using discrete nodes connected in series by Maxwell units (i.e., dashpot and spring

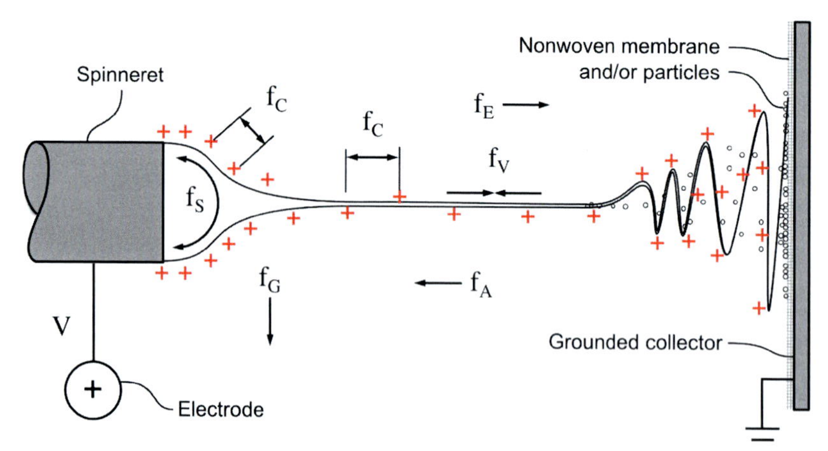


**Fig. 3** Forces governing the electrospinning of a spin dope solution, showing the relationship between Coulombic ($F_C$), electric field ($F_E$), viscoelastic ($F_V$), surface tension ($F_S$), air drag ($F_A$), and gravitational ($F_G$) forces. The arrows depict the directions in which the forces are acting on the jet. Depending on the relative magnitude of $F_V$ and $F_S$, the jet may be laid as a nonwoven or break into liquid droplets due to Rayleigh instability, forming sprayed particles. Schematic not drawn to scale.

**Table 1** Forces governing the electrospinning of spin dope solution and their descriptions.

| **Forces and stress** | **Forces acting on discrete nodes** |
|---|---|
| Coulombic | $F_C = \frac{e^2}{l^2}$ |
| Electric | $F_E = -\frac{eV_0}{h}$ |
| Viscoelastic stress | $\frac{d\sigma_V}{dt} = \frac{G}{l}\frac{dl}{dt} - \frac{G}{\mu}\sigma_V$ |
| Surface tension | $F_S = \frac{\alpha\pi R^2 k}{\sqrt{x_i^2 + y_i^2}}[\mathbf{i}\|x\|sign(x) + \mathbf{j}\|y\|sign(y)]$ |
| Air drag | $F_A = 0.65\pi R\rho_{air}v^2\left(\frac{2vR}{v_{air}}\right)^{-0.81}$ |
| Gravitational | $F_G = \rho g\pi R^2$ |

Where $e$ = charge; $l$ = length of linear jet; $V_o$ = applied voltage; $h$ = distance to collector; $\sigma_V$ = viscoelastic stress; $G$ = elastic modulus; $\mu$ = viscosity; $\alpha$ = surface tension; $k$ = jet curvature; $\rho$ = density; $v$ = kinematic viscosity.

Adapted from Ramakrishna, S., Fujihara, K., Teo, W.-E., Lim, T.-C., & Ma, Z. (2005). *An introduction to electrospinning and nanofibers*. World Scientific; Reneker, D. H., Yarin, A. L., Fong, H., & Koombhongse, S. (2000). Bending instability of electrically charged liquid jets of polymer solutions in electrospinning. *Journal of Applied Physics*, *87*, 4531–4547.

arranged in series) with fluid constitutive properties governed by a series of equations (Table 1). In their model, while the Coulombic and electric field forces aid in jetting, the viscoelastic and surface tension forces tend to counter jet formation. The relative magnitudes of these forces dictate the morphology of the final materials derived from the electrohydrodynamic process. By and large, increasing bulk viscosity and reducing surface tension tend to stabilize the polymer jet through chain entanglements, thereby inhibiting the formation of beads and increasing the fiber diameter. On the other hand, polymer solutions with low viscosity and high surface tension tend to electrospray into droplets or form fibers embedded with beads along their length (i.e., bead-in-fiber). Polymer solutions that electrospray readily are typically those prepared from low-molecular-weight polymers, low solution concentration, and polymers that adopt compact morphology when dispersed in solution (Regev, Vandebril, Zussman, & Clasen, 2010; Yu, Fridrikh, & Rutledge, 2006). During electrospraying, because the aerosol particles are charged, they are self-repulsing, thereby preventing the aggregation of droplets. As a result, electrosprayed particles have a narrower size distribution than those produced using conventional spray drying processes wherein the atomization of fluid is achieved by hydraulic pressure

(Jaworek, 2008; Quispe-Condori, Saldaña, & Temelli, 2011). Deposition efficiency is also higher in the electrospraying process since the majority of the charged particles are being attracted toward the grounded collector. The electrospraying technique can be adapted for spraying of colloidal suspension systems in encapsulation applications (Jaworek & Sobczyk, 2008).

## 2.3 Bending instability

The slender jet emitted from the Taylor cone maintains an initial linear trajectory for a short distance (from a few millimeters to centimeters) before undergoing bending instability characterized by looping (Fig. 4A). This phenomenon causes an increased flight path essential in stretching the polymer jet into ultrafine dimension, as well as providing adequate flight time for solvent evaporation (or cooling in case of molten spin dope) to form solidified nonwoven on the collector. The bending instability phenomenon has been discussed in detail by Reneker, Yarin, and their colleagues in a series of papers (Reneker et al., 2000; Reneker & Yarin, 2008; Yarin et al., 2001). The theory of bending stability mechanism treats the spin dope in Taylor cone as a perfect electric conductor while the bending jet as a perfect

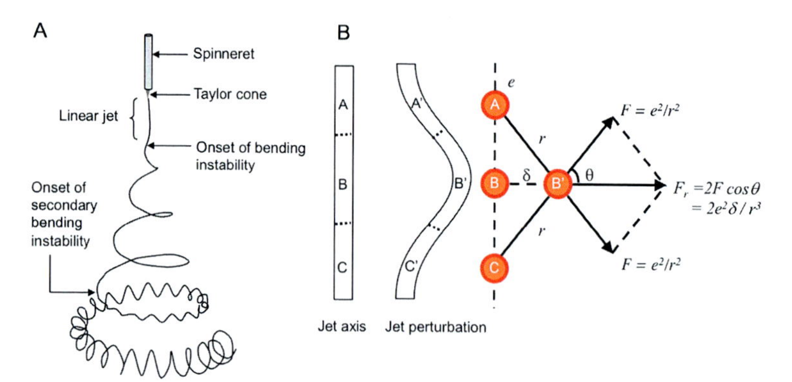


**Fig. 4** (A) Typical trajectory of spin dope exiting from a spinneret during electrospinning, showing the initial linear section of the slender jet ejecting from the Taylor cone, followed by bending instability of polymer jet as it takes flight toward the collector (not shown). The jet transforms into smaller coils along the looping trajectory. (B) Proposed mechanism of bending instability due to perturbation of spin jet during flight, showing three charges (*A*, *B*, and *C*), each with a value of *e*. A deviation of *B* by $\delta$ from the jet axis creates a resultant lateral force, $F_r$ that pushes the jet toward the right. *Adapted from Reneker, D. H., Yarin, A. L., Fong, H., & Koombhongse, S. (2000). Bending instability of electrically charged liquid jets of polymer solutions in electrospinning.* Journal of Applied Physics, 87, *4531–4547; Reneker, D. H., & Yarin, A. L. (2008). Electrospinning jets and polymer nanofibers.* Polymer, 49, *2387–2425.*

dielectric. The latter assumes the conductive current along the jet can be neglected, i.e., the charges are "stuck" in the liquid jet. The initial linear region of the jet is mainly governed by Coulomb's law with a minimal influence from the external field. However, further away from the Taylor cone, the jet is susceptible to environmental perturbation. To illustrate, consider three charges (*A*, *B*, and *C*) each carries a charge of $q$ in a linear segment of the polymer jet (Fig. 4B). Here, two Coulombic forces ($F=e^2/r^2$) are acting on *B* from opposite directions due to *A* and *C*. If a perturbation causes *B* to deviate a distance by $\delta$ from the linear jet axis to $B'$, the net lateral force of $F_r=2F\cos\theta=(2e^2/r^3)\delta$ will push $B'$ away toward the right, which is acting against the viscoelastic resistance and surface tension of the polymer jet. The interplay between these intrinsic factors, along with other extrinsic forces, results in complex flight trajectory during the electrospinning process.

## 2.4 Electrical conductivity and polarity

When the spinneret is connected to a positive electrode, anions and/or negative charged species are attracted to the electrode, causing a buildup of excess cations and/or positive species in the spin dope solution. Similarly, a negative electrode will induce excess anions and/or negative species in the solutions (Ramakrishna et al., 2005). The formation of a jet requires the migration of charge to the surface establishing charge repulsion essential in overcoming the polymer solution surface tension. Highly conductive solutions will cause surface charge to transfer more readily downstream and reduce the surface charge density and electrostatic force required to initiate jetting (Feng, 2002).

Factors that affect excess charge build up at the surface of the solution will determine the magnitude of electrostatic force and electrospinning behavior. Supaphol, Mit-Uppatham, and Nithitanakul (2005a, 2005b) observed that the average diameters of electrospun polyamide-6 (dissolved in 85% formic acid aqueous solution) fibers were significantly smaller when the spin dope solution was positively charged, compared to negatively charged solutions. This observation was attributed to different charged species induced by either positive or negative electrode. When dissolved in formic acid, the polyamide chain ends are protonated, which accumulate on the surface of the Taylor cone when the spinneret is positively charged. Conversely, charging the spin dope with a negative electrode pushes formate anions to the surface. Because the cationic polyamide chains are bulkier than formate anions, a negative polarity results in higher surface charge density of formate ions and hence higher electrostatic force than the positively charged spin dope solution. Experimentally, this effect is manifested as extended initial linear slender jet with increased electrospinning process mass throughput, both of which result in increased fiber diameter due to reduced available flight time for stretching and solvent evaporation (Supaphol et al., 2005a). The same research group further reported an increase in polyamide-6 fiber diameter when NaCl was added to the aqueous formic acid solvent (Mit-uppatham Manit & Supaphol, 2004; Supaphol et al., 2005b). Intuitively, increasing charge density should increase the Coulombic repulsion force acting along the polymer jet and promote stretching/thinning of the fiber. However, this effect was not observed by Supaphol et al. (2005a, 2005b), because of the increased electrical conductivity of the spin dope solution. Similarly, Carroll and Joo (2006) observed that the addition of small amounts of lithium chloride to glycerol dramatically increased the electrical conductivity, hampering the formation of the slender jet (e.g., higher potential difference is required). Carroll and Joo (2006) postulated a "leaky dielectric" model, where the charges present on the surface of highly conductive fluids tend to move freely and slip through the fluid when stretched by the electric field. On the other hand, for low-conductor fluids, the charges are "stuck" on the jet, allowing the force induced by the electric field to transmit effectively into pulling the jet away from the Taylor cone, as well as inducing rapid thinning of the jet through Coulombic repulsion (Carroll & Joo, 2006). For electrospraying process, the dielectric properties of the polymer jet are less critical since fission of a polymer solution to produce fine aerosol droplets is the main goal. On the other extreme, a non-conducting solution will be difficult to electrospin or electrospray due to reduced transfer of charged species to the solution surface.

## 2.5 Solvent selection

In electrohyrodynamic processes, solvents used for the preparation of spin dope solutions can be polar-protic (e.g., water, ethanol, formic acid, acetic acid, 2-propanol, trifluoroethanol, hexafluoro-2-propanol), polar-aprotic (e.g., dimethylsulfoxide, *N*,*N*-dimethylformamide (DMF), acetone, acetonitrile, dichloromethane, ethyl acetate, tetrahydrofuran), and non-polar (e.g., chloroform, toluene, hexane, benzene, 1,4-dioxane). The thermodynamic compatibility of polymer and solvent can be predicted using Hansen solubility parameters (HSPs). The idea is based on the concept that liquid's total cohesive energy of vaporization is made up of energies associated with interatomic dispersion forces, the permanent dipole–dipole interactive forces acting between molecules, and the hydrogen bonding forces due to electron exchange between molecules. Each of these components can be represented $\delta_D$, $\delta_P$, $\delta_H$, respectively, with a unit of $MPa^{0.5}$ (Hansen, 2007). Solubility parameter distance ($R_a$) can be calculated for two materials ($a$, $b$) (Hansen & Skaarup, 1967):

$$R_a = \left[4(\delta D_a - \delta D_b)^2 + (\delta P_a - \delta P_b)^2 + (\delta H_a - \delta H_b)^2\right]^{1/2}$$

The ratio of $R_a$ and interaction radius ($R_o$) is known as relative energy difference (RED $= R_a/R_o$), which is indicative of compatibility of the polymer and the solvent. A RED value of less than unity implies that the solvent is located within the polymer's solubility sphere, and therefore there is a high affinity between the two components and vice versa (Hansen, 2007). The HSP concept has been applied by researchers to select solvents optimal for the preparation of spin dope solutions. Haas, Heinrich, and Greil (2010) applied HSPs to develop binary solvent systems involving the blend of ketones (acetone or methylethylketone) with alcohols (benzyl alcohol or propylene glycol) for the electrospinning of cellulose acetate (Haas et al., 2010). Kurban et al. (2010) applied HSP to select optimal solvents for electrospinning of polystyrene (PS) to encapsulate hydride ammonia borane (AB), using a coaxial electrospinning technique. When immiscible core-shell solutions were used (e.g., shell = 20 wt% PS in 3:1:1 toluene:1,2-dichloroethane:pyridinium formate; core = 10 wt% AB in water), electrospun composite fibers of smooth surface morphologies were produced. However, when semi-miscible (e.g., shell = 18 wt% PS in 7:1.2 toluene:DMF; core = 10 wt% AB in DMSO) or miscible (e.g., shell = 20 wt% PS in DMF; core = 10 wt% AB in DMSO) core-shell solutions were used instead, the resulting fibers exhibited highly porous surface morphologies, which greatly promoted the release of hydride AB. The core inclusions extending from the core through the fiber, exhibiting an ordered radial and longitudinal distribution of nanoscale pores on the fiber surface. Their studies demonstrated that by careful selection of solvents for the core and shell solutions, continuum of coaxial PS-AB fibers with various degrees of porosity can be attained to control the release of the borazine (Kurban et al., 2010). Similarly, Lubasova and Martinova (2011) manipulated the pore size on polyvinyl butyral electrospun fiber, by using the HSPs concept when selecting optimal solvents. Adjusting blend ratio of tetrahydrofuran (THF) and dimethylsulfoxide (DMSO), which has high and low affinity toward the polymer, respectively, resulted in porous electrospun fibers of different pore sizes and morphologies (Lubasova & Martinova, 2011). Jash and Lim (2018) applied HSPs concept to explain the release behavior of hexanal vapor from 1,3-dibenzylethane-2-pentyl imidazolidine (a stable hexanal precursor) encapsulated in electrospun poly(lactic acid) (PLA) fibers and electrosprayed ethylcellulose (EC) particles (Jash & Lim, 2018). On the basis of RED analysis, the higher release rate of hexanal from the electrospun PLA nonwoven than EC was attributed to the greater compatibility of hexanal with PLA than EC.

From a processing standpoint, volatility of the solvent must be taken into consideration to achieve a stable electrodynamic process. The solvent should be volatile enough to produce solidified fibers or particles as they deposit on the collector. Otherwise, the materials will have tendency to fuse together, thereby reducing the porosity of the nonwoven. In cases where a nonvolatile solvent is used due to polymer solubility constraints, the solid collector may be replaced with a grounded coagulation bath to extract the nonvolatile solvent with a compatible liquid wherein the polymer is not soluble. On the other hand, solvents that are highly volatile may cause premature gelling/solidification of the polymer jet during the electrospinning process, thereby preventing further stretching of polymer jet during bending instability that is essential for fiber size reduction. Moreover, solidification of the spin dope solution and dislodging of polymer gel due to hydraulic pressure buildup can induce flow perturbation, producing heterogeneous materials on the collector. To overcome these challenges, a second miscible but less volatile liquid, which may or may not be the solvent for the polymer, could be added to depress the vapor pressure of the main solvent. For instance, DMF has been added to chloroform to depress the vapor pressure of chloroform to enable continuous electrospinning of PLA (212 and 0.49 kPa vapor pressure at 20 °C, respectively) (Jash & Lim, 2018; Zhou & Lim, 2009). The proportion of the binary solvent will need to be optimized to balance the polymer solubility and solvent evaporation rate.

Binary solvent systems are also being used by researchers to manipulate the morphology of the electrospun fibers and electrosprayed particles. Using a blend of chloroform and DMF (9:1, w/w) as the solvent, electrospun PLA fibers with porous skin morphologies were observed (Fig. 5A), which can be useful for enhanced surface functionalization (Jash & Lim, 2018). This porous morphology can be attributed to considerable differences in volatility between the two compatible solvents and different compatibility of PLA in pristine chloroform (soluble) and DMF (limited solubility). During the

**Fig. 5** Effects of solvent on morphology of electrohydrodynamic processes. (A) 10wt% PLA in chloroform:DMF (9:1, w/w); (B) 1:9 (w/w) 1,3-dibenzylethane-2-pentyl imidazolidine in 10wt% PLA in chloroform:DMF (9:1, w/w); (C) 10wt% ethylcellulose in anhydrous ethanol; (D) 1:9 (w/w) 1,3-dibenzylethane-2-pentyl imidazolidine in 10wt% PLA in anhydrous ethanol; (E) 10wt% zein in aqueous ethanol 70wt% aqueous ethanol; and (F) 20wt% zein in aqueous ethanol 70wt% aqueous ethanol. *Adapted from Jash, A., & Lim, L. T. (2018). Triggered release of hexanal from an imidazolidine precursor encapsulated in poly(lactic acid) and ethylcellulose carriers.* Journal of Materials Science, *53(3), 2221–2235; Moomand, K., & Lim, L.-T. (2015). Effects of solvent and n-3 rich fish oil on physicochemical properties of electrospun zein fibres.* Food Hydrocolloids, 46. *https://doi.org/10.1016/j.foodhyd.2014.12.014.*

electrospinning process, chloroform evaporates rapidly from the surface, leaving DMF-rich (PLA-depleted) domains that formed pore imprints on the fiber surface after the remaining solvent was evaporated. On the other hand, the addition of a nonvolatile 1,3-dibenzylethane-2-pentyl imidazolidine to the PLA solution (0.9:1 PLA:imidazolidine, w/w) resulted in fibers with smooth surface morphology (Fig. 5B) due to the compatible nature of the imidazolidine with the both solvents. By contrast, 10wt% EC prepared in anhydrous ethanol resulted in electrosprayed particles, indicating that the surface tension force dominated the viscoelastic stress, causing the polymer jet to break up into droplets. Rapid evaporation of ethanol from the surface of the droplets formed a skin layer that buckled inward (Fig. 5C). The addition of 1,3-dibenzylethane-2-pentyl imidazolidine to the ethanolic EC solution resulted in particles that were less wrinkly than those produced from PLA solutions, due to inclusions of nonvolatile imidazolidine in EC particles (Fig. 5D). At higher magnification, imidazolidine-loaded EC solutions also result in ultrafine fibers connecting the beads stringing among the particles (Jash & Lim, 2018). The aggregated particulate morphologies observed could be due to the vapor depression effect of the imidazolidine on ethanol, causing partial fusion of sprayed particles. Similar phenomena were reported for electrohydrodynamic processing of zein dissolved in binary solvent of 70% aqueous ethanol. At 10 wt% zein, similar imploded polymer particles were observed (Fig. 5E), while at 20 wt% zein concentration, continuous ribbon-shaped fibers were produced (Fig. 5F). The origin of the ribbon-shaped fiber has been discussed by Arinstein and Zussman (2011) and Koombhongse, Liu, and Reneker (2001), driven by the formation of skin layer as a result of rapid solvent evaporation from the jet surface. The solvent evaporation is dictated by the competition between diffusion of solvent into the skin and evaporation of solvent to the surrounding (Arinstein & Zussman, 2011). As solvent continues to evaporate, the tube implodes as a result of the depleting core material and atmospheric pressure (Fig. 6). During this process, the circular cross section becomes elliptical and eventually flattens to form ribbons. In some cases, two small tubes are formed along the edge of the ribbon with a polymer skin connecting them.

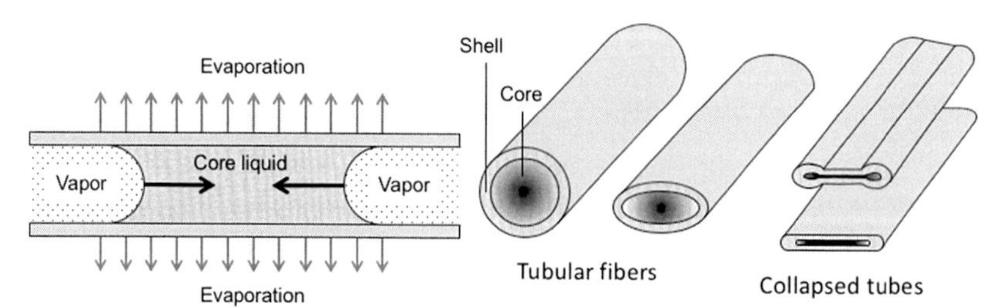


**Fig. 6** Schematic representation of the mechanisms that produce tubular electrospun fibers. For flexible polymer, such as electrospun zein fibers, the collapse of the tubular fiber results in the formation of flat ribbons or ribbons with two tubes along the edges. *Adapted from Arinstein, A., & Zussman, E. (2011). Electrospun polymer nanofibers: Mechanical and thermodynamic perspectives.* Journal of Polymer Science Part B: Polymer Physics, *49, 691–707; Koombhongse, S., Liu, W., & Reneker, D. H. (2001). Flat polymer ribbons and other shapes by electrospinning.* Journal of Polymer Science Part B: Polymer Physics, *39, 2598–2606.*

As the collapse continues, electrical charges tend to flow to the edges of the ribbon, producing a lateral force favoring the flattening of the fiber (Koombhongse et al., 2001).

## 2.6 Viscoelasticity properties

To produce continuous fibers, the spin dope solutions should have optimal viscoelastic properties that are influenced by the molecular weight/morphology of the polymer, polymer–solvent interaction, and concentration. Increasing the molecular weight promotes polymer chain entanglement, stabilizing the polymer jet during electrospinning, and preventing beading along the fibers and/or breaking of the jet into droplets. For proteins, at pH below and above their isoelectric point, they will acquire net positive and negative electrostatic charges on the surface, respectively. The resulting chain repulsion will increase protein–solvent interaction (Vega Lugo & Lim, 2008, 2012). On the other extreme, excessive chain entanglement can result in high yield stress, preventing polymer solutions from jetting (Alborzi, Lim, & Kakuda, 2010).

Although polymer chain entanglement is beneficial for electrospinning, it is not a prerequisite for fiber formation. Boger fluids, such as diluted polymer solutions that exhibit constant (or nearly constant) viscosity as a function of shear rate (James, 2009), can be electrospun into fiber even though their concentration is well below the critical overlap level for chain entanglement, provided if the relaxation time of the fluid is longer than the time of extensional deformation. Yu et al. (2006) prepared a Boger fluid by adding trace quantities of poly(ethylene oxide) (PEO; 0.1–0.2%, w/w; 672–1030 kDa) to poly(ethylene glycol) (PEG; 8–42%, w/w; 10 kDa) dissolved in water. The addition of PEO imparted elasticity to the inelastic PEG solutions, but the relatively high PEG concentrations masked the viscous contribution of PEO, effectively forming a series of Boger fluids that exhibit Newtonian behaviors (i.e., absence of chain entanglement). By using capillary-breakup extensional rheometry, they showed that the Deborah number (De; ratio of fluid relaxation time to the instability growth time) correlated positively with the formation of continuous fibers, i.e., a large fluid relaxation time ($De \gg 1$) tends to arrest Rayleigh instability, promoting the formation of fiber, while a small De value promotes jet breakup, i.e., electrospraying. On the other hand, no correlation was observed between fiber morphology and the Newtonian fluid viscosity. These observations suggest that fluid elasticity, as measured by relaxation time, can be useful for morphology prediction during electrospinning. Similarly, Regev et al. (2010) studied the electrospinning of bovine serum albumin (BSA) solution (10%, w/w in 2,2,2-trifluoroethanol). By reducing disulfide bonds in the globular protein with β-mercaptoethanol, the researchers were able to electrospin the protein solution into continuous fibers. They argued that the denatured protein adopted an unfolded conformation that favors their adsorption onto the jet surface increasing its interfacial moduli (Regev et al., 2010). Conceivably, the relatively volatile solvent used may have also contributed to the formation of viscoelastic skin layer near the apex of the Taylor cone that stabilized the slender jet, which facilitates the solidification of polymer jet to form continuous fibers (Theron, Zussman, & Yarin, 2004).

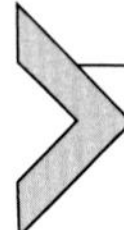

## 3. Electrohydrodynamic processing of biopolymers for food and nutrition applications

Polysaccharides are biopolymers consisting of repeating monomeric units of monosaccharides linked by glycosidic bonds. Interests on applying electrospun and electrosprayed polysaccharides for food and biomedical applications have been increasing due to their abundance, low cost, chemical diversity, and biological activities (García-Moreno, Mendes, Jacobsen, & Chronakis, 2018; Lee, Jeong, Kang, Lee, & Park, 2009; Mendes, Stephansen, & Chronakis, 2017). Feasibility of electrospinning and electrospraying polysaccharides depends on their chemical properties (e.g., molecular weight, functional groups, charges, degree of modification), which dictate the suitable solvent for spin dope solutions (Mendes et al., 2017; Stijnman, Bodnar, & Hans Tromp, 2011). Adjustments of solution properties such as electrical conductivity, viscosity, surface tension, and vapor pressure are needed to manipulate the physical morphologies of the materials from the electrostatic process. In many polysaccharide solution systems, concentration is increased to a level where molecular hydrodynamic radii start overlapping to induce sufficient polymer chain entanglements to stabilize the polymer jet (Shekarforoush, Faralli, Ndoni, Mendes, & Chronakis, 2017; Stijnman et al., 2011). To compensate for the extensive shear-thinning behavior of polysaccharide solutions, other spin aid polymers, such as polyvinyl alcohol (PVA) and PEO, pullulan are often added to the spin dope formulation.

Proteins are polypeptides with unique amino acids sequences that define their structural conformation (e.g., globular or random coil), hydrophobicity, chemical reactivity, biological activity, electrical charge, hydrophobicity, etc. (García-Moreno et al., 2018). Their molecular conformation is sensitive to extrinsic factors (e.g., pH, temperature, ionic strength), changes of which can alter their properties/functionalities (El-salam & El-shibiny, 2016; García-Moreno et al., 2018). Typical challenges encountered during electrohydrodynamic processing of proteins are: (i) high surface tension which can hinder the formation of stable Taylor cone, especially for water-based spin dope solutions; (ii) limited chain entanglements due to low molecular weight, limited chain flexibility, and/or globular structure; and (iii) polyelectrolytic nature of polypeptide chains which weaken charge build-up essential to initiate jetting (García-Moreno et al., 2018). To address these issues, denaturation of protein to unfold the polypeptide chains is essential for sufficient chain entanglement. In addition, the use of organic solvents (e.g., HFP, trifluoroethanol), pH adjustment, application of heat, as well as the incorporation of denaturing agent (e.g., β-mercaptonol, dithiothreitol) and/or surfactants can substantially alter the molecular conformation of proteins rendering them more conducive for electrohydrodynamic processing (Mendes et al., 2017). Blending protein with other electrospinning/electrospraying polymer aid, such as polysaccharides and synthetic polymers, is another strategy commonly adopted by researchers to overcome processing difficulties (Nieuwland et al., 2014).

Nutraceuticals from plants such as flavonoids, hydroxycinnamic acids, anthocyanidins, and carotenoids are health-promoting compounds with antioxidative, anti-inflammatory, antimicrobial, and anticancer properties. However, the incorporation of these bioactives into food products is challenging due to stability (oxidation, photodegradation, reaction with food components, etc.) and sensory (off-color, undesirable taste/odor, etc.) issues. Researchers have exploited electrohydrodynamic processing techniques to encapsulate these bioactive compounds. This section reviews selected polysaccharides and proteins that have successfully been electrospun or electrosprayed, highlighting the polymer solution formulations, processing characteristics, and potential end-use applications related to food, nutrition, and health.

## 3.1 Xanthan

Xanthan gum is an extracellular anionic biopolymer from *Xanthomonas campestris* (Faria et al., 2011; Rosalam & England, 2006). It forms viscous aqueous solutions, which at sufficiently high concentration, exhibits weak gel properties (Chimie, Bercea, Darie, & Morariu, 2013; Lachke, 2004; Mendes, Baran, Pereira, Azevedo, & Reis, 2012; Ungeheuer, Bewersdorff, & Singh, 1989; Zirnsak, Boger, & Tirtaatmadja, 1999). When dissolved in water, the biopolymer lacks entanglements required for electrospinning (Stijnman et al., 2011). However, when dissolved in formic acid, xanthan can be electrospun into continuous fibers, with an average diameter ranging from 128 ± 36.7 to 240 ± 80.7 nm (Fig. 7) (Shekarforoush, Faralli, et al., 2017). At 1% (w/v) polymer concentration, beaded fibers were observed, below which fibers were not formed. Above 1% (w/v) polymer concentration, increases in elastic modulus, apparent viscosity, and first normal stress difference were observed, thereby stabilizing the polymer jet essential for fiber formation. Beyond 2.5% (w/v) level, the spin dope was too viscous to initiate the jetting process.

Curcumin has traditionally been used as a natural food dye that imparts a bright yellow-orange color. It has antioxidative, anti-inflammatory, antimicrobial, and anticancer properties. Shekarforoush, Ajalloueian, Zeng, Mendes, and Chronakis (2018) investigated a electrospun fibrous carrier for curcumin, by blending chitosan with xanthan gum to produce water-stable fibers (Shekarforoush et al., 2018). These fibers released approximately 20% of curcumin after 120 h, at pH 2.2, as compared with approximately 50% release in neutral media (Fig. 8), suggesting that chitosan-xanthan

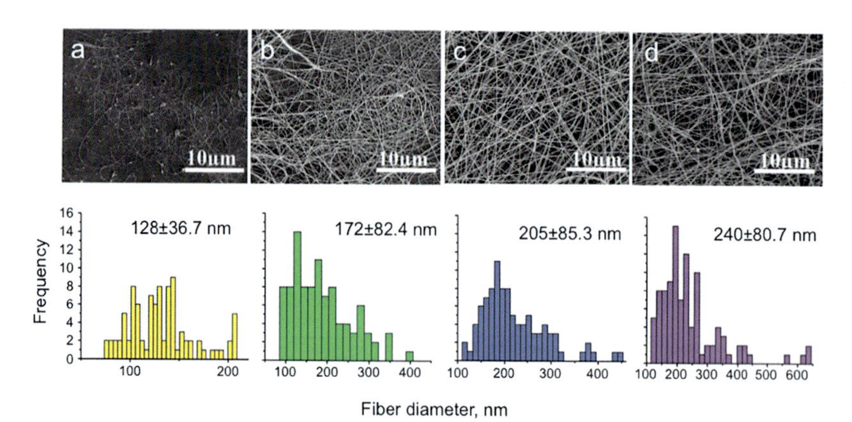


**Fig. 7** SEM micrographs of electrospun xanthan fibers at different polymer concentrations in formic acid: (A) 1.0; (B) 1.5; (C) 2.0; and (D) 2.5% (w/v). The histograms summarize the diameter distribution of the electrospun fibers. *Adapted from Shekarforoush, E., Faralli, A., Ndoni, S., Mendes, A. C., & Chronakis, I. S. (2017). Electrospinning of xanthan polysaccharide.* Macromolecular Materials and Engineering, 302. *https://doi.org/10.1002/mame.201700067.*

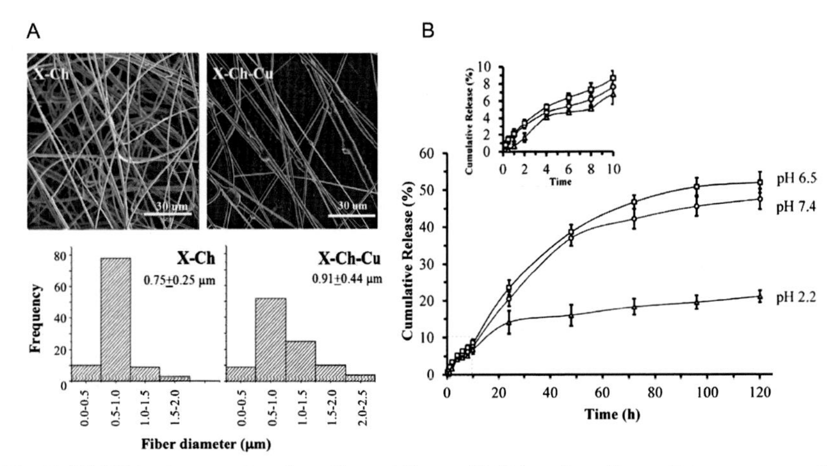


**Fig. 8** (A) SEM micrographs of xanthan-chitosan (X-Ch) and xanthan-chitosan-curcumin (X-Ch-Cu) electrospun fibers and histograms showing their diameter distribution. (B) Release of curcumin from the electrospun fiber carrier. *Adapted from Shekarforoush, E., Ajalloueian, F., Zeng, G., Mendes, A. C., & Chronakis, I. S. (2018). Electrospun xanthan gum-chitosan nanofibers as delivery carrier of hydrophobic bioactives.* Materials Letters, 228, *322–326. https://doi.org/10.1016/j.matlet.2018.06.033.*

nanofibers could be used as a carrier for encapsulated hydrophobic bioactive compounds with pH-dependent release properties. Moreover, when the curcumin-loaded xanthan-chitosan (X-Ch-Cu) fibers were incubated with Caco-2 cells, an enhance in vitro absorption of curcumin across the cell monolayers was observed, with a 3.4-fold increase of curcumin permeability, compared to free curcumin (Faralli, Shekarforoush, Mendes, & Chronakis, 2019). After 24h of incubation, the exposure of Caco-2 cell monolayers to X-Ch-Cu nanofibers resulted in a cell viability of ~80%. The increased in vitro transepithelial permeation of curcumin, without compromising cellular viability, was attributed to the interactions between the electrospun X-Ch fiber carrier and the Caco-2 cells, leading to the opening of the tight junctions (Fig. 9). These results revealed that X-Ch fibers may be promising for oral delivery of poorly water-soluble bioactives.

## 3.2 Alginate

Sodium alginate is an anionic linear polysaccharide extracted from brown seaweed. Its building blocks are β-(1–4)-linked D-mannuronic acid units and α-(1–4) linked L-guluronic acid units, commonly known as M and G blocks, respectively. The polysaccharide can exist in homopolymeric M–M and G–G forms, or alternating sequence of M–G blocks

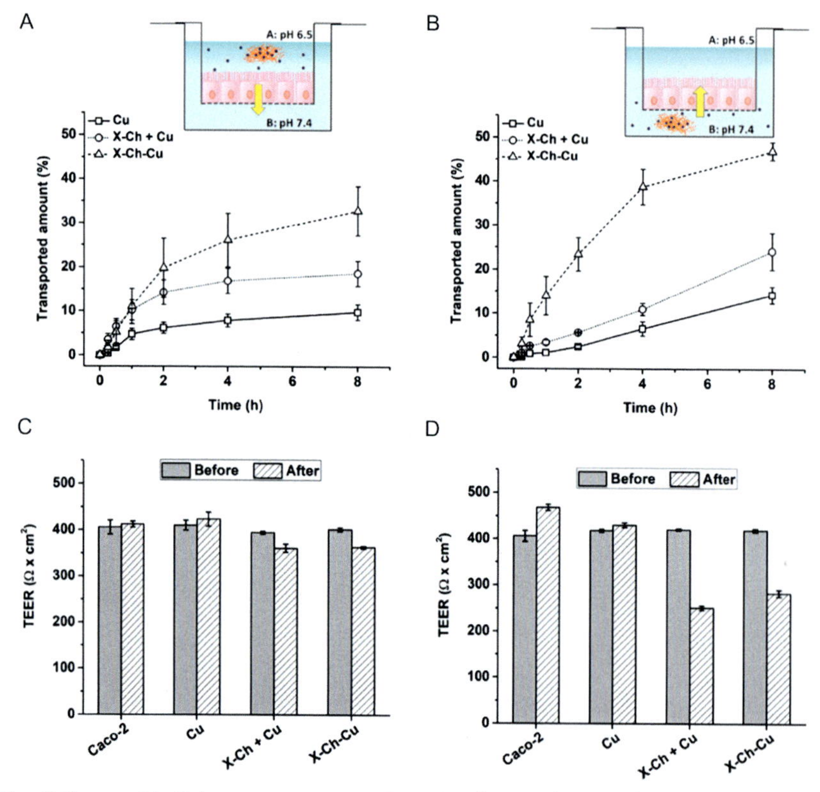


**Fig. 9** Transepithelial transport across Caco-2 cell monolayers of free curcumin (Cu, 150 μM), free curcumin (150 μM) + 3.0 mg xanthan-chitosan nanofibers (X-Ch + Cu), and 9.0 mg curcumin-loaded xanthan-chitosan nanofibers (X-Ch-Cu, fibers amount that corresponded to 150 μM released curcumin) in the donor chamber. Apical-to-basolateral (AB) and basolateral-to-apical (BA) transports were conducted for a time interval of 8 h under a proton gradient (pH 6.5 on the apical side and pH 7.4 on the basolateral chamber). Insets (A) and (B) summarize the percentage of the transported curcumin over time for AB and BA studies, respectively. Insets (C) and (D) are the corresponding transepithelial electrical resistance (TEER) values of the tissue cultures before and after treatments (Faralli et al., 2019).

(Hay, Rehman, Ghafoor, & Rehm, 2010; Reddy & Yang, 2015). Due to the rigid, extended structure of alginate, electrospinning of alginate tends to be challenging. To overcome this issue, Nie et al. (2008) utilized glycerol and water as co-solvent where glycerol improved the flexibility and the entanglement of alginate chains in solution, by disrupting the inter- and intramolecular hydrogen bonds among the alginate chains, allowing the formation of alginate fibers ranging from 120 to 300 nm. Bonino et al. (2011)

electrospun alginate fibers of low (37 kDa) and high (196 kDa) molecular weights, using PEO as a spinning aid and Triton X-100 (*p*-Tertiary-octylphenoxy polyethyl) as a surfactant. Alginate concentrations were 13.5 and 4 wt% for 37 and 196 kDa polymers, with alginate:PEO:surfactant component ratios at 8.0:1.6:2.0 and 2.8:1.2:2.0, respectively. To render the electrospun fibers insoluble, the researchers crosslinked the resulting non-wovens first in ethanol for 1 min followed by a 10 s treatment in 2% $CaCl_2$ in 1:5 ethanol:water. Since PEO is soluble in ethanol, this process selectively removes PEO from the fiber while retaining the integrity of the alginate fiber matrices (Bonino et al., 2011).

Alborzi et al. (2010) reported that a blend of alginate with pectin, at 70:30 (w/w) ratio, could not be electrospun unless PEO (900 kDa; 20–50% (w/w) of total alginate/pectin content) was added to the polymer solution. Depending on polymer concentration (3–5%, w/w), fibers (smooth fibers or beaded fibers) of 39–147 nm were formed. The fiber-promoting properties of PEO were attributed to its electrical conductivity and surface tension lowering effects on the spin dope solutions. Using a similar technique as Bonino et al. (2011), Alborzi, Lim, and Kakuda (2014) subjected folic acid-loaded electrospun alginate/pectin/PEO fiber to 5 min 95% ethanol treatment, followed by 10 min in a solution of 1% $CaCl_2$ in ethanol to crosslink the alginate. The resulting carrier exhibited pH-dependent folic acid release behaviors. Approximately 21% and 97% (w/w) of folic acid were released in 2 h from the electrospun fibers when exposed to water at pH 3 and 7.8, respectively (Alborzi et al., 2014). The lower release rate, under the acidic condition, can be attributed to the protonation of alginate and pectin, resulting in a collapsed network trapping folic acid. On the other hand, the alkaline condition induced a negative charge on the polymers, causing chain-chain repulsion and matrix swelling, favoring the release of the micronutrient.

Electrosprayed alginate particles have been exploited for the encapsulation of viable bacteria. For example, Laelorspoen, Wongsasulak, Yoovidhya, and Devahastin (2014) prepared alginate–zein core-shell microcapsules by an electrospraying process to encapsulate viable *Lactobacillus acidophilus*. The core was made of electrosprayed alginate/glycerol solution (1.4% (w/w) alginate; 8% (w/w) glycerol) collected in an electrically grounded acidic zein solution (7% (w/w) zein, 0.10–0.15% (w/w) citric acid, 75% (v/v) aqueous ethanol, and 1.5% (w/w) $CaCl_2$). The viability of the encapsulated cells was reduced by one log CFU/mL, while the non-encapsulated bacteria exhibited a 5 log CFU/mL reduction after a 2 h-incubation in simulated gastric fluid at pH 1.2 containing pepsin. In another study, an alginate solution (2.5%, w/v) containing 8% (w/v) glycerol and viable *L. acidophilus*, was electrosprayed directly into the gelling bath, containing egg albumen (EA; 8%, w/v), stearic acid (SA)/Tween 40 (1:1.25 ratio), and $CaCl_2$ (1.5%, w/v). The bath was acidified to pH 3.65 using citric acid (Pitigraisorn, Srichaisupakit, Wongpadungkiat, & Wongsasulak, 2017). The resulting capsules were coated with cassava starch granules in fluidized bed dryer using cassava pearls as a drying aid. This method resulted in encapsulation efficiencies of higher than 90%. Moreover, by increasing the amount of SA, the cell viability was further improved, resulting in a minimal loss of viability of only 0.6 log CFU/g. While these matrices are promising to enhance the viability of probiotics in gastrointestinal environments, the beads formed were several hundreds of micrometers in size, due to the close proximity of the spinneret to the grounded gelation bath (6 cm in both studies). Further research in reducing the particle size to nanoscale may present new opportunities for new product innovation.

## 3.3 Starch

Starch is one of the most abundant polysaccharides, it is composed of amylose and amylopectin (Lancuški, Vasilyev, Putaux, & Zussman, 2015). Due to its linearity, amylose can be electrospun into fibers readily. Pure high amylose starches have been successfully electrospun into fibers using dimethylsulfoxide (DMSO) or DMSO/water mixtures as solvents (Kong & Ziegler, 2012, 2013, 2014a, 2014b). Commercial starches have been electrospun and include: Gelose 80, Hylon V, and Hylon VII with amylose contents of ~80%, 70%, and 55%, respectively (Kong & Ziegler, 2012, 2014a, 2014b). The stability of these electrospun fibers in aqueous media can be enhanced by post-spinning treatment in ethanol at 70 °C for 1 h to increase starch crystallinity or crosslinking with 25% (v/v) aqueous glutaraldehyde (Kong & Ziegler, 2014a). Other researchers applied aqueous formic acid (60–100%) to dissolve Hylon VII maize starch (17 wt%) to produce spin dope solution for electrospinning into pure starch fibers of 80–300 nm in diameter (Lancuški et al., 2015).

Although starches with high amylose content are favorable for electrospinning, other starches with lower amylose content have also been electrospun into fibers without the use of spinning aid polymer, through aging in aqueous organic acid solvents. Fonseca, Silva, et al. (2019) and Fonseca, Oliveira, et al. (2019) electrospun a soluble potato starch, with a normal amylose content (32.54 ± 3.65%), into fibers. The starch solution (40 wt%) was prepared using aqueous formic acid (75%) and stirred for 24 h before electrospinning. The electrospun fibers show a morphology with beads and average diameter in the range 128–143 nm. Aging the starch solution for up to 72 h decreased diameter distribution but did not change the fiber morphology significantly (Fonseca, Silva, et al., 2019). Similarly, native and anionic (modified with sodium tripolyphosphate) corn starches were prepared and dissolved in 75% formic acid (v/v) solvent for electrospinning. At elevated concentration (12% (w/v) for regular amylose content starch, 15% (w/v) for high amylose content Hylon VII and V) cannot be electrospun due to high solution viscosity. However, aging the solution up to 72 h resulted in a reduction in viscosity, rendering the spin dope solutions electrospinnable, producing fibers with diameter ranging from 70 to 364 nm (Fonseca, Oliveira, et al., 2019).

To increase the stability in water and enhance the mechanical properties of electrospun starch fiber, other starch derivatives such as starch acetate have been explored. Xu, Yang, and Yang (2009) prepared starch acetates of various degrees of substitution (DS) by treating a 70% amylose starch with acetic anhydride, dissolving the modified polymer in formic acid/water or formic acid/ethanol as solvents, at polymers concentrations ranging from 12 to 24 wt%. They reported that 90% (v/v) formic acid/water solvent resulted in fibers of 50 μm with highest tenacity of 17.9 MPa (Xu et al., 2009). In another study, a modified polysaccharide, hydroxypropyl starch (<7% degree of substitution) was blended with PEO (900 kDa) and electrospun into ultrafine fibers with diameters ranging from 143 to 334 nm, as the PEO increased from 10 to 70 wt% of the total polymer content (fixed at 10 wt%) using water as a solvent (Silva et al., 2013).

Electrospun starch-based, core-sheath composite fibers, used for the encapsulation of viable probiotic cells, were developed by Lancuški et al. (2017) as carriers for *Lactobacillus paracasei* bacteria, where the glycerol fiber core acts as a cell suspension medium. The sheath solution was comprised of Hylon VII high-amylose maize starch (17 wt%) dissolved in formic acid to give starch-formate. Starch-formate is produced due to substitution of hydroxyl groups of the sugar unit with formyl groups. The core-sheath fibers were produced using a coaxial spinneret where the core and sheath spin dopes were pumped into the inner and outer needles, respectively. The resulting core-sheath starch-formate/glycerol composite fibers had mean diameters of 4.13 ± 1.05 μm, capable of retaining bacterial viability when stored at 4 °C and room temperature for up to 21 days

(Lancuški et al., 2017). These fibers are promising delivery vehicles for biotherapeutic applications to enhance the stability of live bacteria.

## 3.4 Cyclodextrin

Cyclodextrins (CDs) are truncated cone shaped cyclic oligosaccharides consisting of 6, 7 or 8 glucose units, respectively, known as α-, β-, γ-CD. CDs are versatile for forming guest–host complexes to stabilize bioactives and control their release (Celebioglu & Uyar, 2010; Fathi, Martín, & McClements, 2014). Initial studies on electrospinning of CD by Uyar's research group demonstrated that high molecular weight was not a prerequisite for electrospinning ultrafine fibers, as long as the intermolecular interactions among the low molecular weight species can result in substantial elasticity to prevent the polymer jet from breaking up into droplets during the electrospinning process. Solvent properties and concentration of the spin dope solution are important factors in promoting intermolecular interactions required during the assembly of CD into supramolecular fiber structures. Celebioglu and Uyar (2013) electrospun α-CD and β-CD fibers of 175–375 nm, using aqueous solutions at 120–160% (w/v) oligosaccharide concentrations. They highlighted the importance of hydrogen bonding and elevated viscosity in forming bead-free CD fibers. Their X-ray diffractometry (XRD) results indicated that CD fibers were highly amorphous, although the as-received CD samples were crystalline in nature (Celebioglu & Uyar, 2013). Methyl-β-cyclodextrin (MβCD) solutions, prepared in either water or *N′N*-dimethylformamide (DMF), have been successfully electrospun at relatively high concentrations (100–160%, w/v), forming fibers ranging from 20 to 1200 nm (Celebioglu & Uyar, 2010). Celebioglu and Uyar (2011) electrospun hydroxypropyl-β-cyclodextrin (HPβCD), without the additional spinning aid polymer, at 160% (w/v) HPβCD concentration. Despite the elevated concentration, the spin dope solutions exhibited Newtonian shear behaviors and possessed adequate elasticity to prevent the breakup of the spinning jet. Besides pristine CDs, researchers have shown that vanillin-CD inclusion complex can be electrospun into fibers without the incorporation of any spinning aid polymer in the spin dope solutions, using water, dimethylacetamide, and DMF as solvents (Celebioglu, Kayaci-Senirmak, Kusku, Durgun, & Uyar, 2016). The researchers reported that the polymer-free vanillin-CD fibers had significantly higher vanillin loading as compared to polymer-based carrier.

While CD complexes can be electrospun into fibers at elevated concentration, the hydrophilic nature of the resulting fibers may not be optimal for certain applicantions, such as controlled release in aqueous media. To address this issue, the CD guest–host complexes can be encapsulated in a higher order structure using other polymeric matrices that are not soluble but instead permeable to water. Kayaci, Ertas, and Uyar (2013) and Kayaci, Umu, Tekinay, and Uyar (2013) incorporated triclosan/CDs in PLA solution using a binary chloroform/DMF (9:1, v/v) solvent. The resulting electrospun fibers, with average diameters ranging from 640 to 940 nm, exhibited higher antibacterial activity against *Staphylococcus aureus* and *Escherichia coli*, as compared with the electrospun PLA containing triclosan alone (i.e., without forming the CD complex), despite the smaller fibers (diameter ~560 nm) of the latter (Kayaci, Umu, et al., 2013). This observation suggests that the added CD might have enhanced the diffusion of triclosan and water in the PLA fiber matrix.

CD-containing electrospun fibers can be further functionalized by the incorporation of other nanoscale bioactive species. Wang, Bai, Li, and Zhang (2012) electrospun polyacrylonitrile (PAN) composite fibers (~200 nm diameter) containing CD and Ag nanoparticles (5–10 nm), using DMF. In this work, the nanoparticles were produced in situ by heating $AgNO_3$/PAN solutions at 90 °C for 10 min to reduce $Ag^+$ into Ag nanoparticles, followed by the addition of β-CD to form the final spin dope. The resulting nonwovens exhibited broad-spectrum antimicrobial properties against Gram-positive *S. aureus* and Gram-negative *Escherichia coli*. The researchers hypothesized that the added CD acted as controlled release carrier for Ag nanoparticles to enhance their antimicrobial properties (Wang et al., 2012). Similarly, vanillin- and eugenol-CD inclusion complexes have been dispersed in electrospun PVA to enhance the thermal and storage stability of the phenolic compounds (Kayaci, Ertas, & Uyar, 2013; Kayaci & Uyar, 2012).

## 3.5 Pullulan

Pullulan is a water-soluble extracellular linear polysaccharide produced by a ubiquitous fungus, *Aureobasidium pullulans*. The pullulan chain is made up of mainly maltotriose units connected via repetitive α-(1,6) linkages that interrupt what would otherwise a typical linear amylose (Fig. 10A), although other structures such as maltotetraose may also present in the pullulan chain (Singh, Saini, & Kennedy, 2008). The alternation of α-(1,6) and α-(1,4)

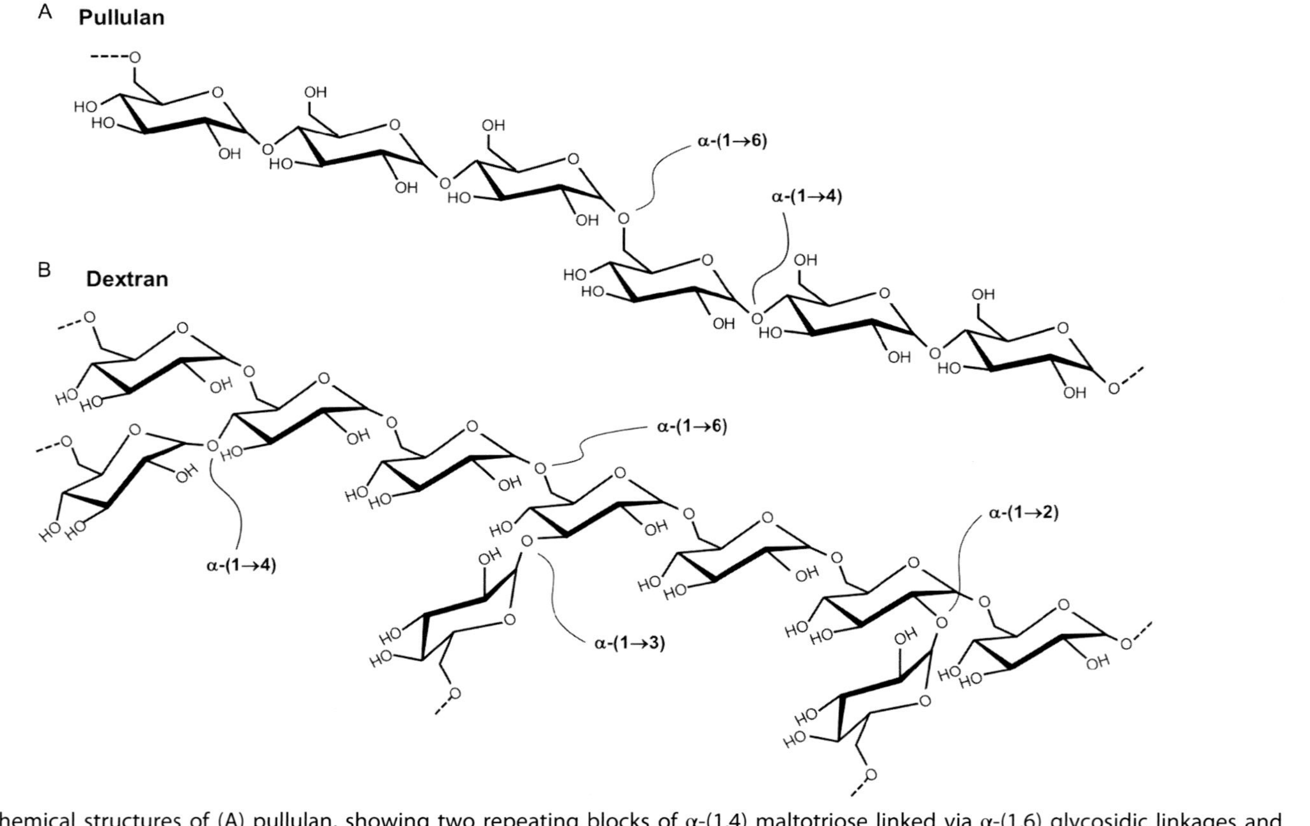


**Fig. 10** Chemical structures of (A) pullulan, showing two repeating blocks of α-(1,4) maltotriose linked via α-(1,6) glycosidic linkages and (B) dextran showing α-(1,6) linkages of the glucose main chain with branching points at 2-, 3-, and 4-positions. *Adapted from Heinze, T., Liebert, T., Heublein, B., & Hornig, S. (2006). Functional polymers based on dextran.* Advances in Polymer Science, 205, *199–291; Leathers, T. D. (2003). Biotechnological production and applications of pullulan.* Applied Microbiology and Biotechnology, 62, *468–473.*

linkages give rise to pullulan's unique properties, i.e., high solubility in water to give polymer solutions with Newtonian flow behavior, even at a relatively high polymer concentration (>10 wt%) as compared to typical polymer solutions (Kong & Ziegler, 2014c; Shingel, 2004; Xiao & Lim, 2018).

Pullulan can be electrospun into uniform fibers using a number of solvent systems. Sun, Jia, Kang, Cheng, and Li (2013) dissolved pullulan in distilled water at concentrations up to 22 wt%, producing fibers ranging from 100 to 700 nm, using a typical electrospinning setup. Kong and Ziegler (2014c) prepared a series of spin dope solutions by dissolving pullulan in DMSO/water of different blend ratios. The solutions were electrospun into ethanol bath to coagulate the fibers. At 12% (w/v) concentration, pullulan forms continuous smooth fibers in DMSO:water blend through the entire 0:100 to 100:0 range. They reported that the pullulan fibers were micron in diameter for spin dope solutions prepared in 100–40% DMSO solvents, whereas those from 20% to 0% DMSO were submicron in size (Kong & Ziegler, 2014c). Aceituno-Medina, Mendoza, Lagaron, and López-Rubio (2013) dissolved blends of amaranth protein isolate (API) and pullulan at different ratios (50:50 to 80:20, w/w) in 95% formic acid, at a constant 20% (w/v) polymer concentration. Moreover, a nonionic surfactant, Tween 80 (polyoxyethylene (20) sorbitan monooleate) was added to the spin dope solutions which facilitated the dispersion of API, thereby improving the fiber morphology consistency. Fibers ranging from average diameters of 227–352 nm were produced. Increasing pullulan content led to an increase in the apparent viscosity and decrease in electrical conductivity, suggesting interactions between API and pullulan. Without Tween 80, heterogeneous morphology with bead defects were observed at elevated API content (API: pullulan 80:20, w/w) (Aceituno-Medina et al., 2013). Recently, Xiao and Lim (2018) electrospun aqueous pullulan/alginate solutions (pullulan:alginate 10:0.0 to 10:2.4 ratios; 10 wt% total polymer concentration), using a free-surface method similar to the setup depicted in Fig. 2B. Pullulan solution alone could be electrospun into beaded fibers of 110 nm in average diameter, while continuous and smooth fibers were formed when small amounts of alginate were added to the spin dope solutions. At 10:0.8 and 10:1.6 pullulan:alginate ratios, the polymer solutions resulted in nanofibers of average diameters of 87 and 82 nm, respectively. The addition of trace amount of $CaCl_2$ (up to 0.045 wt%) increased the charge density of spin dope solutions, forming limited number of intermolecular junction zones between the alginate chains that promoted jet stabilization, forming smooth nanofibers that were significantly smaller in diameter. For instance, at 10:1.6

pullulan:alginate ratio, the average fiber diameter decreased to 69 nm when 0.03 wt% of $CaCl_2$ was added to the spin dope solution. Moreover, the $CaCl_2$-containing fibers had greater thermal stability than those without the addition of $CaCl_2$. Fourier transform infrared (FTIR) spectroscopy analysis revealed that the incorporation of alginate into pullulan increased the hydrogen bonding interaction in the electrospun fibers. The addition of $CaCl_2$ resulted in a blue shift for the symmetric stretching vibrations of $COO^-$ groups from 1409 to 1415 $cm^{-1}$, confirming ionic interaction between $Ca^{2+}$ and the $COO^-$ groups, which contributed to the observed enhanced thermal stability (Xiao & Lim, 2018). Electrospun pullulan fibers (average diameters 50–500 nm) of enhanced thermal stability, as well as increased tensile strength, were reported by Karim et al. (2009), when pristine monomorillonite (MMT) layered silicate was incorporated into aqueous pullulan spin dope solutions (20 wt% polymer concentration). At 3–10 wt% MMT levels, XRD analysis and transmission electron microscopy (TEM) revealed the presence of intercalated MMT layers, and increased electrospun materials crystallinity. Moreover, FTIR analysis suggested MMT–pullulan interaction in the fiber matrices (Karim et al., 2009).

Electrospun fibers prepared using food-compatible solvent systems for the electrospinning of pullulan fibers have been exploited for the encapsulation of bioactives. For instance, electrospun pullulan fibers, loaded with inclusion complex of β-CD and R-(+)-limonene, were prepared and characterized by Fuenmayor et al. (2013). Their study showed that the release of limonene from the pullulan fibers could be activated when exposed above 80% relative humidity. Electrospun pullulan fibers were used as carrier for the encapsulation of omega-3 poly-unsaturated fatty acids rich fish oil by electrospinning emulsions, where the lipid was the dispersed phase and aqueous pullulan was the continuous phase. The oxidative stability of the fish oil was enhanced when natural antioxidants (e.g., tocopherol and rosemary) were incorporated into the dispersed phase of the spin dope formulation (García-Moreno, Damberg, Chronakis, & Jacobsen, 2017). Moreover, these researchers noted that replacing the water solvent of pullulan with formic acid resulted in higher oxidative stability for the fish oil, probably due to substitution of hydroxyl groups on the sugar unit with formyl groups. López-Rubio, Sanchez, Wilkanowicz, Sanz, and Lagaron (2012) encapsulated live *Bifidobacterium animalis* subsp. *lactis* Bb12 cells within pullulan or whey protein concentrate (WPC) capsules, using skim milk as a solvent. Within the pullulan and WPC concentration ranges (15–20 and 30–40 wt%, respectively) investigated, the spin dopes were electrosprayed into droplets (~1170 and 4697 nm average diameters, respectively), instead of forming continuous fibers. Compared with the un-encapsulated freeze-dried cells, probiotics encapsulated in both electrosprayed pullulan and WPI carriers had significantly higher viability at 20 °C, with WPC capsules provided a greater protection than the pullulan counterpart (López-Rubio et al., 2012).

## 3.6 Dextran

Dextran is a water-soluble bacterial polysaccharide produced by Gram-positive, facultative anaerobic cocci, such as those from *Leuconostoc* and *Streptococcus* spp. It is a family of neutral polysaccharides consisting of 97–50% $\alpha$-(1 → 6) glycosidic bonds on the main chain with the remaining $\alpha$-(1 → 2), $\alpha$-(1 → 3) and $\alpha$-(1 → 4) linkages that form branches, depending on the bacteria strains (Fig. 10B). Dextran polymers are soluble in water and various other solvents (e.g., DMSO, formamide), biocompatible, and biodegradable, allowing their applications in medical and biomedical areas (Heinze, Liebert, Heublein, & Hornig, 2006).

Dextran produced by *Leuconostoc mesenteroides* (64–76 kDa) was electrospun by Jiang, Fang, Hsiao, Chu, and Chen (2004) into continuous fibers of several microns in diameter at 0.75 and 1.0 g/mL polymer concentrations, in water. At lower polymer concentration (0.5 and 0.65 g/mL), electrospray beads were observed along with ultrafine fibers of ~100 nm in diameter. At 0.75 g/mL dextran concentration, up to 10 wt% BSA could be loaded into the electrospun dextran fibers without affecting the fiber morphology, although the fiber diameter significantly decreased from 2500 to 500 nm, attributed to the increased net charge of the spinning jet due to the addition of the amphoteric BSA. While dextran is soluble in DMSO, electrospinning of the polymer in the solvent was unsuccessful due to the low volatility of DMSO, resulting in fused materials on the collector. However, the addition of DMF produced solidified continuous fibers (Jiang et al., 2004).

The as-spun dextran fibers are highly soluble in water, limiting their use in certain applications where water-resistance is important. To crosslink the polymer, Jiang et al. (2004) further methacrylated the dextran polymer by dissolving 5 g of dextran in 45 mL of DMSO, followed by the addition of 1.0 g of 4-(*N*,*N*-dimethylamino)pyridine and 1.32 g of glycidyl methacrylate (GMA). After reacting at room temperature for 48 h, the reaction was terminated by adding 0.67 mL of concentrated HCl (37%) solution.

Precipitation was carried out in 500 mL of acetone, followed by washing with acetone and then drying. The treatment resulted a DS value of 25% (molar ratio of methacrylate groups to glucopyranose residues in dextran) GMA. The resulting methacrylated dextran was re-dissolved in 6:4 DMSO:DMF (v/v) to form the spin dope solution, added with 1% (w/w to the methacrylated dextran) of 2,2-dimethoxy-2-phenylacetophenone as a photoinitiator. The electrospun membranes composed of methacrylated dextran and 1% of DMPA were crosslinked by UV irradiation (365 nm, 8 W) for 2 days. The resulting fibers swelled to form hydrogels when exposed to water, but retained their shape for more than 1 month (Jiang et al., 2004). Alternatively, chemical crosslinking of dextran fibers can be achieved by the addition of glutaraldehyde as the crosslinking agent and $MgCl_2$ as the catalyst into the spin dope solution, followed by electrospinning. The electrospun nonwoven is then subjected to heating 70–90 °C in vacuum oven for up to 48 h to induce crosslinking (Ritcharoen et al., 2008).

Electrosprayed dextran particles have been explored by Pérez-Masiá, Lagaron, and Lopez-Rubio (2015) for the encapsulation of lycopene. They produced an oil-in-water emulsion of 16% (w/v) lycopene/soybean oil (1.25%, w/v) in water with Tween 20 (10%, w/v; polyoxyethylene (20) sorbitan monolaurate) as a surfactant. After sonication into particle size of around 150 nm, 20% (w/v) of dextran was added to form the final homogeneous emulsion. For comparison, the researchers also replaced the dextran with 30% (w/v) of whey protein concentrate (WPC). SEM micrographs revealed that the electrosprayed dextran-based solution led to spherical capsules with an average 0.7 μm diameter, while WPC resulted in more irregular electrosprayed particles of 1.8 μm average diameter. However, the electrosprayed dextran capsules had a lower encapsulation efficiency (26%) than that of WPC capsules (73%), due to the less stable dextran-based emulsion than that of the WPC system (Pérez-Masiá, Lagaron, et al., 2015).

## 3.7 Modified celluloses

Cellulose can be electrospun into fibers, but requires strong organic solvents, such as *N*-methylmorpholine-*N*-oxide/water (NMMO/$H_2O$), lithium chloride/dimethylacetamide (DMAc), ionic liquids, ethylene diamine/salt, and so on (Frey, 2008; Lindman, Karlstrom, & Stigsson, 2010). Because many of these solvents are toxic, the use of electrospun cellulose fibers for food applications is limited. However, cellulose derivatives with enhanced solubility in aqueous and food-compatible solvents have received considerable research interests (Frey, 2008). Cellulose derivatives and their solvents (selected examples in parentheses) adopted by researchers for electrospinning include cellulose acetate (acetone, formic acid, DMAc, acetone/water, acetone/DMAc, acetic acid/water), carboxymethylcellulose (CMC; water) cellulose triacetate (methylene chloride/ethanol), methyl cellulose (ethanol/water), EC (ethanol, tetrahydrofuran (THF), THF/DMAc), ethyl-cyanoethyl cellulose (THF) and hydroxypropyl cellulose (ethanol, 2-propanol), and so on (Cai, Niu, Yu, Xiong, & Lin, 2017; Frenot, Henriksson, & Walkenstrom, 2007; Frey, 2008; Lee et al., 2009; Wu, Wang, Yu, & Huang, 2005). Post-electrospinning treatments are often applied by researchers to further enhance the mechanical properties of the resulting nonwovens. For example, Cai et al. (2017) electrospun 37 wt% bamboo cellulose acetate (2.48 DS and 110 degree of polymerization (DP)) prepared in a 2:1 acetone/DMAc binary solvent. Treatment of the resulting nonwovens in 95:5 ethanol/acetone solvent resulted in significant increases in tensile strength, modulus, strain at break, and thermal decomposition temperature as compared to the untreated samples, which is correlated to increases in crystallinity and glass transition temperatures (Cai et al., 2017). Frenot et al. (2007) applied a post-electrospinning treatment to remove PEO (spinning aid) from electrospun CMC fibers, produced from water-based spin dope solutions comprised of 1:1 CMC:PEO at 8 wt% total polymer concentration. Depending on the DS (0.72–1.24) and molecular weight (120–350 kDa) of the CMC polymers, the removal of PEO resulted in CMC structures of varying morphologies, indicating that the distributions of PEO in the CMC matrices varied with DS and molecular weight (Frenot et al., 2007).

Modified celluloses derived from electrospinning process have been exploited for a number of applications. Taepaiboon, Rungsardthong, and Supaphol (2007) electrospun cellulose acetate (CA) ultrafine fibers with diameters ranging from 247 to 265 nm, using a blend of 2:1 acetone/DMAc as solvent, at 17% (w/v) polymer concentration. Vitamins A and E were added to the spin dope solutions, resulting in electrospun CA fibers with loading capacities of 45% and 83%, respectively. The releases of these vitamins were evaluated in acetate buffers at pH 5.5 to simulate the human skin pH condition. The nonwovens exhibited gradual and monotonous increases of the vitamins over a 24 h period for vitamin E and 6 h for vitamin A. By contrast, burst release kinetics were observed for the corresponding cast the film carrier matrix, suggesting that the electrospun CA nonwovens could be promising for the controlled release of hydrophobic bioactives (Taepaiboon et al., 2007). Another study exploited electrospun CA fibers with 200 nm nominal diameter for covalent immobilization of lipase from *Candida rugosa* to form a biocatalyst with high activity retention (Huang et al., 2011). CA polymer of about 30% acetyl content was dissolved in 3:2 acetone:DMAc, followed by electrospinning using a typical spinneret setup. The CA nonwovens were deacetylate using KOH in ethanol for 3 h to convert the fiber surface into regenerated cellulose (RC), followed by oxidation reaction using $NaIO_4$. At optimal oxidation conditions of $NaIO_4$ (4.2 mg/mL), reaction time (6.8 h), temperature (31 °C), and pH (6.1), the reaction resulted in 13.7 wt% aldehyde groups (theoretical max. 18.1 wt%). The aldehyde groups generated on the RC fiber surface coupled with the lipase, resulting in a biocatalyst with an enzymatic activity of 29.6 U/g. The electrospun cellulose carrier significantly enhanced the thermal stability and durability of the lipase as compared to the free enzyme (Huang et al., 2011).

## 3.8 Chitosan

Chitosan is one of the few positively charged polysaccharides with unique biological properties, such as biocompatibility, biodegradability, hemostatic activity, antibacterial, and antimycotic (Balan & Verestiuc, 2014; Jayakumar, Menon, Manzoor, Nair, & Tamura, 2010; Luo & Wang, 2014). Electrospun chitosan fibers can be produced by using trifluoroacetic acid (TFA) as a solvent (Ohkawa, Cha, Kim, Nishida, & Yamamoto, 2004), although the resulting fibers are soluble in neutral and weak basic aqueous solvents due to the high solubility of the TFA-chitosan salt residues (Pakravan, Heuzey, & Ajji, 2012; Sun & Li, 2011). Alternatively, chitosan fibers can be electrospun using aqueous acetic acid, which lower surface tension and increase charge density of the spin dope solution (Geng, Kwon, & jang, 2005; Homayoni, Ravandi, & Valizadeh, 2009). Other solvents used include: lactic acid, formic acid, ethanol, acetone, DMSO, dichloromethane, and so on (Table 2). To enhance fiber stability in aqueous media, researchers have neutralized electrospun chitosan nonwovens with saturated aqueous $Na_2CO_3$ (Gudjónsdóttir et al., 2015; Sangsanoh & Supaphol, 2006), as well as crosslinking the biopolymer with various agents, including glutaraldehyde (Schiffman & Schauer, 2007), genipin, hexamethylene-1,6-diaminocarboxysulphonate (HDACS), and epi-chlorohydrin (ECH) (Austero, Donius, Wegst, & Schauer, 2012).

In food and nutrition applications, hybrid chitosan/phospholipids electrospun fibers with diameters ranging from 250 to 600 nm were

**Table 2** Spin dope formulations for chitosan and its co-spinning polymers for fiber electrospinning and bead electrospraying.

| Polymers | Spin dope solutions | Fiber morphology | References |
|---|---|---|---|
| CTS | – >30 wt% acetic acid<br>– 7 wt% CTS | – Smooth fibers @ 130 nm for 9% acetic acid<br>– Beaded fibers at lower acetic acid concentrations | Geng et al. (2005) |
| CTS | – 70–90% acetic acid<br>– 4–7 wt% CTS | – Smooth and beaded fibers<br>– 140–284 nm | Homayoni et al. (2009) |
| CTS | – TFA or TFA:DCM (90:10)<br>– 6–8 wt% CTS | – Smooth and beaded fibers<br>– 490 nm | Ohkawa et al. (2004) |
| CTS | – 90% acetic acid<br>– 2 wt% CTS | – Beads<br>– 520 nm | Arya, Chakraborty, Dube, and Katti (2009) |
| CTS | – 1 vol% lactic acid<br>– 1% (w/v) CTS<br>– Electrosprayed into 5% (w/v) TPP | – Beads<br>– 221–569 nm | Songsurang, Praphairaksit, Siraleartmukul, and Muangsin (2011) |
| CTS | – 30 vol% acetic acid in ethanol<br>– 0.5–1.0% (w/v) CTS | – Beads<br>– 261–848 nm | Sreekumar, Lemke, Moerschbacher, Torres-Giner, and Lagaron (2017) |
| CTS/PEO | – 75–90% acetic acid<br>– 1.33 wt% CTS:PEO (90:10) | – Smooth fibers<br>– 88–132 nm diameter | Desai et al. (2009) |
| CTS/PEO | – 0.5 M acetic acid, 10% DMSO, 0.3%Triton X-100™<br>– 1–2 wt% CTS:PEO (90:10) | – Smooth fibers<br>– Microns to 40 nm | Bhattarai, Edmondson, Veiseh, Matsen, and Zhang (2005) |

*Continued*

**Table 2** Spin dope formulations for chitosan and its co-spinning polymers for fiber electrospinning and bead electrospraying.—cont'd

| Polymers | Spin dope solutions | Fiber morphology | References |
|---|---|---|---|
| CTS/PEO | – Coaxial spinneret<br>– 4 wt% CTS as shell solution in 50% acetic acid<br>– up to 4% PEO as core solution in 50% acetic acid | – Smooth PEO core—Ct shell composite fibers<br>– 150–190 nm | Pakravan et al. (2012) |
| CTS/PVA | – Blending CTS in TFA:DCM (12:9) with PVA in water | – Smooth fibers<br>– 94–110 nm | Jeannie Tan and Zhang (2011) |
| CTS /PVA | – 2% (v/v) aqueous acetic acid<br>– 3–6 wt% polymer concentration up to 50:50 CTS:PVA ratio | – Smooth and beaded fibers<br>– 20–100 nm | Li and Hsieh (2006) |
| CTS/PVA | – Blend of 7 wt% CTS in formic acid with 9 wt% PVA in water | – CTS:PVA 90:10 formed beads<br>– CTS:PVA 70:30 formed beaded fibers<br>– CTS:PVA 50:50 formed smooth fibers; 120 nm | Ohkawa et al. (2004) |
| CTS/PLA | – TFA:DCM (70:30)<br>– 5 wt% CTS:PLA (50:50) | – Smooth fibers<br>– 840 nm | Ignatova, Manolova, Markova, and Rashkov (2009) |
| CTS/ cellulose | – TFA:acetic acid (3–4 wt% acetic acid)<br>– 4.5–5.0 wt% CTS:cellulose (10:0 to 0:10) | – Smooth fibers<br>– 60–75 nm | Devarayan et al. (2013) |
| CTS/HA | – 1 wt% HA in 75% formic acid<br>– 7 wt% CTS in 80% formic acid<br>– 9:1 to 6:4 HA:CTS (v/v) | – Smooth fibers at 9:1 and 8:2 HA:CTS ratios<br>– Beads and fibers at 7:3 and 6:4 HA: CTS ratios | Ma et al. (2012) |

**Table 2** Spin dope formulations for chitosan and its co-spinning polymers for fiber electrospinning and bead electrospraying.—cont'd

| Polymers | Spin dope solutions | Fiber morphology | References |
|---|---|---|---|
| CST/PCL | – Formic acid:acetone (70:30)<br>– 0.5–2 wt% CTS<br>– 4–10 wt% PCL | – Smooth and beaded fibers<br>– 102–147 nm | Shalumon et al. (2010) |
| CST/Nylon | – HFP:formic acid (90:10, v/v)<br>– 6 wt% Nylon:CTS (85:15, w/w) | – Smooth fibers<br>– 80–310 nm | Zhang et al. (2010) |
| CST/collagen | – HFP:TFA (90:10)<br>– 6–10 wt% CST:collagen (50:50) | – Smooth fibers<br>– 300–500 nm | Chen, Mo, and Qing (2007) |
| CST/SF | – 12 wt% SF in formic acid<br>– 3.6 wt% CST in formic acid<br>– Blending SF:CS solutions (100/0 to 50/50) | – Smooth and beaded fibers<br>– 130–450 nm | Park, Jeong, Yoo, and Hudson (2004) |
| CST/zein | – Ethanol:TFA (2:1, w/w)<br>– 25 wt% zein:CTS (99:1, 97:3, 95:5, 90:10, w/w) | – Smooth fibers; 129–192 nm<br>– Beads 598 nm | Torres-Giner, Ocio, and Lagaron (2009) |

*CTS*, chitosan; *DCM*, dichloromethane; *PVA*, poly(vinyl alcohol); *PCL*, polycaprolactone; *PLA*, poly(lactic acid); *PEO*, poly(ethylene oxide); *TFA*, trifluoroacetic acid; *HA*, hyaluronate acid; *SF*, silk fibroin; *HFP*, 1,1,1,3,3,3-hexafluoro-2-propanol; *DMSO*, dimethylsulfoxide; *TPP*, sodium tripolyphosphate.

produced to encapsulate vitamin B12, diclofenac, and curcumin. The releases of these bioactives from the chitosan/phospholipid fibers were mainly governed by the solubility of the bioactives in water, with the release of vitamin B12 being the quickest, while curcumin the slowest. Increasing phospholipid content increased the hydrophilicity of the matrix which facilitated the diffusion of vitamin B12 molecules to the media. FTIR analysis revealed electrostatic interactions between the positively charged amine groups of chitosan with the positively charged phospholipid. In vitro release of curcumin, diclofenac, and vitamin B12 from the fibers demonstrated their potential use as a transdermal drug delivery carrier (Mendes, Gorzelanny, Halter, Schneider, & Chronakis, 2016). Morphological, mechanical, and mucoadhesive properties of the electrospun chitosan/phospholipid fibers were affected by the phospholipid content, chitosan molecular weights and its degree of acetylation (DA) (Fig. 11). The average diameter of the nanofibers increased with increasing molecular weight and DA of chitosan, as well as increasing phospholipid content. Enhanced elastic and adhesive properties were observed when higher molecular weight and lower DA chitosan polymers were used (Mendes et al., 2018).

To facilitate the electrospinning process, synthetic polymers have been added to the spin dope as a spinning aid, including PEO (Bhattarai et al., 2005; Desai et al., 2009; Pakravan et al., 2012), PVA (Jeannie Tan & Zhang, 2011; Li & Hsieh, 2006), PLA (Ignatova et al., 2009), polycaprolactone (PCL) (Shalumon et al., 2010), and nylon (Zhang et al., 2010). Chitosan is compatible with polysaccharides and proteins when prepared in optimal solvent systems for electrospinning into composite fibers. For examples, chitosan has been blended with cellulose and hyaluronic acid using aqueous acetic and formic acids as co-solvents, for electrospinning into continuous fibers of submicron in diameters (Devarayan et al., 2013; Ma et al., 2012). Proteins, including collagen (Chen et al., 2007), silk fibroin (Park et al., 2004), and zein (Torres-Giner et al., 2009) have been dissolved in acidic organic media compatible with chitosan for the development of composite fibers. Nonwovens derived from these fibers are promising for a broad range of applications, including enzyme/catalyst immobilization, wound dressing, tissue engineering, filtration, bioactive delivery, active packaging, and so on (Gudjónsdóttir et al., 2015; Hu et al., 2014; Ignatova, Manolova, & Rashkov, 2013; Reddy & Yang, 2015; Sun & Li, 2011; Torres-Giner, Ocio, & Lagaron, 2008).

Chitosan polymers have been electrosprayed into particles by optimizing solution parameters. Chitosan with low DA (below 10%), DP ranging from

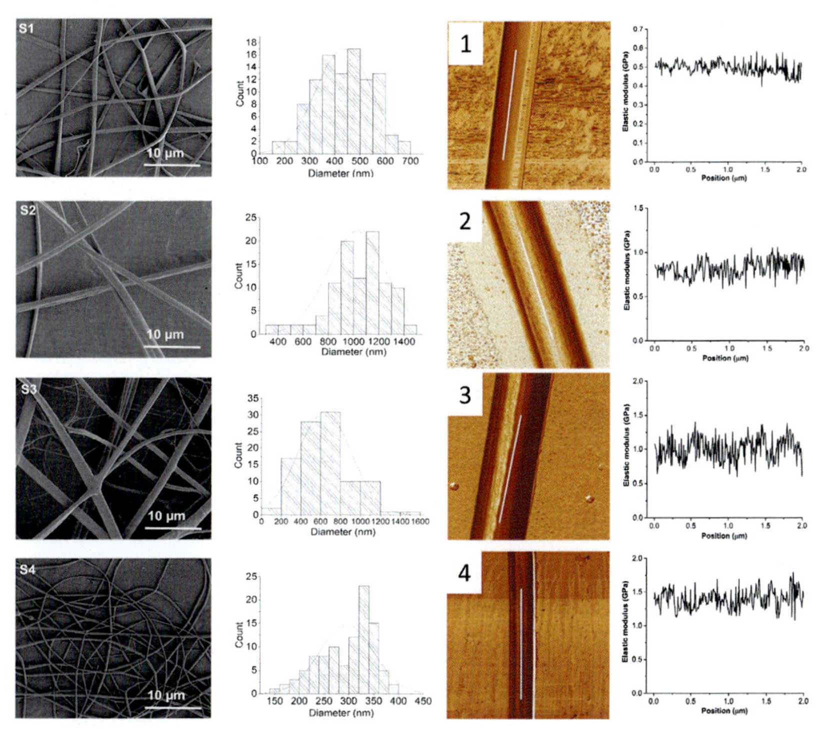


**Fig. 11** SEM micrographs of chitosan/phospholipid (CTS/P) fibers, diameter distribution, and Derjaguin–Muller–Toporov (DMT) modulus: (S1) 1:1 CTS:P, 13% DA, 211 kDa; (S2) 1:3 CTS:P, 13% DA, 211 kDa; (S3) 1:1 CTS:P, 12% DA, 287 kDa; and (S4) 1:1 CTS:P, 6% DA, 276 kDa. DMT modulus values are derived from atomic force microscopy analysis, in the regions as indicated by a 2 μm line profile along the major axes of fibers. *Adapted from Mendes, A., Sevilla Moreno, J., Hanif, M., E. L. Douglas, T., Chen, M., & Chronakis, I. (2018). Morphological, mechanical and mucoadhesive properties of electrospun chitosan/phospholipid hybrid nanofibers.* International Journal of Molecular Sciences, 19*(8), 2266. https://doi.org/10.3390/ijms19082266.*

500 to 1500, spin dope solution of 1–2% (w/w) polymer concentration, with acetic acid/water/ethanol as a solvent, are conducive for the formation of electrosprayed particles (Moreno et al., 2018; Sreekumar et al., 2017). Gómez-Mascaraque et al. investigated the effect of the molecular weight on the electrospraying of chitosan and tested the potential of these particles to encapsulate and release (−)-epigallocatechin gallate (EGCG). The combination of low molecular weight (25 kDa) and higher concentration (5%, w/v) allowed the electrospraying of chitosan into capsules with about 80% EGCG encapsulation efficiency. The EGCG-loaded chitosan capsules elicited a more prolonged antiviral activity against the murine norovirus

than the free compound under simulated physiological conditions, suggesting that the chitosan capsules conferred protective effect upon the phenolic compound (Gómez-Mascaraque, Sanchez, & López-Rubio, 2016). Due to its biocompatibility, mucoadhesive, and nontoxic properties, chitosan has attracted considerable interest as a carrier polymer for drug delivery, especially in the forms micro/nanoparticles, which can be tailored to reduce the side effect of systemic drug administration, as well as to facilitate cross-membrane transport of drugs (Arya et al., 2009; Haley & Frenkel, 2008). To this end, researchers have electrosprayed chitosan into particulate vehicles for medicinal agents, including chemotherapeutic doxorubicin (Songsurang et al., 2011), anti-inflammatory indomethacin (Thien, Hsiao, & Ho, 2012), antibiotic ampicillin (Arya et al., 2009), deoxyribonucleic acid (Sreekumar et al., 2017), and so on.

Table 2 summarizes the formulations of chitosan solutions, along with other polymers commonly added, adopted by researchers for electrospinning and electrospraying of pristine and composite chitosan fibers and particles.

## 3.9 Zein

By classification, prolamins are proteins from plants that are soluble in 70% ethanol. Along with glutelins (soluble in dilute acid and base), prolamins are major storage proteins essential for seed germination (Anderson & Lamsal, 2011; Osborne, 1924). Zein prolamin from corn is insoluble in water due to the presence of large fractions of hydrophobic amino acid residues (e.g., proline, glutamine, leucine, and alanine), but low in charged amino acids. There are several fractions of prolamins present in corn (i.e., $\alpha$, $\beta$, $\gamma$, or $\delta$), with $\alpha$-zein represents about 80% of the total prolamine content (Shukla & Cheryan, 2001). The relatively hydrophobic property of zein is desirable for the encapsulation and controlled release of bioactives in aqueous environment.

The typical solvent used for the electrospinning of zein is aqueous alcohol. Similar to other biopolymers, the composition of the solvent will dictate the morphology of the materials. Moomand and Lim (2015) compared the electrospinning behaviors of zein solution prepared in 70 wt% ethanol or isopropanol aqueous solvents. Both spin dope solutions exhibited shear-thinning behavior, with the ethanolic solution having a higher apparent viscosity than the isopropanol-based zein solution, which was correlated with the continuous ribbon-shaped and beaded electrospun fibers, respectively. FTIR analysis revealed that aqueous ethanol promoted the formation of α-helix secondary structure while isopropanol increased intermolecular and intramolecular β-sheet protein structure (Moomand & Lim, 2015). Protein concentration, ethanol/water ratio, and pH are the mainly factors that influence the morphology of zein during the electrohydrodynamic process. Li, Lim, and Kakuda (2009) reported that at 20 wt% zein concentration, increasing the ethanol concentration from 60 to 90 wt% in water increased the apparent fiber diameter from 369 to 683 nm, with concomitant increased in materials heterogeneity and formation of beads at 80 wt% ethanol and above (Li et al., 2009). This effect was attributed to the increased volatility of the solvent, causing the polymer to solidify sooner which prevented fiber drawing. Torres-Giner, Ocio, et al. (2008) and Torres-Giner, Gimenez, and Lagaron (2008) adjusted the pH of ethanolic zein solution with sodium hydroxide or acetic acid. The zein solution produced ribbon, tubular, and particulate zein at pH at 3.9, 6.0. and 11.3, respectively. The fibrous morphology at low pH was correlated with the relatively high spin dope solution viscosity (243 cP), as compared with the particulate zein observed at pH 11.3 (36 cP) (Torres-Giner, Gimenez, et al., 2008). From a processing feasibility standpoint, ethanolic zein solutions are prone to gelling due to solvent evaporation, which can clog the spinneret. To overcome this issue, blanketing the Taylor cone with the solvent (e.g., ethanol) vapor can effectively prevent the premature solidification of the zein spin dope solution at the spinneret, allowing interrupted electrospinning processing (Li et al., 2009; Moomand & Lim, 2015).

While zein is not soluble in water, long-term exposure of electrospun zein to aqueous environment can cause substantial swelling. Jiang, Reddy, and Yang (2010) applied a pre-electrospinning crosslinking treatment, wherein 50 wt% of zein was dissolved in 70 wt% ethanol solution containing up to 9 wt% of citric acid (by weight of zein) which acted as a crosslinking agent. Before the dispersion of zein, the pH of the crosslinking solution was adjusted to 4.9 by adding sodium hydroxide. The zein solution was aged for 48 h at room temperature to allow for the expansion of zein molecular chains and the formation of crosslinks. The solution was then diluted from 50 to 26 wt% with 70 wt% ethanol in order to obtain the final spin dope solution for electrospinning. Reportedly, with the 9 wt% citric acid pre-crosslinking treatment, the electrospun fibers retained their ultrafine fibrous structure even after immersion in PBS at 37 °C for up to 15 days. Moreover, the citric acid crosslinked electrospun zein scaffolds showed better attachment, spreading and proliferation of fibroblast cells than uncross-linked electrospun zein fibers (Jiang et al., 2010).

Electrospun zein fibers have been used as carriers for various bioactive compounds. Moomand and Lim (2014) encapsulated an omega-3 fatty acid-rich fish oil in electrospun zein fiber with 30 wt% loading level and at >90% encapsulation efficiency. They demonstrated that the electrospun zein fibers provided a greater oxidative stability than their un-encapsulated counterpart. Their TEM and fluorescence analyses revealed the distribution of fish oil along the entire lengths of zein fibers when ethanol was used as a solvent, but the lipid phase was mostly concentrated in beaded areas when isopropanol was used as a solvent instead (Moomand & Lim, 2014). β-Carotene, a food grade colorant and antioxidant, has been encapsulated in electrospun zein fibers of micron and submicron in diameter, resulting in a significant increase in stability when exposed to UV–vis irradiation, albeit the encapsulation efficiency was relative low at 25% due to the poor solubility of the carotenoid in the ethanol/water solvent used (Fernandez, Torres-Giner, & Lagaron, 2009). α-Tocopherol was entrapped within electrospun composite zein fiber (zein:PEO:tocopherol 6:5:3, w/w) with or without the addition of soluble dietary fibers (SDF; 3%, w/w) derived from purpose rice bran (Li, Chotiko, Narcisse, & Sathivel, 2016). The solvent used was 75 wt% ethanol in water containing Tween 80 (0.5 wt%). The addition of SDF in the fibers retained higher amounts of α-tocopherol after the exposure to heat, UV radiation, and storage at room temperature, as compared with the electrospun nonwoven without the SDF. In another study, α-tocopherol was encapsulated in electrospun zein:PEO:chitosan (87.5:10:2.5, w/w) fibers (450 nm average diameter), at 20 wt% loading (Wongsasulak, Pathumban, & Yoovidhya, 2014). The inclusion of α-tocopherol did not affect fiber morphology but enhanced mucoadhesion properties of the fiber matrix. The release of α-tocopherol in simulated gastric fluid at pH 1.2, in the presence of pepsin (simulate digestion), was triggered by matrix erosion, whereas at pH 2 without pepsin (simulate fasting) was driven by swelling and diffusion of the fiber matrices. Antunes et al. (2017) prepared CD-eucalyptus essential oil inclusion complex by a co-precipitation technique and then added to zein polymer solution prepared using aqueous 70% ethanol as a solvent. At 30% (w/v) zein concentration and 24% (w/v) inclusion complex loading, the electrospun zein fibers resulted in 29 and 24% reductions in Gram-positive *Listeria monocytogenes* and *Staphylococcus aureus*, respectively. Whereas minimal impacts inhibition effects were observed for Gram-negative *Escherichia coli* and *Salmonella Typhimurium* (Antunes et al., 2017).

Besides these hydrophobic compounds, electrospun zein fibers have been explored as carriers for other hydrophilic agents. Li et al. (2009) dispersed EGCG from tea at 20 wt% polymer concentration in an aqueous ethanol solvent, and electrospun the spin dope continuous fiber. The zein fibers resisted solubilization in water, although swelling and plasticization were apparent after the water treatment. An 82% recovery of EGCG was observed when freshly spun fibers were submerged in water for up to 1 h. On the other hand, aging the EGCG-loaded fiber at 0% relative humidity for at least 1 day resulted in >98% EGCG recovery. FTIR analysis revealed that the drying and aging treatment facilitated the subsequent changes in zein secondary structures that enhanced the EGCG retention within the zein fiber (Li et al., 2009). Similarly, Neo et al. (2013) encapsulated gallic acid in zein-based electrospun fibers with an encapsulation efficiency of close to 100%. Calorimetry and FTIR analyses suggested interactions between gallic acid and zein, as well as oligomerization or dimerization of gallic acid molecules. There was no significant change in gallic acid 1,10-diphenyl-2-picrylhydrazyl (DPPH) scavenging properties before and after the electrospinning, indicating that the phenolic acid had retained its antioxidant activity during the electrohydrodynamic process (Neo et al., 2013).

In non-edible applications, electrospun zein nonwovens have also been investigated for sensor and detection purposes. Saithongdee, Praphairaksit, and Imyim (2014) encapsulated curcumin in citric acid crosslinked electrospun zein nonwoven, at 5% (w/w) (curcumin/zein) level, for colorimetric detection of $Fe^{3+}$ ions in aqueous solution, due to curcumin-$Fe^{3+}$ complexation. The co-existance of other cations tested (i.e., $Ag^{+}$, $Cd^{2+}$, $Co^{2+}$, $Cr^{3+}$, $Cu^{2+}$, $Fe^{2+}$, $Hg^{2+}$, $Mn^{2+}$, $Ni^{2+}$, $Pb^{2+}$, $Zn^{2+}$, $Ba^{2+}$, $Mg^{2+}$, $Ca^{2+}$, $Na^{+}$, $K^{+}$, and $As^{5+}$) did were not significantly interfered with the color measurement. Optical detection limit of $Fe^{3+}$ was 0.4 mg/L (Saithongdee et al., 2014). Electrospun zein nonwoven was being exploited as a carrier for anthocyanins extracted form red cabbage (*Brassica oleracea* L.) to form a pH indicator derived from biobased materials (Prietto et al., 2018). This type of indicators is promising for intelligent food packaging applications in providing indication on food product quality.

## 3.10 Wheat gluten

80–90% of the protein in wheat flour is the gluten, consisting of prolamins gliadins and glutenins. Gliadins are the main prolamin in wheat with sulfur amino rich (S-rich; α/β- and γ-gliadin monomers, 30–45 kDa), sulfur amino poor (S-poor; ω-gliadin monomer, 30–75 kDa), and high molecular weight gliadins (HMW-gliadins; 100–500 kDa). On the other hand, glutenins are classified into high (HMW-GS; 67–88 kDa) and low (LMW-GS; 30–45 kDa) molecular weight subunits (Ortolan & Steel, 2017). Wheat gluten can be electrospun into continuous fibers, without the use of spinning aid polymer, by using strong organic solvent, such as HFP (Woerdeman et al., 2005). However, the preparation of aqueous polymer solution optimal for electrospinning requires extra treatments to denature the proteins.

Wheat glutenin is inherently stable in water and other solvents because of intramolecular disulfide bridges. Using alkaline and reductive condition (pH 10.5 and 8 M urea), Xu, Cai, Sellers, and Yang (2014) disrupted the disulfide bridges in glutenin, followed by neutralization and washing. The denatured protein was dissolved and heated in 0.3 M sodium carbonate–bicarbonate buffer to form transparent wheat gluten spin dope solution that could be electrospun into continuous fibers. The nonwoven materials were stable in PBS buffer, suggesting they may be promising as tissue scaffolds in biomedical applications. In this approach, the alkaline pH favored the thiol-exchange reaction, during which the thiol groups of cysteine can be deprotonated and attack other disulfide bonds. On the other hand, the 8 M urea solution disrupted inter- and intramolecular hydrogen bonds between the glutenin subunits, thereby unfolding the polypeptide chains (Xu et al., 2014). A similar technique was adopted by Han and Chen (2013) to dissolve glutenin in water via sodium dodecyl sulfate and β-mercaptoethanol, which breaks the disulfide linkages of the glutenin subunits. They then added the denatured glutenin to PVA spin dope solution to produce plasticized electrospun PVA/glutenin composite fibers of about 100 nm in thickness. Enhanced mechanical strength and increased water absorption were observed, which were attributed to the formation of S–S linkages during electrospinning and building up of hydrogen bonds between the side chain of amino acid and hydroxyl group of PVA (Han & Chen, 2013). Woerdeman, Shenoy, and Breger (2007) reported a similar reinforcement in tensile strength of electrospun composite nonwoven of PVA and gluten protein, using HFP as a solvent. They observed that the presence of gluten protein inhibited the recrystallization of PVA, indicating the two polymers were miscible (Woerdeman et al., 2007). Electrospun wheat protein fibers investigated for controlled release applications examined electrospun wheat gluten, at 8% (w/v) concentration level, using ethanol/2-mercaptoethanol as a solvent (Castro-Enríquez et al., 2012).

The resulting nonwoven membranes were loaded with urea to form pastilles for sustained-release applications in agricultural crops. Wang et al. (2015) prepared a composite electrospun membrane, consisting of PVA, wheat gluten, and zirconia, as a carrier for the delivery of an antimicrobial peptide, nisin. The release of the peptide from the nisin-loaded PVA/WG/$ZrO_2$ membranes exhibited Fickian diffusion behavior, effectively inhibiting the growth of *Staphylococcus aureus*. The composite carrier may be useful for drug delivery, would dressing, and active packaging.

## 3.11 Whey proteins

β-Lactoglobulin and α-lactalbumin are proteins in dairy produced during cheese production. Similarly to other proteins, the electrospinning of whey protein isolate (WPI) and whey protein concentrate (WPC) is challenging due to the globular structure of the proteins and high electrical conductivity (Drosou, Krokida, & Biliaderis, 2018). In order to form continuous electrospun fibers, these proteins need to be denatured and unfolded. Moreover, the addition of spinning aid polymer, such as PEO, is essential to initiate jetting from the Taylor cone. Vega-Lugo and Lim (2012) investigated the effects of PEO on the electrospinning of WPI solution (10 wt% WPI and 0.4% (w/w) 900 kDa PEO) using acetic acid (pH 1) or water (pH 7 and 12) as solvents. A substantial increase in apparent viscosity was observed for the acidic polymer solution, which is attributed to the electrostatic interaction between the positively charged protein with the ether oxygen on the PEO chain. The addition of PEO resulted in smaller increases and negligible effects on the alkaline and neutral WPI solutions, respectively. These observations can be correlated with morphology of the materials from the electrohydrodynamic processes. The acidic solution produced smooth fibers (707 nm average diameter), while the alkaline solution produced fibers (191 nm) embedded with spindle-like beads (1.0 μm). On the other hand, electrosprayed particles of an average diameter of 2 μm were observed, inter-dispersed with ultrafine fibers (138 nm). Changes in pH also altered the secondary structures of the proteins. Increased random coil and α-helix secondary structures in WPI, as revealed by FTIR and $^{13}C$ NMR analyses were the main contributors to the formation of bead-less electrospun fibers for the acidic solution (Vega-Lugo & Lim, 2012). Sullivan, Tang, Kennedy, Talwar, and Khan (2014) enabled the electrospinning of WPI in aqueous solution, albeit at a higher PEO concentration (4 wt%; 600 kDa). Similarly, they reported acidic polymer solutions (pH 2–3) favored the formation of fibers (312–690 nm). By heating the WPI-PEO nonwoven at 100 °C for 44 h, the researchers reported that the electrospun fibers resisted solubilization after several days of soaking in water, although substantial fiber swelling was observed (Sullivan et al., 2014). The post-electrospinning heating treatment can be useful for crosslinking electrospun WPI fibers to render them insoluble in aqueous medium, although leaching of PEO still occurs. Colín-Orozco, Zapata-Torres, Rodríguez-Gattorno, and Pedroza-Islas (2015) prepared solutions by combining different proportion of WPI (10 wt%) and PEO (9 wt%) stock solution, to produce electrospun carrier membranes for rosemary extract. They observed that the release kinetics of rosemary extract was controlled by the leaching of the PEO (Colín-Orozco et al., 2015). Recently, Drosou et al. (2018) reported the addition of pullulan to WPI solutions (20 wt% total polymer concentration; WPI: pullulan 80:20 to 20:80, w/w), using water as a solvent, resulted in increased viscosity and lowered electrical conductivity beneficial for promoting the formation of uniform fibers with submicron diameter (~160 to 250 nm). Infrared spectroscopy revealed intermolecular interactions between WPI and pullulan, which explained the increased thermal stability and decreased $T_g$ of the composite fibers as compared to pure WPI (Drosou et al., 2018). Other researchers enabled electrospinning of WPI by adding gelatin, at 1:1 (w/w) proportion, to the spin dope solution (Nieuwland et al., 2014). Although PEO has been widely used in the biomedical and pharmaceutical applications, the use of pullulan and gelatin as a spinning aid for WPI is more attractive than PEO for food application.

Electrospraying aqueous whey protein solution is achieved by addition of low molecular weight strong polar co-solvent. López-Rubio and Lagaron (2012) successfully electrosprayed WPC into spherical submicron capsules by incorporation of glycerol (10–40 wt% of WPC content) into the aqueous WPC solution (40 wt%). Considering that glycerol is an excellent vehicle for β-carotene, these researchers pre-dispersed the carotenoid into glycerol at 0.005 g/mL concentration, followed by adding the mixture into the aqueous WPC solution and homogenized to give a final β-carotene concentration of 20 wt%. They reported that the WPC capsules provided an encapsulation efficiency of ~90%, and dramatically stabilized it against photo-oxidization (López-Rubio & Lagaron, 2012). Based on a similar approach, Fabra, López-Rubio, and Lagaron (2016) electrosprayed an aqueous solution consisting of 30 wt% WPI, 10 wt% glycerol, 5 wt% Tween 80, and 15 wt% α-tocopherol onto an edible film for direct delivery of the antioxidant in food contact applications (Fabra et al., 2016). Pérez-Masiá, Lagaron, et al. (2015) compared the morphology and stability of lycopene micro- and nanocapsules produced through either electrospraying (single or coaxial spinnerets) or spray drying, using WPC or other polymers (chitosan and dextran). The spray drying process, due to the high temperature involved, reduced lycopene stability and resulted in poor encapsulation efficiencies. WPC provided a higher encapsulation efficiency (about 75%) and stronger protection against moisture and thermal degradation for lycopene than the other two biopolymers, which was attributed to protein–lycopene interaction. Similarly, folic acid was encapsulated in WPC or resistant starch matrices, by using electrospraying and spraying drying techniques (Pérez-Masiá, López-Nicolás, et al., 2015). Electrospraying produced spherical nano-, submicro-, and microcapsules with smaller diameters of narrower size distribution than spray drying. Greater encapsulation efficiency was observed in WPC (>80%) compared to resistant starch ($44 \pm 5\%$) matrices, due to interactions between the protein and folic acid. Although both materials and encapsulation techniques resulted in improved folic acid stability, the WPC also conferred higher protection on folic acid than resistant starch capsules when exposed to aqueous and dry conditions.

The encapsulation of probiotics aims to create a micro-environment to protect the cells from the deleterious factors during processing and storage, while allowing their release at the targeted sites in the digestive tract (Chávarri, Marañón, & Villarán, 2012). Electrospraying has been used for this purpose due to its non-thermal process characteristic and high encapsulation efficiency (Pitigraisorn et al., 2017). Researchers have demonstrated that live bacteria can survive the typical shear and electric field of electrohydrodynamic processes. For example, Gomez-Mascaraque, Morfin, Pérez-Masiá, Sanchez, and Lopez-Rubio (2016) encapsulated a model probiotic microorganism, *Lactobacillus plantarum*, within electrosprayed WPC matrices, using fresh bacterial culture. The electrospraying solution was made up of WPC at 0.3 g/mL protein concentration, along with 9 wt% (with respect to WPC content) of Tween 20 and up to 20% (with respect to WPC content) of Fibersol® resistant starch. Viability losses of less than one $\log_{10}$ CFU were achieved with the final bacterial count of $>8.5 \log_{10}$ CFU/g, as well as enhanced protection during in vitro digestion and storage (Gomez-Mascaraque et al., 2016). Using a similar approach, Librán, Castro, and Lagaron (2017) encapsulated a probiotic, *Bifidobacterium longum* subsp. *infantis*, within electrosprayed particles of 2–3 μm in diameter, consisting of WPC, resistant starch, or polyvinylpyrrolidone matrices. Cell viability of over six $\log_{10}$ CFU/g for 600 days at 23% RH was reported, while the unprotected cells were all nonviable (Librán et al., 2017).

## 3.12 Soy proteins

Soy protein consists of a complex mixture of globular proteins of different subunits (e.g., 2S, 7S, 11S, and 15S fractions) ranging from 200 to 600 kDa (Horan, 1974; McClements, 2014; Wolf, 1970). Due to its globular structure, soy protein needs to be denatured to induce polymer chain entanglements for electrospinning. To this end, researchers have employed combined alkaline and thermal treatments. Similar to whey proteins, in aqueous solution, denatured soy protein isolate (SPI) could not be electrospun into fiber. However, the addition of 0.8 wt% PEO to SPI solution (15 wt%) resulted electrospun fibers of 200–260 nm in diameter (Vega Lugo & Lim, 2008). Electrospun SPI/PEO nonwovens, prepared using this approach, have been evaluated as a carrier to encapsulate β-cyclodextrin loaded with allyl isothiocyanate (AITC)—a potent antimicrobial for active packaging application. Due to the hydrophilic nature of the cyclodextrin host–guest complex and the SPI/PEO fibers, the release of AITC can be triggered when exposed to elevated relative humidity, such as within the headspace of food package (Vega-Lugo & Lim, 2009). Alkaline and heat-treated aqueous SPI solution (10 wt%), doped with 0.5 wt% of PEO, has been exploited by researchers to encapsulate an anthocyanin-rich red raspberry (*Rubus strigosus*) extract. The resulting fibers exhibited antibacterial activity against *Staphylococcus epidermidis* (Wang, Marcone, Barbut, & Lim, 2013). Salas, Ago, Lucia, and Rojas (2014) electrospun soy protein fibers, fortified with lignin in aqueous alkaline solution. Again, the addition of PEO (400 kDa; 10 wt % of the total polymer content) was required to produce continuous fibers with diameter ranging from 124 to 400 nm (Salas et al., 2014). Without subjecting SPI to a denaturation treatment, Fabra et al. (2016) prepared SPI solution (10 wt%) in water added with guar gum (1 wt%) and Tween 20 (5 wt%). Due to the lack of polymer unfolding, the aqueous solution was electrosprayed into heterogeneous particles, instead of forming continuous fibers (Fabra et al., 2016). Even by using a strong organic polar solvent (1,1,1,3,3,3-hexafluoro-2-propanol; HFIP), Xu et al. reported an addition of PEO (SPI:PEO 16:1, 10:1, 6:1, and 2:1, w/w) was still needed to enable the electrospinning of SPI (Xu, Jiang, Zhou, Wu, & Wang, 2012). Based on the information available in the literature, it is fair to conclude that the neat SPI alone cannot be electrospun readily into fiber. The incorporation of a spinning aid polymer into the spin dope solution is essential.

## 3.13 Gelatin

Gelatin is extracted from animal collagen via acidic or alkaline hydrolysis processes, producing type A and type B gelatin, respectively. It is a heterogeneous mixture of polypeptides with gel and bloom properties that are highly dependent on the degree of hydrolysis. Gelatin with weight-average-molecular weight of <20 kDa do not form gel upon cooling from hot protein solution (Damodaran, 2008). At higher weight-average-molecular weight, below its gelling temperature (e.g., 10–25 °C for pig and cow gelatin, 0–5 °C for fish gelatin), the protein forms an elastic gel (McClements, 2014). The formation of gel upon cooling is due to the development of triple-helical crosslinks via β-turn to form short double helix, in which another polypeptide strand is being wind around it, forming a triple-helical intermolecular crosslink (Walstra & Vliet, 2008). While gelatin dissolves readily in hot water, the resulting solution cannot be electrospun nor electrosprayed without further spin dope reformulation and modification. Despite the several collagen hydrolysis during the manufacturing process, aqueous gelatin solution is noticeable viscous, suggesting strong polymer–polymer interaction. This property, along with high surface tension of the protein solution, might have prevented the jetting process (Weiss, Kanjanapongkul, Wongsasulak, & Yoovidhya, 2012). Gelatin, as well as other water-soluble proteins, is polyelectrolytic. In aqueous solution, gelatin can be considered a polyion carrying ionizable amine and carboxylic groups which act as counterions that hampered the buildup of electrostatic charge (Huang, Zhang, Ramakrishna, & Lim, 2004). Although gelatin, when dissolved in strong organic solvents, such as 2,2,2-trifluoroethanol (TFE) and HFP can be effectively electrospun into fibers (Huang et al., 2004; Weiss et al., 2012), these solvents are not food compatible.

Li, He, Zheng, and Han (2006) demonstrated that gelatin spin dope solution at 15% (w/v) concentration, using ethanol (9:1 water:ethanol, v/v) as a solvent, can be electrospun into fibers of ~200 nm without the use of spinning aids, provided if the electrohydrodynamic process was carried out at 40 °C—a condition essential in preventing the gelation of the polymer solution, as well as weakening the molecular hydrogen bonds. The electrospinning of gelatin was further enhanced when hyaluronic acid was added to the spin dope solution, attributable to the increased viscosity and electrical conductivity of the spin dope solutions (Li et al., 2006). Hyaluronic acid is a biocompatible and resorbable biopolymer found in vertebrates and bacteria. It has repeated disaccharide units of D-glucuronic acid and *N*-acetylglucosamine linked by β-(1–4) and β-(1–3) glycosidic bonds (Salwowska, Bebenek, Azdło, & Wcisło-Dziadecka, 2016). Steyaert, Rahier, Van Vlierberghe, Olijve, and De Clerck (2016) reported that 13 wt% protein dissolved in acetic acid/water (90:10 acetic acid:water) can be electrospun into uniform fibers, without causing any significant hydrolysis degradation within 48 h. They observed an intriguing phenomenon that the electrospun gelatin fibers were readily soluble in cold demineralized water (19 °C) to form a hydrogel, but not the unprocessed gelatin powder (300 bloom). Their differential scanning calorimetry analysis indicated the presence of helix-to-coil transition for all the electrospin gelatin fibers with dissociation enthalpies of around 4 J/g, which was about 1/3 of the magnitude as compared to that for the original powder, indicating that electrospun fibers were not totally amorphous. The authors postulated that cold-water solubility of electrospun fibers cannot be attributable to amorphous fraction, but rather large surface area-to-volume ratio of the electrospun materials (Steyaert et al. (2016)). Similarly, Yang, Li, and Nie (2007) were able to electrospin neat gelatin into fibers of 133 nm, using 88 wt% aqueous formic acid as a solvent. The addition of polyvinyl alcohol (PVA) further strengthened the mechanical properties of electrospun nonwovens.

Electrospun gelatin fibers (diameter lower than 100 nm) have been used to encapsulate an ethanol extract from *Moringa oleifera*, which was rich in polyphenolic compounds, with encapsulation efficiency values of 80–85% (Hani, Torkamani, Azarian, Mahmood, & Ngalim, 2017). Deng et al. encapsulated curcumin in gelatin nano-microfibers. They added CTAB (cationic cetyltrimethylammonium bromide) to enhance the release of curcumin in aqueous solvents leading to higher radical scavenging activity and reducing power as well as stronger antimicrobial activity (Deng, Kang, Liu, Feng, & Zhang, 2017). Yang et al. (2007) electrospun gelatin/PVA composite fiber as a carrier for controlled release of raspberry ketone (4-(4-hydroxyphenyl) butan-2-one)—a major aromatic compounds of red raspberry with anti-inflammatory and lipid metabolism modulation activities (Bredsdorff, Wedebye, Nikolov, Hallas-Møller, & Pilegaard, 2015; Morimotoa et al., 2005).

Electrosprayed gelatin capsules have been developed for the encapsulation of nutraceuticals. Gómez-Estaca, Gavara, and Hernández-Muñoz (2015) encapsulated curcumin in electrosprayed gelatin microspheres to enhance the bioaccessibility and antioxidative properties. Moreover, the electrosprayed gelatin microcapsules provided a more homogenous distribution of color when dispersed into a gelled fish model food system. They observed that the in vitro digestibility of curcumin in the fish gel product was similar to the commercial encapsulated curcumin (Gómez-Estaca et al., 2015). Gómez-Mascaraque and López-Rubio (2016) compared the efficacy of gelatin, WPC, and SPI micron–submicron capsules, prepared by electrospraying of respective oil-in-water emulsions containing α-linolenic acid dispersed phase. They reported that WPC and SPI capsules provided higher encapsulation efficiencies and stronger protection than gelatin capsules. Gómez-Mascaraque et al. (2017) encapsulated EGCG-rich green tea extract within electrosprayed gelatin, as well as zein capsules, with high encapsulation efficiency (~90%). While both protein matrices were effective in preserving 85–90 g/100 g of their initial content in catechins during a 180 °C thermal treatment (compared with 40 g/100 g was lost in un-encapsulated green tea extract), no significant stability enhancements were observed between the encapsulated and free extract when tested in a biscuit model system (Gómez-Mascaraque et al., 2017). This observation highlighted the importance of evaluating the performance of encapsulant carriers in real food systems, which are more complex than the test media used in laboratory settings.

## 3.14 Other proteins

Proteins from other sources have been converted into fibers using similar spin dope formulations and electrospinning techniques. Selected examples are presented in this section. Protein concentrate derived from *Botryococcus braunii* microalgae residual biomass—a by-product from oil extraction is rich in protein, was electrospun by Verdugo, Lim, and Rubilar (2014) into fibers of average diameters ranging from 192 to 770 nm, depending on the protein and PEO concentrations, as well as the pH of the spin dope solutions. The incorporation of PEO in aqueous spin dope solutions was needed to initiate the jetting process, regardless of the solvents (distilled water, aqueous sodium hydroxide 1% solution, or glacial acetic acid) and pH conditions (Verdugo et al., 2014). Recently, Moreira et al. (2018) electrospun protein concentrate (81 wt% protein content) extracted from spirulina microalga, using a free-surface wire electrospinning technique (Fig. 2B). Uniform fibers at 5–10 wt% protein concentration were observed to have average diameters ranging from 118 to 452 nm when acetic acid:water (3:1) was used as a solvent. On the other hand, 1 wt% NaOH solvent tended to form spherical beads and beaded fibers. They reported that the addition of a trace quantity of PEO (0.5 wt%) was instrumental in enabling the electrohydrodynamic process (Moreira et al., 2018).

Fish muscle proteins made up of 25–30% fish sarcoplasmic proteins (FSP) are heterogeneous mixtures of proteins with molecular weights up to 200 kDa. Fish protein hydrolysates are known to elicit health benefits such as inhibition against dipeptidylpeptidase-4 (DPP-IV) which has been implicated to Type 2 diabetes (Li-Chan, Hunag, Jao, Ho, & Hsu, 2012). Dietary cod proteins have also been shown to improve insulin sensitivity (Ouellet et al., 2008; Ouellet, Marois, Weisnagel, & Jacques, 2007). These and other health-promoting properties of fish hydrolysates (Ewart et al., 2009; Girgih et al., 2015; Himaya, Ngo, Ryu, & Kim, 2012; Pilon et al., 2011; Sabeena Farvin et al., 2014), have prompted researchers to utilize FSP as a polymeric carrier for drug delivery. Stephansen, Chronakis, and Jessen (2014) and Stephansen, García-Díaz, Jessen, Chronakis, and Nielsen (2015, 2016) produced electrospun fish sarcoplasmic proteins (FSP) from cod (*Gadus morhua*) using HFP as a solvent. The fibers were insoluble in water (despite the water-soluble nature of FSP), but were hydrolyzable by proteolytic enzymes (Stephansen et al., 2014). Electrospun FSP fibers have been used as a carrier for drug delivery including dipeptide Ala-Trp, rhodamine B, and insulin and studied in various biorelevant media (Stephansen et al., 2015, 2016).

Amaranth protein isolate (API) from the grain of *Amaranthus hypochondriacus*, has been electrospun into fibers by blending with pullulan, for encapsulation of folic acid (Aceituno-Medina, Mendoza, Lagaron, & Lopez-Rubio, 2014), ferulic acid, and quercetin (Aceituno-Medina, Mendoza, Rodríguez, Lagaron, & López-Rubio, 2015). API/pullulan nanofibers increased the thermal- and UV-stability of folic acid (Aceituno-Medina et al., 2014), revealing the potential of electrospun fibers for protecting photosensitive vitamins (Aceituno-Medina et al., 2014). Furthermore, quercetin and ferulic acid were released in a sustained manner from API/pullulan nanofibers, and antioxidative properties were better preserved (Aceituno-Medina et al., 2015). Similarly, Blanco-Padilla, López-Rubio, Loarca-Piña, Gómez-Mascaraque, and Mendoza (2015) encapsulated curcumin in electrospun API:pullulan (1:1, w/w) fibers of average diameters of 225–249 nm, with curcumin encapsulation efficiencies of 73–93%. A higher antioxidant activity was observed for the encapsulated curcumin as compared to the free counterpart during the in vitro digestion process (Blanco-Padilla et al., 2015).

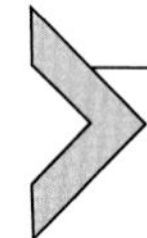

## 4. Electrohydrodynamic processing of functional phospholipid ultrafine fibers

Phospholipids are made up of two hydrophobic "tails" and a hydrophilic negatively charged phosphate "head". The amphiphilic characteristics of phospholipids are essential in the formation of phospholipid bilayer in cell membranes, as well as cellular signaling and regulation (Kim, Shanta, Zhou, & Kim, 2010). Although not a supramolecular structure, phospholipids have been shown to be electrospinnable into nano- and microfibers. McKee, Layman, Cashion, and Long (2006) demonstrated that lecithin solution of above 35% (w/w), prepared in DMF:chloroform (3:2, v/v) could be electrospun into continuous fibers of ~3.3 μm in diameter. The ability for phospholipids to form fiber was attributed to the transition from single molecules to spherical micelles, and then to cylindrical "wormlike" micelles, as the concentration of lecithin increased in the non-polar solvent. As the concentration increased, the cylindrical micelles eventually reached a critical length causing overlap, resulting in "chain entanglement" that enables the electrospinning of the non-polymeric compound (Fig. 12).

Asolectin is a mixture of phospholipids made up of approximately 25–33% of lecithin, cephalin, and phosphatidylinositol, respectively, along with other minor quantities of phospholipid and polar lipids. Fatty acids in asolectin are about 24% saturated, 14% mono-unsaturated, and 62% poly-unsaturated (Johns, Morris, Edwards, & Quirino, 2015). Asolectin extracted from soy has been electrospun into continuous fibers by Jørgensen, Qvortrup, and Chronakis (2015), using chloroform:DMF (3:2, v/v), isooctane, cyclohexane or limonene as solvents, resulting in phospholipid fibers with average diameters of 2.6, 3–8, 4–5, and 14 μm, respectively (Fig. 13). When a coaxial spinneret was used, wherein the outer needle contains the pure solvent and inner needle the asolectin solution, there was a drastic reduction in fiber diameter (0.38, 0.66, 1.01, and 1.54 μm for DMF, chloroform, isooctane, and cyclohexane, respectively). The phenomenon was attributed to the delayed solidification of the polymer jet and increased dielectric constant of the polymer jet that enhanced its bending instability (Jørgensen et al., 2015). In another study, mechanical properties and stability of electrospun asolectin fibers were evaluated by Mendes, Nikogeorgos, Lee, and Chronakis (2015) using an atomic force microscope (AFM) nanoindentation technique. They reported that the phospholipid fibers were stable in ambient conditions, preserving the modulus of elasticity

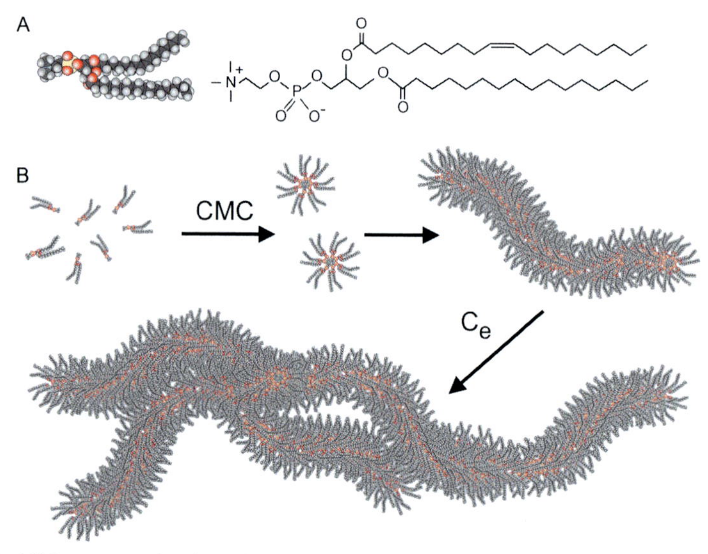


**Fig. 12** (A) Structure of a phospholipid with a polar phosphate head and two hydrophobic tails. (B) Transition of the phospholipid molecules to form reverse micelles above the critical micelle concentration (CMC) in non-polar solvents. Increasing the phospholipid concentration causes a transition from the spherical to cylindrical "wormlike" micelles. As the concentration continues to increase above the entanglement concentration ($C_e$; ~35%), entanglement couplings of the cylindrical micelles eventually occur, forming a "polymeric" network essential for electrospinning. *Adapted from McKee, M. G., Layman, J. M., Cashion, M. P., & Long, T. E. (2006). Phospholipid nonwoven electrospun membranes.* Science (New York, N.Y.), 311*(5759), 353–355. https://doi.org/10.1126/science.1119790.*

up to 24h. A substantial adhesion was observed during the unloading step due to the interaction between the hydrophilic AFM probe (silanol groups) with the hydrophilic inner core of the phospholipid fibers, which might have caused the high adhesion hysteresis observed (Mendes et al., 2015). The average elastic modulus of the assembled asolectin fiber was approximately 17.2 ± 1 MPa, which was higher than the elastic moduli of natural cholesterol and phosphatidylcholine (0.27 and 0.03 MPa, respectively) (Crowley, 1973).

The electrospun asolectin fibers have been exploited as a carrier for curcumin and vanillin (Shekarforoush, Mendes, Baj, Beeren, & Chronakis, 2017). The researchers reported that these phenolic compounds remained stable during a 15-day storage under different temperature (refrigerated, ambient) and pressure (vacuum, ambient) conditions, as reflected by the total antioxidant capacity (TAC), total phenolic content, and $^1$H NMR

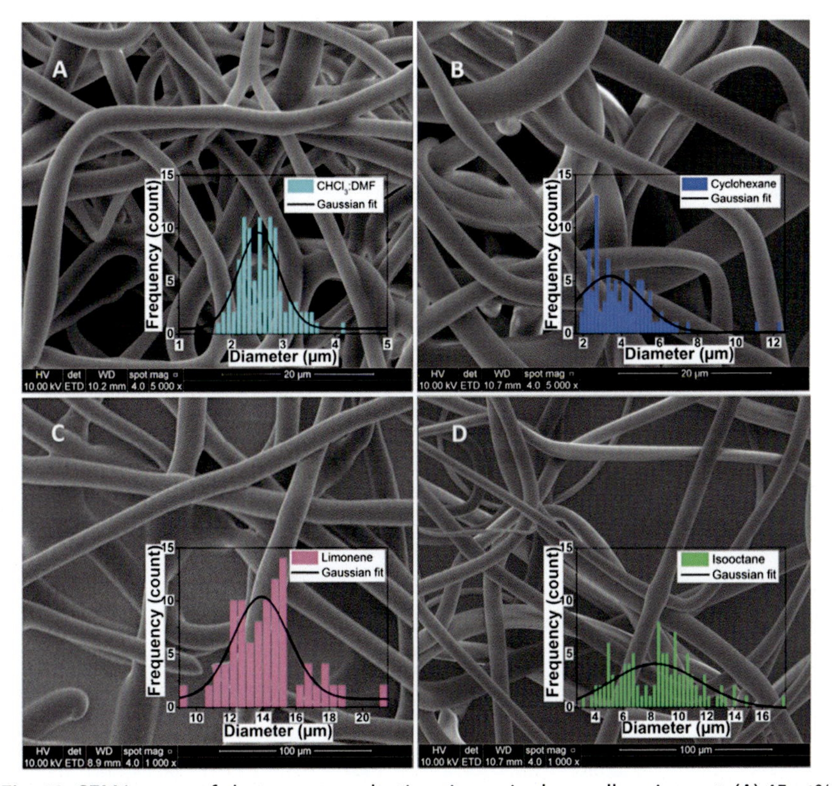


**Fig. 13** SEM images of electrospun asolectin using a single needle spinneret: (A) 45 wt% in chloroform:DMF (3:2, v/v), scale bar 20 μm; (B) 50 wt% in cyclohexane, scale bar 20 μm; (C) 60 wt% in limonene, scale bar 100 μm; and (D) 60 wt% isooctane, scale bar 100 μm. Insets are histogram of the diameter distribution calculated from 100 random fiber measurements (Jørgensen et al., 2015).

analyses. By contrast, considerable losses in TAC were observed under the same storage conditions. Release studies in aqueous media revealed that the phenolics were released mainly due to swelling of phospholipid fiber matrix over time. Moreover, Shekarforoush, Mendes, et al. (2017) observed an enhanced TAC for the curcumin and vanillin loaded into the electrospun phospholipid compared with the free phenolics, suggesting their synergistic antioxidant activity with phospholipids. This phenomenon is in accordance with another study where an increased TAC was observed due to synergistic interactions between lecithins with other phenolic compounds, such as tocopherols (Judde, Villeneuve, Rossignol-Castera, & Guillou, 2003). These studies confirmed the efficacy of electrospun phospholipid fibers for carriers of antioxidants.

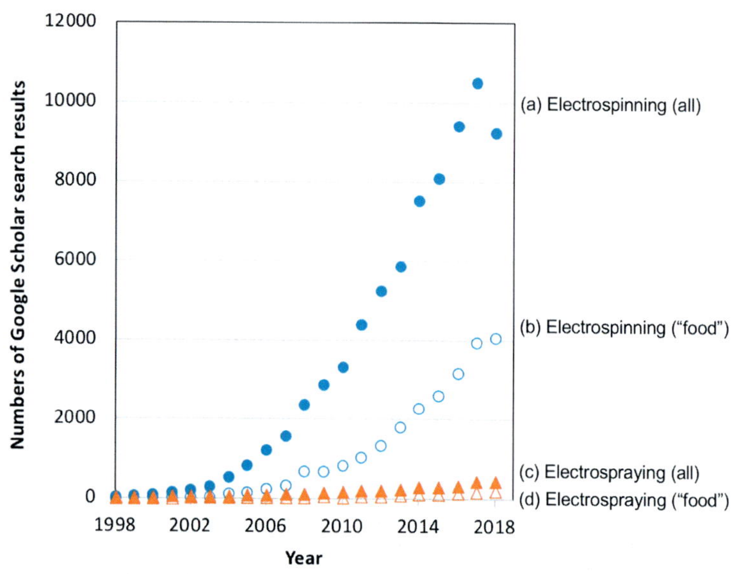


**Fig. 14** Estimated number of published works based on Google scholar. (A) All electrospinning (search keywords: +electrospinning +electrospun); (B) electrospinning related to food (search keywords: +electrospinning +electrospun +food); (C) all electrospraying (keywords: +electrospraying +electrosprayed −ionization); and (D) electrospraying related to food (keywords: +electrospraying +electrosprayed −ionization +food). "−ionization" was used during search for (C) and (D) to avoid irrelevant hits related to "electrospray ionization mass spectrometry" (accessed on Dec 6, 2018).

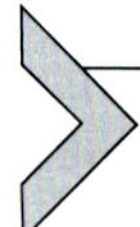

## 5. Conclusions: Challenges, opportunities, and prospects

Electrospinning and electrospraying are versatile techniques for forming ultrafine fibrous and particulate materials that can be engineered for various applications. As revealed by Google Scholar search results (Fig. 14), research and development activities related to these electrohydrodynamic processes continue to increase over the past 2 decades, with the numbers of publication related to electrospinning being substantially higher than those of electrospraying. Search results including the keyword "food" show that there is an increasing reference to food in electrospinning-related literature, suggesting a sustained interest of applying electrospinning technologies to food. Most published works are focusing on spinneret approach which is low in processing rate, in the order of milliliters of spin dope solution per hour.

Future research and development are expected to focus more on other high-throughput techniques, such as those based on free-surface and blow-assist electrohydrodynamic processes.

Although many solvent systems have been developed to solubilize polymers for electrospinning, a large proportion of these solvents are not food compatible. Water, ethanol, and organic acid-based solvent systems have been successfully applied for the electrospinning of food biopolymers (proteins and polysaccharides). However, these solvents are not always optimal due to difficulty in building up electrostatic charge density, lack of chain entanglement caused by intramolecular interactions (e.g., globular proteins), intermolecular aggregation, and so on. The incorporation of edible spinning aid polymers and additives into the spin dope solutions will continue to be a main strategy to overcome processing obstacles. As demonstrated in a number of studies, polymer chain entanglement is not a prerequisite for the formation of electrospun fibers for certain spin dope solution systems. Small molecules, such as phospholipid and cyclodextrins, can be electrospun into uniform nonwoven, due to their propensity to self-assemble into supramolecular structure that prevents the breakup of polymer jet during electrospinning. Similarly, Newtonian fluids, such as aqueous pullulan solutions which maintain at a relatively low viscosity at high polymer concentration (>10 wt%), are capable of forming fiber due to the fluid elasticity. These materials are useful in maximizing the loading capacity of bioactive compounds into the carriers.

When deployed as a carrier of bioactive compounds in food products, evaluation of electrospun and electrosprayed materials in real food systems is critical, considering commercial products are more complex and exposed to more aggressive processing conditions than experimental food models used in the laboratory. From a processing standpoint, the incorporation of bioactive loaded electrospun fibers into food products is challenging since fibers tend to remain entangled and/or fractured when sheared. Methodologies to produce fibers of discrete length will be useful to overcome dispersion challenges. Alternatively, the use of electrosprayed particles may be more desirable than fibrous materials. Successful dispersion of electrospun fibers in food matrices may open up new opportunities in the development of innovative nanostructured food products to mimic fibrous textures in meat tissues. These areas are relatively unexplored. Finally, in active and intelligent applications, research and development innovations will continue to exploit electrospun nonwovens for improving food product shelf-life and safety. Commercial use of electrospun nonwovens as the primary packaging materials for food is unlikely in the near future due to their substantially higher production cost than the existing packaging materials manufactured with the highly efficient converting processes. However, the deployment of functionalized electrospun nonwovens, in the form of an adjunct to food package (e.g., sachet, label, absorbent) is conceivable, such as for controlled release of preservatives (antimicrobial, antioxidant), removal of undesirable components from the headspace, and real-time detection of food spoilage.

## References

Aceituno-Medina, M., Mendoza, S., Lagaron, J. M., & Lopez-Rubio, A. (2014). Photoprotection of folic acid upon encapsulation in food-grade amaranth (*Amaranthus hypochondriacus* L.) protein isolate—Pullulan electrospun fibers. *LWT—Food Science and Technology*, *62*(2), 970–975. https://doi.org/10.1016/j.lwt.2015.02.025.

Aceituno-Medina, M., Mendoza, S., Lagaron, J. M., & López-Rubio, A. (2013). Development and characterization of food-grade electrospun fibers from amaranth protein and pullulan blends. *Food Research International*, *54*(1), 667–674. https://doi.org/10.1016/j.foodres.2013.07.055.

Aceituno-Medina, M., Mendoza, S., Rodríguez, B. A., Lagaron, J. M., & López-Rubio, A. (2015). Improved antioxidant capacity of quercetin and ferulic acid during in-vitro digestion through encapsulation within food-grade electrospun fibers. *Journal of Functional Foods*, *12*, 332–341. https://doi.org/10.1016/j.jff.2014.11.028.

Alborzi, S., Lim, L.-T., & Kakuda, Y. (2014). Release of folic acid from sodium alginate-pectin-poly(ethylene oxide) electrospun fibers under in vitro conditions. *LWT—Food Science and Technology*, *59*(1), 383–388. https://doi.org/10.1016/j.lwt.2014.06.008.

Alborzi, S., Lim, L. T., & Kakuda, Y. (2010). Electrospinning of sodium alginate-pectin ultrafine fibers. *Journal of Food Science*, *75*(1), 100–107. https://doi.org/10.1111/j.1750-3841.2009.01437.x.

Anderson, T. J., & Lamsal, B. P. (2011). Zein extraction from corn, corn products, and coproducts and modifications for various applications: A review. *Cereal Chemistry*, *88*, 159–173.

Antunes, M. D., Dannenberg, G. da S., Fiorentini, Â. M., Pinto, V. Z., Lim, L.-T., Zavareze, E. da R., et al. (2017). Antimicrobial electrospun ultrafine fibers from zein containing eucalyptus essential oil/cyclodextrin inclusion complex. *Internal Journal of Biological Macromolecules*, *104*, 874–882. https://doi.org/10.1016/j.ijbiomac.2017.06.095.

Arinstein, A., & Zussman, E. (2011). Electrospun polymer nanofibers: Mechanical and thermodynamic perspectives. *Journal of Polymer Science Part B: Polymer Physics*, *49*, 691–707.

Arya, N., Chakraborty, S., Dube, N., & Katti, D. S. (2009). Electrospraying: A facile technique for synthesis of chitosan-based micro/nanospheres for drug delivery applications. *Journal of Biomedical Materials Research Part B: Applied Biomaterials*, *88B*(1), 17–31. https://doi.org/10.1002/jbm.b.31085.

Austero, M. S., Donius, A. E., Wegst, U. G. K., & Schauer, C. L. (2012). New crosslinkers for electrospun chitosan fibre mats. I. Chemical analysis. *Journal of the Royal Society, Interface/the Royal Society*, *9*(75), 2551–2562. https://doi.org/10.1098/rsif.2012.0241.

Balan, V., & Verestiuc, L. (2014). Strategies to improve chitosan hemocompatibility: A review. *European Polymer Journal*, *53*, 171–188. https://doi.org/10.1016/j.eurpolymj.2014.01.033.

Baumgarten, P. K. (1971). Electrostatic spinning of acrylic microfibers. *Journal of Colloid and Interface Science*, *36*(1), 71–79.

Bhattarai, N., Edmondson, D., Veiseh, O., Matsen, F. a., & Zhang, M. (2005). Electrospun chitosan-based nanofibers and their cellular compatibility. *Biomaterials*, *26*, 6176–6184. https://doi.org/10.1016/j.biomaterials.2005.03.027.

Blanco-Padilla, A., López-Rubio, A., Loarca-Piña, G., Gómez-Mascaraque, L. G., & Mendoza, S. (2015). Characterization, release and antioxidant activity of curcumin-loaded amaranth-pullulan electrospun fibers. *LWT—Food Science and Technology*, *63*, 1137–1144. https://doi.org/10.1016/j.lwt.2015.03.081.

Bonino, C. A., Krebs, M. D., Saquing, C. D., Jeong, S. I., Shearer, K. L., Alsberg, E., et al. (2011). Electrospinning alginate-based nanofibers: From blends to crosslinked low molecular weight alginate-only systems. *Carbohydrate Polymers*, *85*(1), 111–119. https://doi.org/10.1016/j.carbpol.2011.02.002.

Braghirolli, D. I., Steffens, D., & Pranke, P. (2014). Electrospinning for regenerative medicine: A review of the main topics. *Drug Discovery Today*, *19*, 743–753.

Bredsdorff, L., Wedebye, E. B., Nikolov, N. G., Hallas-Møller, T., & Pilegaard, K. (2015). Raspberry ketone in food supplements—High intake, few toxicity data—A cause for safety concern? *Regulatory Toxicity and Pharmacology*, *73*, 196–200.

Cai, J., Niu, H., Yu, Y., Xiong, H., & Lin, T. (2017). Effect of solvent treatment on morphology, crystallinity and tensile properties of cellulose acetate nanofiber mats. *The Journal of the Textile Institute*, *108*, 555–561.

Carroll, C. P., & Joo, Y. L. (2006). Electrospinning of viscoelastic Boger fluids: Modeling and experments. *Physics of Fluids*, *18*(053102), 1–14.

Castro-Enríquez, D., Rodríguez-Félix, F., Ramírez-Wong, B., Torres-Chávez, P., Castillo-Ortega, M., Rodríguez-Félix, D., et al. (2012). Preparation, characterization and release of urea from wheat gluten electrospun membranes. *Materials*, *5*(12), 2903–2916. https://doi.org/10.3390/ma5122903.

Celebioglu, A., Kayaci-Senirmak, F., Kusku, S. İ., Durgun, E., & Uyar, T. (2016). Polymer-free nanofibers from vanillin/cyclodextrin inclusion complexes: High thermal stability, enhanced solubility and antioxidant property. *Food and Function*, 7, 3141–3153.

Celebioglu, A., & Uyar, T. (2010). Cyclodextrin nanofibers by electrospinning. *Chemical Communications (Cambridge, England)*, *46*(37), 6903–6905. https://doi.org/10.1039/c0cc01484b.

Celebioglu, A., & Uyar, T. (2011). Electrospinning of polymer-free nanofibers from cyclodextrin inclusion complexes. *Langmuir*, *27*(10), 6218–6226. https://doi.org/10.1021/la1050223.

Celebioglu, A., & Uyar, T. (2013). Electrospinning of nanofibers from non-polymeric systems: Electrospun nanofibers from native cyclodextrins. *Journal of Colloid and Interface Science*, *404*, 1–7. https://doi.org/10.1016/j.jcis.2013.04.034.

Cengiz, F., Krucińska, I., Gliścińsk, E., Chrzanowski, M., & Göktepe, F. (2009). Comparative analysis of various alectrospinning methods of nanofibre formation. *Fibers & Textiles in Eastern Europe*, *17*(1), 13–19.

Chávarri, M., Marañón, I., & Villarán, M. C. (2012). *Encapsulation technology to protect probiotic bacteria. In Probiotics* (pp. 501–540). Riejka: InTech Open Access. https://doi.org/10.5772/50046.

Chen, Z., Mo, X., & Qing, F. (2007). Electrospinning of collagen-chitosan complex. *Materials Letters*, *61*(16), 3490–3494. https://doi.org/10.1016/j.matlet.2006.11.104.

Chimie, R. R., Bercea, M., Darie, R. N., & Morariu, S. (2013). Rheological investigation of xanthan/Pluronic F127 hydrogels. *Revue Roumaine de Chimie*, *58*, 189–196.

Colín-Orozco, J., Zapata-Torres, M., Rodríguez-Gattorno, G., & Pedroza-Islas, R. (2015). Properties of poly (ethylene oxide)/whey protein isolate nanofibers prepared by electrospinning. *Food Biophysics*, *10*(2), 134–144. https://doi.org/10.1007/s11483-014-9372-1.

Crowley, J. M. (1973). Bimolecular lipid membranes as an electromechanical instability. *Biophysical Journal*, *13*, 711–724.

Damodaran, S. (2008). Chapter 5: Amino acids, peptides, and proteins. In S. Damodaran, K. L. Parkin, & O. R. Fennema (Eds.), *Fennema's food chemistry* (4th ed.). Boca Raton, FL: CRC Press.

Deng, L., Kang, X., Liu, Y., Feng, F., & Zhang, H. (2017). Effects of surfactants on the formation of gelatin nanofibres for controlled release of curcumin. *Food Chemistry*, *231*, 70–77. https://doi.org/10.1016/j.foodchem.2017.03.027.

Desai, K., Kit, K., Li, J., Michael Davidson, P., Zivanovic, S., & Meyer, H. (2009). Nanofibrous chitosan non-wovens for filtration applications. *Polymer*, *50*(15), 3661–3669. https://doi.org/10.1016/j.polymer.2009.05.058.

Devarayan, K., Hanaoka, H., Hachisu, M., Araki, J., Ohguchi, M., Behera, B. K., et al. (2013). Direct electrospinning of cellulose-chitosan composite nanofiber. *Macromolecular Materials and Engineering*, *298*, 1059–1064. https://doi.org/10.1002/mame.201200337.

Drosou, C., Krokida, M., & Biliaderis, C. G. (2018). Composite pullulan-whey protein nanofibers made by electrospinning: Impact of process parameters on fiber morphology and physical properties. *Food Hydrocolloids*, *77*, 726–735. https://doi.org/10.1016/j.foodhyd.2017.11.014.

El-salam, M. H. A., & El-shibiny, S. (2016). *Natural biopolymers as encapsulations*. Elsevier. https://doi.org/10.1016/B978-0-12-804307-3/00019-3.

Ewart, H. S., Dennis, D., Potvin, M., Tiller, C., Fang, L. H., Zhang, R., et al. (2009). Development of a salmon protein hydrolysate that lowers blood pressure. *European Food Research and Technology*, *229*(4), 561–569. https://doi.org/10.1007/s00217-009-1083-3.

Fabra, M. J., López-Rubio, A., & Lagaron, J. M. (2016). Use of the electrohydrodynamic process to develop active/bioactive bilayer films for food packaging applications. *Food Hydrocolloids*, *55*, 11–18. https://doi.org/10.1016/j.foodhyd.2015.10.026.

Faralli, A., Shekarforoush, E., Mendes, A. C., & Chronakis, I. S. (2019). In vitro permeability enhancement of curcumin across Caco-2 cells monolayers using electrospun xanthan-chitosan nanofibers. *Carbohydrate Polymers*, *206*, 38–47.

Faria, S., De Oliveira Petkowicz, C. L., De Morais, S. A. L., Terrones, M. G. H., De Resende, M. M., De Frana, F. P., et al. (2011). Characterization of xanthan gum produced from sugar cane broth. *Carbohydrate Polymers*, *86*(2), 469–476. https://doi.org/10.1016/j.carbpol.2011.04.063.

Fathi, M., Martín, Á., & McClements, D. J. (2014). Nanoencapsulation of food ingredients using carbohydrate based delivery systems. *Trends in Food Science and Technology*, *39*(1), 18–39. https://doi.org/10.1016/j.tifs.2014.06.007.

Feng, J. J. (2002). The stretching of an electrified non-Newtonian jet—A model for electrospinning. *Physics of Fluids*, *14*(11), 3912–3926.

Feng, J. J. (2003). Stretching of a straight electrically charged viscoelastic jet. *Journal of Non-Newtonian Fluid Mechanics*, *116*, 55–70.

Fernandez, A., Torres-Giner, S., & Lagaron, J. M. (2009). Novel route to stabilization of bioactive antioxidants by encapsulation in electrospun fibers of zein prolamine. *Food Hydrocolloids*, *23*, 1427–1432. https://doi.org/10.1016/j.foodhyd.2008.10.011.

Fonseca, L. M., Oliveira, J. P., Oliveira, P. D., Zavareze, E. R., Dias, A. R. G., & Lim, L.-T. (2019). Electrospinning of native and anionic corn starch fibers with different amylose contents. *Food Research International*, *116*, 1318–1326. https://doi.org/10.1016/j.foodres.2018.10.021.

Fonseca, L. M., Silva da, F. T., Antunes, M. D., Halal El, S. L. M., Lim, L.-T. ., & Dias, A. R. G. (2019). Aging time of soluble potato starch solutions for ultrafine fibers formation by electrospinning. *Starch*, *71*, 1800089. doi.org/10.1002/star.201800089.

Frenot, A., Henriksson, M. W., & Walkenstrom, P. (2007). Electrospinning of cellose-based nanofibers. *Journal of Applied Polymer Science, 103*, 1473–1482.
Frey, M. W. (2008). Electrospinning cellulose and cellulose derivatives. *Polymer Reviews, 48*(2), 378–391. https://doi.org/10.1080/15583720802022281.
Fuenmayor, C. A., Mascheroni, E., Cosio, M. S., Piergiovanni, L., Benedetti, S., Ortenzi, M., et al. (2013). Encapsulation of R-(+)-limonene in edible electrospun nanofibers. *Chemical Engineering Transactions, 32*, 1771–1776. https://doi.org/10.3303/CET1332296.
García-Moreno, P. J., Damberg, C., Chronakis, I. S., & Jacobsen, C. (2017). Oxidative stability of pullulan electrospun fibers containing fish oil: Effect of oil content and natural antioxidants addition. *European Journal of Lipid Science and Technology, 119*, 1–11. 1600305. https://doi.org/10.1002/ejlt.201600305.
García-Moreno, P. J., Mendes, A. C., Jacobsen, C., & Chronakis, I. S. (2018). Biopolymers for the nano-microencapsulation of bioactive ingredients by electrohydrodynamic processing. *Polymers for Food Applications*, 447–479. https://doi.org/10.1007/978-3-319-94625-2_17.
Geng, X., Kwon, O.-H., & jang, J. (2005). Electrospinning of chitosan dissolved in concentrated acetic acid solution. *Biomaterials, 26*, 5427–5432.
Girgih, A. T., He, R., Hasan, F. M., Udenigwe, C. C., Gill, T. A., & Aluko, R. E. (2015). Evaluation of the in vitro antioxidant properties of a cod (Gadus morhua) protein hydrolysate and peptide fractions. *Food Chemistry, 173*, 652–659. https://doi.org/10.1016/j.foodchem.2014.10.079.
Gómez-Estaca, J., Gavara, R., & Hernández-Muñoz, P. (2015). Encapsulation of curcumin in electrosprayed gelatin microspheres enhances its bioaccessibility and widens its uses in food applications. *Innovative Food Science and Emerging Technologies, 29*, 302–307. https://doi.org/10.1016/j.ifset.2015.03.004.
Gómez-Mascaraque, L. G., Hernández-Rojas, M., Tarancón, P., Tenon, M., Feuillère, N., Ruiz, J. F. V., et al. (2017). Impact of microencapsulation within electrosprayed proteins on the formulation of green tea extract-enriched biscuits. *LWT—Food Science and Technology, 81*, 77–86. https://doi.org/10.1016/j.lwt.2017.03.041.
Gómez-Mascaraque, L. G., & López-Rubio, A. (2016). Protein-based emulsion electrosprayed micro- and submicroparticles for the encapsulation and stabilization of thermosensitive hydrophobic bioactives. *Journal of Colloid and Interface Science, 465*, 259–270. https://doi.org/10.1016/j.jcis.2015.11.061.
Gomez-Mascaraque, L. G., Morfin, R. C., Pérez-Masiá, R., Sanchez, G., & Lopez-Rubio,-A. (2016). Optimization of electrospraying conditions for the microencapsulation of probiotics and evaluation of their resistance during storage and in-vitro digestion. *LWT—Food Science and Technology, 69*, 438–446. https://doi.org/10.1016/j.lwt.2016.01.071.
Gómez-Mascaraque, L. G., Sanchez, G., & López-Rubio, A. (2016). Impact of molecular weight on the formation of electrosprayed chitosan microcapsules as delivery vehicles for bioactive compounds. *Carbohydrate Polymers, 150*, 121–130. https://doi.org/10.1016/j.carbpol.2016.05.012.
Greiner, A., & Wendorff, J. H. (2008). Functinal self-assembled nanofibers by eletrospinning. *Advances in Polymer Science, 219*, 107–171.
Gudjónsdóttir, M., Gacutan, M. D., Mendes, A. C., Chronakis, I. S., Jespersen, L., & Karlsson, A. H. (2015). Effects of electrospun chitosan wrapping for dry-ageing of beef, as studied by microbiological, physicochemical and low-field nuclear magnetic resonance analysis. *Food Chemistry, 184*, 167–175. https://doi.org/10.1016/j.foodchem.2015.03.088.
Haas, D., Heinrich, S., & Greil, P. (2010). Solvent control of cellulose acetate nanofibre felt structure produced by electrospinning. *Journal of Materials Scienceaterials Science, 45*, 1299–1306.

Haley, B., & Frenkel, E. (2008). Nanoparticles for drug delivery in cancer treatment. *Urologic Oncology*, *26*, 57–64.
Han, Y., & Chen, H. (2013). Enhancement of nanofiber elasticity by using wheat glutenin as an addition. *Polymer Science, Series A*, *55*(5), 320–326. https://doi.org/10.1134/S0965545X13050076.
Hani, N. M., Torkamani, A. E., Azarian, M. H., Mahmood, K. W. A., & Ngalim, S. H. (2017). Characterisation of electrospun gelatine nanofibres encapsulated with Moringa oleifera bioactive extract. *Journal of the Science of Food and Agriculture*, *97*, 3348–3358. https://doi.org/10.1002/jsfa.8185.
Hansen, C. (2007). *Hansen solubility parameters: A user's handbook* (2nd ed.). . CRC Press.
Hansen, C. M., & Skaarup, K. (1967). The three dimensional solubility parameter—Key to paint component affinities III. *Journal of Paint Technology*, *39*, 511–514.
Hay, I. D., Rehman, Z. U., Ghafoor, A., & Rehm, B. H. A. (2010). Bacterial biosynthesis of alginates. *Journal of Chemical Technology and Biotechnology*, *85*, 752–759.
Heinze, T., Liebert, T., Heublein, B., & Hornig, S. (2006). Functional polymers based on dextran. *Advances in Polymer Science*, *205*, 199–291.
Himaya, S. W. A., Ngo, D. H., Ryu, B., & Kim, S. K. (2012). An active peptide purified from gastrointestinal enzyme hydrolysate of Pacific cod skin gelatin attenuates angiotensin-1 converting enzyme (ACE) activity and cellular oxidative stress. In *Vol. 132. Food chemistry* (pp. 1872–1882). Elsevier. https://doi.org/10.1016/j.foodchem.2011.12.020.
Homayoni, H., Ravandi, S. A. H., & Valizadeh, M. (2009). Electrospinning of chitosan nanofibers: Processing optimization. *Carbohydrate Polymers*, *77*, 656–661.
Horan, F. E. (1974). Soy protein products and their production. *Journal of the American Oil Chemists Society*, *51*, 67–73.
Hou, H., & Reneker, D. H. (2004). Carbon nanotubes on carbon nanofibers—A novel structure based on electrospun polymer nanofibers. *Advanced Materials*, *16*(1), 69–73.
Hu, X., Liu, S., Zhou, G., Huang, Y., Xie, Z., & Jing, X. (2014). Electrospinning of polymeric nanofibers for drug delivery applications. *Journal of Controlled Release*, *185*(1), 12–21. https://doi.org/10.1016/j.jconrel.2014.04.018.
Huang, X. J., Chen, P. C., Huang, F., Ou, Y., Chen, M. R., & Xu, Z. K. (2011). Immobilization of Candida rugosa lipase on electrospun cellulose nanofiber membrane. *Journal of Molecular Catalysis B: Enzymatic*, *70*(3–4), 95–100. https://doi.org/10.1016/j.molcatb.2011.02.010.
Huang, Z.-M., Zhang, Y. Z., Ramakrishna, S., & Lim, C. T. (2004). Electrospinning and mechanical characterization of gelatin nanofibers. *Polymer*, *45*, 5361–5368.
Ignatova, M., Manolova, N., Markova, N., & Rashkov, I. (2009). Electrospun non-woven nanofibrous hybrid mats based on chitosan and PLA for wound-dressing applications. *Macromolecular Bioscience*, *9*(1), 102–111. https://doi.org/10.1002/mabi.200800189.
Ignatova, M., Manolova, N., & Rashkov, I. (2013). Electrospun antibacterial chitosan-based fibers. *Macromolecular Bioscience*, *13*(7), 860–872. https://doi.org/10.1002/mabi.201300058.
James, D. F. (2009). Boger fluids. *Annual Review of Fluid Mechanics*, *41*(1), 129–142.
Jash, A., & Lim, L. T. (2018). Triggered release of hexanal from an imidazolidine precursor encapsulated in poly(lactic acid) and ethylcellulose carriers. *Journal of Materials Science*, *53*(3), 2221–2235.
Jaworek, A. (2008). Electrostatic micro- and nanoencapsulation and electroemulsification: A brief review. *Journal of Microencapsulation*, *25*, 443–468.
Jaworek, A., & Sobczyk, A. T. (2008). Electrospraying route to nanotechnology: An overview. *Journal of Electrostatics*, *66*, 197–219.
Jayakumar, R., Menon, D., Manzoor, K., Nair, S. V., & Tamura, H. (2010). Biomedical applications of chitin and chitosan based nanomaterials—A short review. *Carbohydrate Polymers*, *82*(2), 227–232. https://doi.org/10.1016/j.carbpol.2010.04.074.

Jeannie Tan, Z. Y., & Zhang, X. W. (2011). Influence of chitosan on electrospun PVA nanofiber mat. *Advanced Materials Research*, *311–313*(311–313), 1763–1768. https://doi.org/10.4028/www.scientific.net/AMR.311-313.1763.

Jiang, H., Fang, D., Hsiao, B. S., Chu, B., & Chen, W. (2004). Optimization and characterization of dextran membranes prepared by electrospinning. *Biomacromolecules*, *5*(2), 326–333.

Jiang, Q., Reddy, N., & Yang, Y. (2010). Cytocompatible cross-linking of electrospun zein fibers for the development of water-stable tissue engineering scaffolds. *Acta Biomaterialia*, *6*, 4042–4051.

Johns, A., Morris, S., Edwards, K., & Quirino, R. L. (2015). Asolectin from soybeans as a natural compatibilizer for cellulose-reinforced biocomposites from tung oil. *Journal of Applied Polymer Science*, *41833*, 1–9.

Jørgensen, L., Qvortrup, K., & Chronakis, I. S. (2015). Phospholipid electrospun nanofibers: Effect of solvents and co-axial processing on morphology and fiber diameter. *RSC Advances*, *5*(66), 53644–53652. https://doi.org/10.1039/C5RA10498J.

Judde, A., Villeneuve, P., Rossignol-Castera, A., & Guillou, A. (2003). Antioxidant effect of soy lecithins on vegetable oil stability and their synergism with tocopherols. *Journal of the American Oil Chemists' Society*, *80*(12), 1209–1215. https://doi.org/10.1007/s11746-003-0844-4.

Karim, M. R., Lee, H. W., Kim, R., Ji, B. C., Cho, J. W., Son, T. W., et al. (2009). Preparation and characterization of electrospun pullulan/montmorillonite nanofiber mats in aqueous solution. *Carbohydrate Polymers*, *78*(2), 336–342. https://doi.org/10.1016/j.carbpol.2009.04.024.

Kayaci, F., Ertas, Y., & Uyar, T. (2013). Enhanced thermal stability of eugenol by cyclodextrin inclusion complex encapsulated in electrospun polymeric nanofibers. *Journal of Agricultural and Food Chemistry*, *61*(34), 8156–8165. https://doi.org/10.1021/jf402923c.

Kayaci, F., Umu, O. C. O., Tekinay, T., & Uyar, T. (2013). Antibacterial electrospun poly(lactic acid) (PLA) nano fi brous webs incorporating triclosan/cyclodextrin inclusion complexes. *Journal of Agricultural and Food Chemistry*, *61*, 3901–3908. https://doi.org/10.1021/jf400440b.

Kayaci, F., & Uyar, T. (2012). Encapsulation of vanillin/cyclodextrin inclusion complex in electrospun polyvinyl alcohol (PVA) nanowebs: Prolonged shelf-life and high temperature stability of vanillin. *Food Chemistry*, *133*(3), 641–649. https://doi.org/10.1016/j.foodchem.2012.01.040.

Kim, Y., Shanta, S. R., Zhou, L.-H., & Kim, K. P. (2010). Mass spectrometry based cellular phosphoinositides profiling and phospholipid analysis: A brief review. *Experimental and Molecular Medicine*, *42*(1), 11.

Ko, F. K. (2004). Formation of nanofibers and nanotubes production. In S. Guceri, Y. G. Gogotsi, & V. Kuznetsov (Eds.), *NATO science series: Vol. 169. Nanoengineered nanofibrous materials*. Dordrecht: Springer.

Kong, L., & Ziegler, G. R. (2012). Role of molecular entanglements in starch fiber formation by electrospinning. *Biomacromolecules*, *13*(8), 2247–2253. https://doi.org/10.1021/bm300396j.

Kong, L., & Ziegler, G. R. (2013). Quantitative relationship between electrospinning parameters and starch fiber diameter. *Carbohydrate Polymers*, *92*(2), 1416–1422. https://doi.org/10.1016/j.carbpol.2012.09.026.

Kong, L., & Ziegler, G. R. (2014a). Fabrication of pure starch fibers by electrospinning. *Food Hydrocolloids*, *36*, 20–25. https://doi.org/10.1016/j.foodhyd.2013.08.021.

Kong, L., & Ziegler, G. R. (2014b). Formation of starch-guest inclusion complexes in electrospun starch fibers. *Food Hydrocolloids*, *38*, 211–219. https://doi.org/10.1016/j.foodhyd.2013.12.018.

Kong, L., & Ziegler, G. R. (2014c). Rheological aspects in fabricating pullulan fibers by electro-wet-spinning. *Food Hydrocolloids*, *38*, 220–226. https://doi.org/10.1016/j.foodhyd.2013.12.016.

Koombhongse, S., Liu, W., & Reneker, D. H. (2001). Flat polymer ribbons and other shapes by electrospinning. *Journal of Polymer Science Part B: Polymer Physics*, *39*, 2598–2606.

Kurban, Z., Lovell, A., Bennington, S. M., Jenkins, D. W. K., Ryan, K. R., Jones, M. O., et al. (2010). A solution selection model for coaxial electrospinning and its application to nanostructured hydrogen storage materials. *Journal of Physical Chemistry C*, *114*, 21201–21213.

Lachke, A. R. (2004). Xanthan—A versatile gum. *Resonance*, *9*, 25–33. https://doi.org/10.1007/BF02834866.

Laelorspoen, N., Wongsasulak, S., Yoovidhya, T., & Devahastin, S. (2014). Microencapsulation of Lactobacillus acidophilus in zein-alginate core-shell microcapsules via electrospraying. *Journal of Functional Foods*, 7(1), 342–349. https://doi.org/10.1016/j.jff.2014.01.026.

Lancuški, A., Abu Ammar, A., Avrahami, R., Vilensky, R., Vasilyev, G., & Zussman, E. (2017). Design of starch-formate compound fibers as encapsulation platform for biotherapeutics. *Carbohydrate Polymers*, *158*, 68–76. https://doi.org/10.1016/j.carbpol.2016.12.003.

Lancuški, A., Vasilyev, G., Putaux, J. L., & Zussman, E. (2015). Rheological properties and electrospinnability of high-amylose starch in formic acid. *Biomacromolecules*, *16*(8), 2529–2536. https://doi.org/10.1021/acs.biomac.5b00817.

Lee, K. Y., Jeong, L., Kang, Y. O. O., Lee, S. J., & Park, W. H. (2009). Electrospinning of polysaccharides for regenerative medicine. *Advanced Drug Delivery Reviews*, *61*(12), 1020–1032. https://doi.org/10.1016/j.addr.2009.07.006.

Li-Chan, E. C. Y., Hunag, S.-L., Jao, C.-L., Ho, K.-P., & Hsu, K.-C. (2012). Peptides derived from atlantic salmon skin gelatin as dipeptidyl-peptidase IV inhibitors. *Journal of Agricultural and Food Chemistry*, *60*, 973–978.

Li, D., & Xia, Y. (2004). Electrospinning of nanofibers: Reinventing the wheel? *Advanced Materials*, *16*(14), 1151–1170.

Li, J., Chotiko, A., Narcisse, D. A., & Sathivel, S. (2016). Evaluation of alpha-tocopherol stability in soluble dietary fiber based nanofiber. *LWT—Food Science and Technology*, *68*, 485–490. https://doi.org/10.1016/j.lwt.2015.12.042.

Li, J., He, A., Zheng, J., & Han, C. C. (2006). Gelatin and gelatin—Hyaluronic acid nanofibrous membranes produced by electrospinning of their aqueous solutions. *Biomacromolecules*, 7, 2243–2247. https://doi.org/10.1021/bm0603342.

Li, L., & Hsieh, Y. L. (2006). Chitosan bicomponent nanofibers and nanoporous fibers. *Carbohydrate Research*, *341*(3), 374–381. https://doi.org/10.1016/j.carres.2005.11.028.

Li, Y., Lim, L.-T., & Kakuda, Y. (2009). Electrospun zein fibers as carriers to stabilize (-)-epigallocatechin gallate. *Journal of Food Science*, *74*(3), C233–C240.

Librán, C. M., Castro, S., & Lagaron, J. M. (2017). Encapsulation by electrospray coating atomization of probiotic strains. *Innovative Food Science and Emerging Technologies*, *39*, 216–222. https://doi.org/10.1016/j.ifset.2016.12.013.

Lindman, B., Karlstrom, G., & Stigsson, L. (2010). On the mechanism of dissolution of cellulose. *Journal of Molecular Liquids*, *156*(1), 76–81. https://doi.org/10.1016/j.molliq.2010.04.016.

López-Rubio, A., & Lagaron, J. M. (2012). Whey protein capsules obtained through electrospraying for the encapsulation of bioactives. *Innovative Food Science and Emerging Technologies*, *13*, 200–206. https://doi.org/10.1016/j.ifset.2011.10.012.

López-Rubio, A., Sanchez, E., Wilkanowicz, S., Sanz, Y., & Lagaron, J. M. (2012). Electrospinning as a useful technique for the encapsulation of living bifidobacteria in food hydrocolloids. *Food Hydrocolloids*, *28*(1), 159–167. https://doi.org/10.1016/j.foodhyd.2011.12.008.

Lubasova, D., & Martinova, L. (2011). Controlled morphology of porous polyvinyl butyral nanofibers. *Journal of Nanomaterials, 2011*, 292516.

Luo, Y., & Wang, Q. (2014). Recent development of chitosan-based polyelectrolyte complexes with natural polysaccharides for drug delivery. *International Journal of Biological Macromolecules, 64*, 353–367. https://doi.org/10.1016/j.ijbiomac.2013.12.017.

Ma, G., Liu, Y., Fang, D., Chen, J., Peng, C., Fei, X., et al. (2012). Hyaluronic acid/chitosan polyelectrolyte complexes nanofibers prepared by electrospinning. *Materials Letters, 74*, 78–80. https://doi.org/10.1016/j.matlet.2012.01.012.

McClements, D. (2014). *Nanoparticle- and microparticle-based delivery systems*. CRC Press. *https://doi.org/10.1201/b17280.*

McKee, M. G., Layman, J. M., Cashion, M. P., & Long, T. E. (2006). Phospholipid nonwoven electrospun membranes. *Science (New York, N.Y.), 311*(5759), 353–355. https://doi.org/10.1126/science.1119790.

Mendes, A. C., Baran, E. T., Pereira, R. C., Azevedo, H. S., & Reis, R. L. (2012). Encapsulation and survival of a chondrocyte cell line within xanthan gum derivative. *Macromolecular Bioscience, 12*(3), 350–359. https://doi.org/10.1002/mabi.201100304.

Mendes, A. C., Gorzelanny, C., Halter, N., Schneider, S. W., & Chronakis, I. S. (2016). Hybrid electrospun chitosan-phospholipids nanofibers for transdermal drug delivery. *International Journal of Pharmaceutics, 510*(1), 48–56. https://doi.org/10.1016/j.ijpharm.2016.06.016.

Mendes, A. C., Nikogeorgos, N., Lee, S., & Chronakis, I. S. (2015). Nanomechanics of electrospun phospholipid fiber. *Applied Physics Letters, 106*(22), 223108-4. https://doi.org/10.1063/1.4922283.

Mendes, A. C., Stephansen, K., & Chronakis, I. S. (2017). Electrospinning of food proteins and polysaccharides. *Food Hydrocolloids, 68*, 53–68. https://doi.org/10.1016/j.foodhyd.2016.10.022.

Mendes, A., Sevilla Moreno, J., Hanif, M., E. L. Douglas, T., Chen, M., & Chronakis, I. S. (2018). Morphological, mechanical and mucoadhesive properties of electrospun chitosan/phospholipid hybrid nanofibers. *International Journal of Molecular Sciences, 19*(8), 2266. https://doi.org/10.3390/ijms19082266.

Mit-uppatham Manit, C. N., & Supaphol, P. (2004). Ultrafine electrospun polyamide-6 fibers: Effect of solution conditions on morphology and average fiber diameter. *Macromolecular Chemistry and Physics, 205*, 2327–2338.

Moomand, K., & Lim, L.-T. (2014). Oxidative stability of encapsulated fish oil in electrospun zein fibres. *Food Research International, 62*, 523–532.

Moomand, K., & Lim, L.-T. (2015). Effects of solvent and n-3 rich fish oil on physicochemical properties of electrospun zein fibres. *Food Hydrocolloids, 46*, 191–200. https://doi.org/10.1016/j.foodhyd.2014.12.014.

Moreira, J. B., Lim, L.-T., Zavareze, E. d. R., Dias, A. R. G., Costa, J. A. V., & Morais, M. G. d. (2018). Microalgae protein heating in acid/basic solution for nanofibers production by free surface electrospinning. *Journal of Food Engineering, 230*, 49–54.

Moreno, J. A. S., Mendes, A. C., Stephansen, K., Engwer, C., Goycoolea, F. M., Boisen, A., et al. (2018). Development of electrosprayed mucoadhesive chitosan microparticles. *Carbohydrate Polymers, 190*, 240–247. https://doi.org/10.1016/j.carbpol.2018.02.062.

Morimotoa, C., Satohb, Y., Hara, M., Inoue, S., Tsujita, T., & Okuda, H. (2005). Anti-obese action of raspberry ketone. *Life Sciences, 77*, 194–204.

Neo, Y. P., Ray, S., Jin, J., Gizdavic-Nikolaidis, M., Nieuwoudt, M. K., Liu, D., et al. (2013). Encapsulation of food grade antioxidant in natural biopolymer by electrospinning technique: A physicochemical study based on zein-gallic acid system. *Food Chemistry, 136*, 1013–1021. https://doi.org/10.1016/j.foodchem.2012.09.010.

Nie, H., He, A., Zheng, J., Xu, S., Li, J., & Han, C. C. (2008). Effects of chain conformation and entanglement on the electrospinning of pure alginate. *Biomacromolecules*, *9*(5), 1362–1365. https://doi.org/10.1021/bm701349j.

Nieuwland, M., Geerdink, P., Brier, P., Eijnden Van Den, P., Henket, J. T. M. M., Langelaan, M. L. P., et al. (2014). Food-grade electrospinning of proteins. *Innovative Food Science and Emerging Technologies*, *24*, 138–144.

Ogata, N., Yamaguchi, S., Shimada, N., Lu, G., Iwata, T., Nakane, K., et al. (2007). Poly(lactide) nanofibers produced by a melt-electrospinning system with a laser melting device. *Journal of Applied Polymer Science*, *104*, 1640–1645.

Ohkawa, K., Cha, D., Kim, H., Nishida, A., & Yamamoto, H. (2004). Electrospinning of chitosan. *Macromolecular Rapid Communications*, *25*(18), 1600–1605. https://doi.org/10.1002/marc.200400253.

Ortolan, F., & Steel, C. J. (2017). Protein characteristics that affect the quality of vital wheat gluten to be used in baking: A review. *Comprehensive Reviews in Food Science and Food Safety*, *16*, 369–381.

Osborne, T. B. (1924). *Classification of vegetable proteins*. New Work: Longmans.

Ouellet, V., Marois, J., Weisnagel, S. J., & Jacques, H. (2007). Dietary cod protein improves insulin sensitivity in insulin-resistant men and women. *Diabetes Care*, *30*, 2816–2821.

Ouellet, V., Weisnagel, S. J., Marois, J., Bergeron, J., Julien, P., Gougeon, R., et al. (2008). Dietary cod protein reduces plasma C-reactive protein in insulin-resistant men and women. *Journal of Nutrition*, *138*, 2386–2391.

Pakravan, M., Heuzey, M.-C., & Ajji, A. (2012). Core-shell structured PEO-chitosan nanofibers by coaxial electrospinning. *Biomacromolecules*, *13*(2), 412–421. https://doi.org/10.1021/bm201444v.

Park, W. H., Jeong, L., Yoo, D. I., & Hudson, S. (2004). Effect of chitosan on morphology and conformation of electrospun silk fibroin nanofibers. *Polymer*, *45*(21), 7151–7157. https://doi.org/10.1016/j.polymer.2004.08.045.

Pérez-Masiá, R., Lagaron, J. M., & Lopez-Rubio, A. (2015). Morphology and stability of edible lycopene-containing micro- and nanocapsules produced through electrospraying and spray drying. *Food and Bioprocess Technology*, *8*, 459–470. https://doi.org/10.1007/s11947-014-1422-7.

Pérez-Masiá, R., López-Nicolás, R., Periago, M. J., Ros, G., Lagaron, J. M., & López-Rubio, A. (2015). Encapsulation of folic acid in food hydrocolloids through nanospray drying and electrospraying for nutraceutical applications. *Food Chemistry*, *168*, 124–133. https://doi.org/10.1016/j.foodchem.2014.07.051.

Pilon, G., Ruzzin, J., Rioux, L. E., Lavigne, C., White, P. J., Frøyland, L., et al. (2011). Differential effects of various fish proteins in altering body weight, adiposity, inflammatory status, and insulin sensitivity in high-fat-fed rats. *Metabolism, Clinical and Experimental*, *60*(8), 1122–1130. https://doi.org/10.1016/j.metabol.2010.12.005.

Pitigraisorn, P., Srichaisupakit, K., Wongpadungkiat, N., & Wongsasulak, S. (2017). Encapsulation of Lactobacillus acidophilus in moist-heat-resistant multilayered microcapsules. *Journal of Food Engineering*, *192*, 11–18. https://doi.org/10.1016/j.jfoodeng.2016.07.022.

Prietto, L., Pinto, V. Z., Halal El, S. L. M., Morais de, M. G., Costa, J. A. V., Lim, L.-T., et al. (2018). Ultrafine fibers of zein and anthocyanins as natural pH indicator. *Journal of the Science of Food and Agriculture*, *98*, 2735–2741.

Quispe-Condori, S., Saldaña, M. D. A., & Temelli, F. (2011). Microencapsulation of flax oil with zein using spray and freeze drying. *LWT—Food Science and Technology*, *44*(9), 1880–1887.

Ramakrishna, S., Fujihara, K., Teo, W.-E., Lim, T.-C., & Ma, Z. (2005). *An introduction to electrospinning and nanofibers*. World Scientific.

Reddy, N., & Yang, Y. (2015). *Innovative biofibers from renewable resources*. Springer New York.

Regev, O., Vandebril, S., Zussman, E., & Clasen, C. (2010). The role of interfacial viscoelasticity in the stabilization of an electrospun jet. *Polymer*, *51*, 2611–2620. https://doi.org/10.1016/j.polymer.2010.03.061.

Reneker, D. H., & Yarin, A. L. (2008). Electrospinning jets and polymer nanofibers. *Polymer*, *49*, 2387–2425.

Reneker, D. H., Yarin, A. L., Fong, H., & Koombhongse, S. (2000). Bending instability of electrically charged liquid jets of polymer solutions in electrospinning. *Journal of Applied Physics*, *87*, 4531–4547.

Ritcharoen, W., Thaiying, Y., Saejeng, Y., Jangchud, I., Rangkupan, R., Meechaisue, C., et al. (2008). Electrospun dextran fibrous membranes. *Cellulose*, *15*(3), 435–444. https://doi.org/10.1007/s10570-008-9199-3.

Rosalam, S., & England, R. (2006). Review of xanthan gum production from unmodified starches by Xanthomonas comprestris sp. *Enzyme and Microbial Technology*, *39*(2), 197–207. https://doi.org/10.1016/j.enzmictec.2005.10.019.

Rutledge, G. C., & Fridrikh, S. V. (2007). Formation of fibers by electrospinning. *Advanced Drug Delivery Reviews*, *59*, 1384–1391.

Sabeena Farvin, K. H., Andersen, L. L., Nielsen, H. H., Jacobsen, C., Jakobsen, G., Johansson, I., et al. (2014). Antioxidant activity of Cod (Gadus morhua) protein hydrolysates: In vitro assays and evaluation in 5% fish oil-in-water emulsion. *Food Chemistry*, *149*, 326–334. https://doi.org/10.1016/j.foodchem.2013.03.075.

Saithongdee, A., Praphairaksit, N., & Imyim, A. (2014). Electrospun curcumin-loaded zein membrane for iron(III) ions sensing. *Sensors and Actuators B: Chemical*, *202*, 935–940.

Salas, C., Ago, M., Lucia, L. A., & Rojas, O. J. (2014). Synthesis of soy protein-lignin nanofibers by solution electrospinning. *Reactive and Functional Polymers*, *85*, 221–227.

Salwowska, N. M., Bebenek, K. A., Azdło, D. A., & Wcisło-Dziadecka, D. (2016). Physiochemical properties and application of hyaluronic acid: A systematic review. *Journal of Cosmetic Dermatology*, *15*, 520–526.

Sangsanoh, P., & Supaphol, P. (2006). Stability improvement of electrospun chitosan nanofibrous membranes in neutral or weak basic aqueous solutions. *Biomacromolecules*, 7(10), 2710–2714.

Schiffman, J. D., & Schauer, C. L. (2007). One-step electrospinning of cross-linked chitosan fibers. *Biomacromolecules*, *8*(9), 2665–2667. https://doi.org/10.1021/bm7006983.

Shalumon, K. T., Anulekha, K. H., Girish, C. M., Prasanth, R., Nair, S. V., & Jayakumar, R. (2010). Single step electrospinning of chitosan/poly(caprolactone) nanofibers using formic acid/acetone solvent mixture. *Carbohydrate Polymers*, *80*(2), 414–420. https://doi.org/10.1016/j.carbpol.2009.11.039.

Shekarforoush, E., Ajalloueian, F., Zeng, G., Mendes, A. C., & Chronakis, I. S. (2018). Electrospun xanthan gum-chitosan nanofibers as delivery carrier of hydrophobic bioactives. *Materials Letters*, *228*, 322–326. https://doi.org/10.1016/j.matlet.2018.06.033.

Shekarforoush, E., Faralli, A., Ndoni, S., Mendes, A. C., & Chronakis, I. S. (2017). Electrospinning of xanthan polysaccharide. *Macromolecular Materials and Engineering*, *302*(8), 1700067. https://doi.org/10.1002/mame.201700067.

Shekarforoush, E., Mendes, A., Baj, V., Beeren, S., & Chronakis, I. S. (2017). Electrospun phospholipid fibers as micro-encapsulation and antioxidant matrices. *Molecules*, *22*(10), 1708. https://doi.org/10.3390/molecules22101708.

Shingel, K. I. (2004). Current knowledge on biosynthesis, biological activity, and chemical modification of the exopolysaccharide, pullulan. *Carbohydrate Research*, *339*, 447–460.

Shukla, R., & Cheryan, M. (2001). Zein: The industrial protein from corn. *Industrial Crops and Products*, *13*, 171–192.

Silva, I., Gurruchaga, M., Goni, I., Fernandez-Gutierrez, M., Vazquez, B., & Roman, J. S. (2013). Scaffolds based on hydroxypropyl starch: Processing, morphology, characterization, and biological behavior. *Journal of Applied Polymer Science*, *127*(3), 1475–1484. https://doi.org/10.1002/app.37551.

Singh, R. S., Saini, G. K., & Kennedy, J. F. (2008). Pullulan: Microbial sources, production and applications. *Carbohydrate Polymers*, *73*, 515–531.

Songsurang, K., Praphairaksit, N., Siraleartmukul, K., & Muangsin, N. (2011). Electrospray fabrication of doxorubicin-chitosan-tripolyphosphate nanoparticles for delivery of doxorubicin. *Archives of Pharmacal Research*, *34*(4), 583–592. https://doi.org/10.1007/s12272-011-0408-5.

Sreekumar, S., Lemke, P., Moerschbacher, B. M., Torres-Giner, S., & Lagaron, J. M. (2017). Preparation and optimization of submicron chitosan capsules by water-based electrospraying for food and bioactive packaging applications. *Food Additives and Contaminants—Part A Chemistry, Analysis, Control, Exposure and Risk Assessment*, *34*(10), 1795–1806. https://doi.org/10.1080/19440049.2017.1347284.

Stephansen, K., Chronakis, I. S., & Jessen, F. (2014). Bioactive electrospun fish sarcoplasmic proteins as a drug delivery system. *Colloids and Surfaces B: Biointerfaces*, *122*, 158–165. https://doi.org/10.1016/j.colsurfb.2014.06.053.

Stephansen, K., García-Díaz, M., Jessen, F., Chronakis, I. S., & Nielsen, H. M. (2015). Bioactive protein-based nanofibers interact with intestinal biological components resulting in transepithelial permeation of a therapeutic protein. *International Journal of Pharmaceutics*, *495*(1), 58–66. https://doi.org/10.1016/j.ijpharm.2015.08.076.

Stephansen, K., García-Díaz, M., Jessen, F., Chronakis, I. S., & Nielsen, H. M. (2016). Interactions between surfactants in solution and electrospun protein fibers: Effects on release behavior and fiber properties. *Molecular Pharmaceutics*, *13*(3), 748–755. https://doi.org/10.1021/acs.molpharmaceut.5b00614.

Steyaert, I., Rahier, H., Vlierberghe Van, S., Olijve, J., & Clerck De, K. (2016). Gelatin nanofibers: Analysis of triple helix dissociation temperature and cold-water-solubility. *Food Hydrocolloids*, *57*, 200–208.

Stijnman, A. C., Bodnar, I., & Hans Tromp, R. (2011). Electrospinning of food-grade polysaccharides. *Food Hydrocolloids*, *25*(5), 1393–1398. https://doi.org/10.1016/j.foodhyd.2011.01.005.

Sullivan, S. T., Tang, C., Kennedy, A., Talwar, S., & Khan, S. a. (2014). Electrospinning and heat treatment of whey protein nanofibers. *Food Hydrocolloids*, *35*, 36–50. https://doi.org/10.1016/j.foodhyd.2013.07.023.

Sun, K., & Li, Z. H. (2011). Preparations, properties and applications of chitosan based nanofibers fabricated by electrospinning. *Express Polymer Letters*, *5*(4), 342–361. https://doi.org/10.3144/expresspolymlett.2011.34.

Sun, X. B., Jia, D., Kang, W. M., Cheng, B. W., & Li, Y. B. (2013). *Research on electrospinning process of pullulan nanofibers. In Materials, mechanical engineering and manufacture, Pts. 1–3* (pp. 198–201). 268–270. (pp. 198–201). https://doi.org/DOI https://doi.org/10.4028/www.scientific.net/AMM.268-270.198.

Supaphol, P., Mit-Uppatham, C., & Nithitanakul, M. (2005a). Ultrafine electrospun polyamide-6 fibers—Effects of solvent system and emitting electrode polarity on morphology and average fiber diameter. *Macromolecular Materials and Engineering*, *290*, 933–942.

Supaphol, P., Mit-Uppatham, C., & Nithitanakul, M. (2005b). Ultrafine electrospun polyamide-6 fibers: Effect of emitting electrode polarity on morphology and average fiber diameter. *Jounal of Polymer Science Part B: Polymer Physics*, *43*, 3699–3712.

Taepaiboon, P., Rungsardthong, U., & Supaphol, P. (2007). Vitamin-loaded electrospun cellulose acetate nanofiber mats as transdermal and dermal therapeutic agents of vitamin A acid and vitamin E. *European Journal of Pharmaceutics and Biopharmaceutics: Official Journal of Arbeitsgemeinschaft Für Pharmazeutische Verfahrenstechnik e.V*, *67*(2), 387–397. https://doi.org/10.1016/j.ejpb.2007.03.018.

Taylor, G. (1966). The force exerted by an electric field on a long cylindrical conductor. *Proceedings of the Royal Society A: Mathematical, Physical and Engineering Sciences*, *291*, 145–158.

Taylor, G. (1969). Electrically driven jets. *Proceedings of the Royal Society A: Mathematical, Physical and Engineering Sciences*, *313*, 453–475.

Teo, W. E., & Ramakrishna, S. (2006). A review on electrospinning design and nanofiber assemblies. *Nanotechnology*, *17*, R89–R106.

Theron, S. A., Zussman, E., & Yarin, A. L. (2004). Experimental investigation of the governing parameters in the electrospinning of polymer solutions. *Polymer*, *45*, 2017–2030.

Thien, D. V. H., Hsiao, S. W., & Ho, M. H. (2012). Synthesis of electrosprayed chitosan nanoparticles for drug sustained release. *NanoLife*, *02*(01), 1250003. https://doi.org/10.1142/S1793984411000360.

Torres-Giner, S., Gimenez, E., & Lagaron, J. M. (2008). Characterization of the morphology and thermal properties of zein prolamine nanostructures obtained by electrospinning. *Food Hydrocolloids*, *22*, 601–614.

Torres-Giner, S., Ocio, M. J., & Lagaron, J. M. (2009). Novel antimicrobial ultrathin structures of zein/chitosan blends obtained by electrospinning. *Carbohydrate Polymers*, *77*(2), 261–266. https://doi.org/10.1016/j.carbpol.2008.12.035.

Torres-Giner, S., Ocio, M. J., & Lagaron, J. M. (2008). Development of active antimicrobial fiber-based chitosan polysaccharide nanostructures using electrospinning. *Engineering in Life Sciences*, *8*(3), 303–314. https://doi.org/10.1002/elsc.200700066.

Ungeheuer, S., Bewersdorff, H., & Singh, R. P. (1989). Turbulent drag effectiveness and shear stability of xanthan-gum-based graft copolymers. *Journal of Applied Polymer Science*, *37*, 2933–2948.

Vega-Lugo, A.-C., & Lim, L.-T. (2009). Controlled release of allyl isothiocyanate using soy protein and poly(lactic acid) electrospun fibers. *Food Research International*, *42*(8), 933–940.

Vega-Lugo, A.-C., & Lim, L.-T. (2012). Effects of poly(ethylene oxide) and pH on the electrospinning of whey protein isolate. *Journal of Polymer Science Part B: Polymer Physics*, *50*(16), 1188–1197.

Vega Lugo, A.-C., & Lim, L.-T. (2008). Electrospinning of soy protein isolate nanofibers. *Journal of Biobased Materials and Bioenergy*, *2*, 223–230.

Verdugo, M., Lim, L.-T., & Rubilar, M. (2014). Electrospun protein concentrate fibers from microalgae residual biomass. *Journal of Polymers and the Environment*, *22*(3), 373–383. https://doi.org/10.1007/s10924-014-0678-3.

Walstra, P., & Vliet, T. v. (2008). Chapter 13: Dispersed systems—Basic consideration. In S. Damodaran, K. L. Parkin, & O. R. Fennema (Eds.), *Fennema's food chemistry* (4th ed.). Boca Raton, FL: CRC Press.

Wang, H., She, Y., Chu, C., Liu, H., Jiang, S., Sun, M., et al. (2015). Preparation, antimicrobial and release behaviors of nisin-poly (vinyl alcohol)/wheat gluten/$ZrO_2$ nanofibrous membranes. *Journal of Materials Science*, *50*(14), 5068–5078. https://doi.org/10.1007/s10853-015-9059-0.

Wang, S., Bai, J., Li, C., & Zhang, J. (2012). Functionalization of electrospun B-cyclodextrin/polyacrylonitrile (PAN) with silver nanoparticles: Broad-spectrum antibacterial property. *Applied Surface Science*, *261*, 499–503. https://doi.org/10.1016/j.apsusc.2012.08.044.

Wang, S., Marcone, M. F., Barbut, S., & Lim, L.-T. (2013). Electrospun soy protein isolate-based fiber fortified with anthocyanin-rich red raspberry (Rubus strigosus) extracts. *Food Research International*, *52*, 467–472.

Weiss, J., Kanjanapongkul, K., Wongsasulak, S., & Yoovidhya, T. (2012). *13—Electrospun fibers: fabrication, functionalities and potential food industry applications. In Nanotechnology in the food, beverage and nutraceutical industries.* Woodhead Publishing Limited. https://doi.org/10.1533/9780857095657.2.362.

Woerdeman, D. L., Shenoy, S., & Breger, D. (2007). Role of chain entanglements in the electrospinning of wheat protein-poly(vinyl alcohol) blends. *The Journal of Adhesion, 83*, 785–798. https://doi.org/10.1080/00218460701588398.

Woerdeman, D. L., Ye, P., Shenoy, S., Parnas, R. S., Wnek, G. E., & Trofimova, O. (2005). Electrospun fibers from wheat protein: Investigation of the interplay between molecular structure and the fluid dynamics of the electrospinning process. *Biomacromolecules, 6*, 707–712.

Wolf, W. J. (1970). Soybean proteins: Their functional, chemical, and physical properties. *Journal of Agricultural and Food Chemistry, 18*, 969–976.

Wongsasulak, S., Pathumban, S., & Yoovidhya, T. (2014). Effect of entrapped α-tocopherol on mucoadhesivity and evaluation of the release, degradation, and swelling characteristics of zein-chitosan composite electrospun fibers. *Journal of Food Engineering, 120*(1), 110–117. https://doi.org/10.1016/j.jfoodeng.2013.07.028.

Wu, X., Wang, L., Yu, H., & Huang, Y. (2005). Effect of solvent on morphology of electrospinning ethyl cellulose fibers. *Journal of Applied Polymer Science, 97*(3), 1292–1297. https://doi.org/10.1002/app.21818.

Xiao, Q., & Lim, L.-T. (2018). Pullulan-alginate fibers produced using free surface electrospinning. *International Journal of Biological Macromolecules, 112*, 809–817.

Xu, H., Cai, S., Sellers, A., & Yang, Y. (2014). Electrospun ultrafine fibrous wheat glutenin scaffolds with three-dimensionally random organization and water stability for soft tissue engineering. *Journal of Biotechnology, 184*, 179–186. https://doi.org/10.1016/j.jbiotec.2014.05.011.

Xu, W., Yang, W., & Yang, Y. (2009). Electrospun starch acetate nanofibers: Development, properties, and potential application in drug delivery. *Biotechnology Progress, 25*(6), 1788–1795. https://doi.org/10.1002/btpr.242.

Xu, X., Jiang, L., Zhou, Z., Wu, X., & Wang, Y. (2012). Preparation and properties of electrospun soy protein isolate/polyethylene oxide nanofiber membranes. *Applied Materials & Interfaces, 4*, 4331–4337.

Yang, D., Li, Y., & Nie, J. (2007). Preparation of gelatin/PVA nanofibers and their potential application in controlled release of drugs. *Carbohydrate Polymers, 69*(3), 538–543. https://doi.org/10.1016/j.carbpol.2007.01.008.

Yarin, A. L., Koombhongse, S., & Reneker, D. H. (2001). Taylor cone and jetting from liquid droplets in electrospinning of nanofibers. *Journal of Applied Physics, 90*, 4836–4846.

Yu, J. H., Fridrikh, S. V., & Rutledge, G. C. (2006). The role of elasticity in the formation of electrospu fibers. *Polymer, 47*, 4789–4797.

Zhang, C. L., & Yu, S. H. (2014). Nanoparticles meet electrospinning: Recent advances and future prospects. *Chemical Society Reviews, 43*, 4423–4448.

Zhang, H., Wu, C., Zhang, Y., White, C. J. B., Xue, Y., Nie, H., et al. (2010). Elaboration, characterization and study of a novel affinity membrane made from electrospun hybrid chitosan/nylon-6 nanofibers for papain purification. *Journal of Materials Science, 45*(9), 2296–2304. https://doi.org/10.1007/s10853-009-4191-3.

Zhou, F.-L., Gong, R.-H., & Porat, I. (2009). Mass production of nanofibre assemblies by electrostatic spinning. *Polymer International, 58*(4), 331–342.

Zhou, H., Green, T. B., & Joo, Y. L. (2006). The thermal effects on electrospinning of poly-lactic acid melts. *Polymer, 47*, 7497–7505.

Zhou, Y., & Lim, L.-T. (2009). Activation of lactoperoxidase system in milk by glucose oxidase immobilized in electrospun polylactide microfibers. *Journal of Food Science, 74*(2), C170–C176.

Zirnsak, M. a., Boger, D. V., & Tirtaatmadja, V. (1999). Steady shear and dynamic rheological properties of xanthan gum solutions in viscous solvents. *Journal of Rheology, 43*(3), 627. https://doi.org/10.1122/1.551007.

CHAPTER SIX

# Bioavailability of nanotechnology-based bioactives and nutraceuticals

**Dena Jones, Sarah Caballero, Gabriel Davidov-Pardo***
Nutrition and Food Science Department, California State Polytechnic University, Pomona, CA, United States
*Corresponding author: e-mail address: gdavidov@cpp.edu

## Contents



## Abstract

Bioaccessibility and bioavailability of some hydrophobic bioactives (e.g., carotenoids, polyphenols, fat-soluble vitamins, phytosterols and fatty acids) are limited due to their low water solubility, and in some instances low chemical stability. Nanotechnology involving nanometric ($r < 500$ nm) delivery systems, can be used to improve the solubility and thus enhance the bioaccessibility and bioavailability of hydrophobic compounds. Nanometric delivery systems, derived from food grade phospholipids and biopolymers adopt many forms, including liposomes, micelles, micro/nanoemulsions,

*Advances in Food and Nutrition Research*, Volume 88
ISSN 1043-4526
https://doi.org/10.1016/bs.afnr.2019.02.014

particles, polyelectrolyte complexes, and hydrogels. The small particle sizes and customized materials used to create delivery systems confer their unique properties such as higher stability and/or resistance to enzymatic activity in the gastrointestinal tract. This chapter provides an overview of bioaccessibility and bioavailability of different classes of hydrophobic bioactive compounds, focusing on nanometric delivery systems and methods of evaluation.

## 1. Introduction

A number of nanometric delivery systems have been designed for the food, supplements, and pharmaceutical industries to control the digestion, release, and absorption of lipophilic bioactives essential for physiological functions in the human body or host disease-prevention. However, incorporation of bioactives into food is challenging and absorption by the human body is often limited due to poor chemical stability, poor solubility in water, or low bioavailability (McClements, 2010). In order for the body to utilize a specific bioactive, the compound must be released from the food matrix and absorbed through the epithelial layer in the gastrointestinal tract (GIT). Two coexisting mechanisms are involved in the absorption of bioactives through the epithelial cells of the GIT (Fig. 1). Passive transcellular transport happens

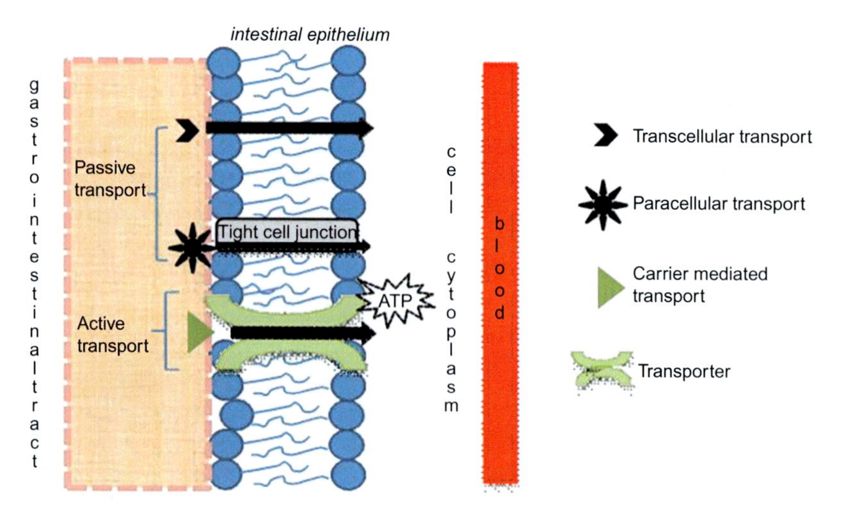


**Fig. 1** Intestinal bioactives transport mechanisms. *Reprinted with permission from Renukuntla, J., Vadlapudi, A. D., Patel, A., Boddu, S. H. S., & Mitra, A. K. (2013). Approaches for enhancing oral bioavailability of peptides and proteins.* International Journal of Pharmaceutics, *447(1–2), 75–93. https://doi.org/10.1016/j.ijpharm.2013.02.030. Copyright (2013) Elsevier.*

due to a concentration gradient from high concentration to low concentration. The mass transport is governed by Fick's law of diffusion (Renukuntla, Vadlapudi, Patel, Boddu, & Mitra, 2013). On the other hand, active transport involves the use of transmembrane proteins and the expenditure of ATP, wherein the bioactive component is transported against the concentration gradient (Renukuntla et al., 2013). The extent in which these processes occur for a particular bioactive is described by bioaccessibility and bioavailability. Bioavailability refers to the portion of the bioactive compound that is absorbed in the body, enters systemic circulation, and performs functions in the human body (Saini, Nile, & Park, 2015). In contrast, bioaccessibility refers to the portion of the compound released from the food matrix, incorporated into micelles in the small intestine, and made available for absorption. In general, *in vitro* digestion models best predict bioaccessibility, whereas blood plasma must be analyzed to determine bioavailability. *In vitro* bioaccessibility often correlate with bioavailability values, although the strength of correlation depends on food processing and delivery matrix (Reboul et al., 2006).

Nanometric colloidal delivery systems, consisting of particles or droplets ($r < 500\,nm$), have been utilized in the food industry for encapsulation, protection, and controlled release of bioactives (Joye & McClements, 2013). This book chapter surveys the challenges involved with incorporating lipophilic compounds into foods and beverages. The impacts of various nanotechnology-based delivery systems on the bioaccessibility and/or bioavailability of these compounds are discussed.

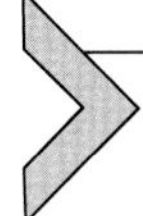

## 2. Classification of bioactive compounds

### 2.1 Polyphenols

Phenolic compounds have at least one phenolic ring with hydroxyl groups present. These are secondary metabolites, produced to defend against predators and diseases, as well they are essential for growth and reproduction in plants. Polyphenols have been associated with a variety of health-promoting activities in humans such as anti-carcinogenic, -diabetes, -oxidant, -inflammatory, -viral, -microbial, and cardioprotective (Hu, Liu, Zhang, & Zeng, 2017; Krishnaswamy, Orsat, & Thangavel, 2012; Massounga Bora, Ma, Li, & Liu, 2018). Examples of bioactive phenolic compounds include flavonoids, stilbenes, phenolic acids and curcuminoids (Fig. 2) (Hu et al., 2017; Lall, Syed, Adhami, Khan, & Mukhtar, 2015; Massounga Bora et al., 2018).

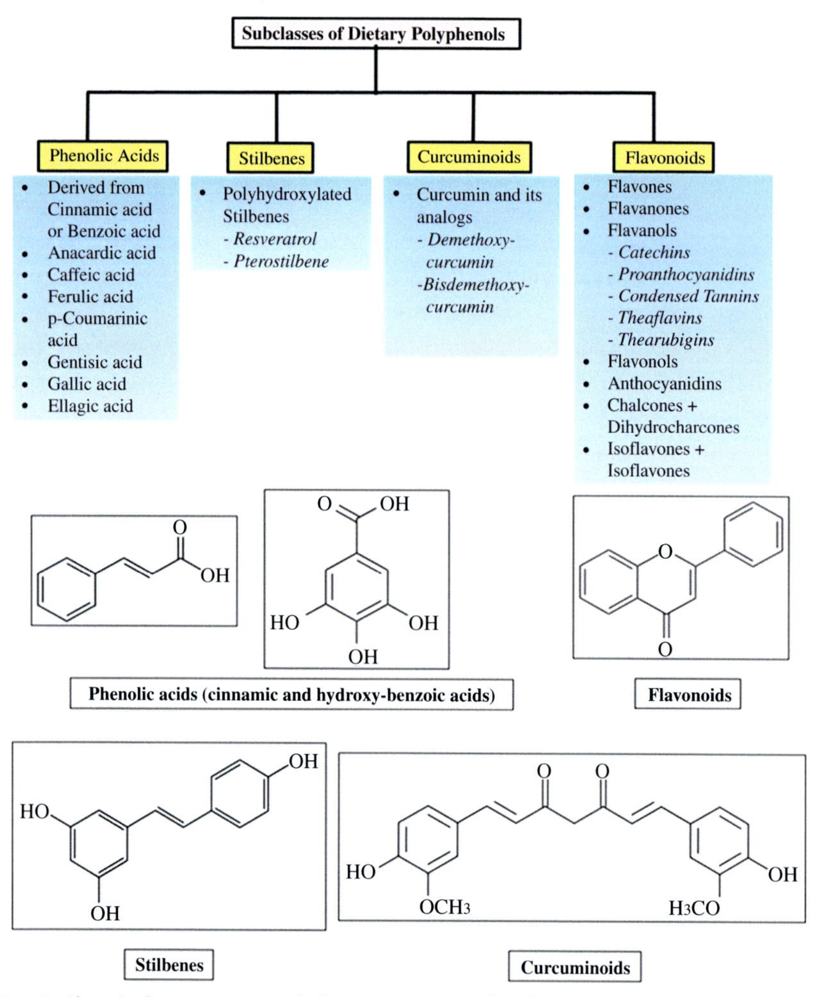


**Fig. 2** Chemical structures and dietary sources of different groups of polyphenols. *Reprinted from Lall, R. K., Syed, D. N., Adhami, V. M., Khan, M. I., & Mukhtar, H. (2015). Dietary polyphenols in prevention and treatment of prostate cancer.* International Journal of Molecular Sciences, 16*(2), 3350–3376. https://doi.org/10.3390/ijms16023350. Open access.*

Flavonoids are identified by their 15-carbon skeleton, which can be further divided into $C_6$-$C_3$-$C_6$. Flavonoids are classified as flavones, flavanols, flavanones, isoflavones, flavanols and anthocyinidins; although flavanones, flavones, flavanols, and anthocyanins are the most commonly occurring in nature. Dietary flavonoids are obtained from a variety of plant-based sources, such as fruits, vegetables, wine, and cocoa (Pérez-Abril et al., 2017).

Stilbene, another class of phenolic, includes *trans*-resveratrol isolated from grape skins, which has been associated with anti-oxidant, -cancer, -atherosclerosis, and -cardiovascular disease effects (Sessa et al., 2014). Curcuminoid is another class of polyphenol mainly found in turmeric, of which curcumin is the most abundant and powerful. Curcumin has been reported to have anti-carcinogenic, -inflammatory and -oxidant properties (Lall et al., 2015). Finally, simple phenols are either derivatives of cinnamic or benzoic acids, for instance, caffeic and coumaric acids.

Following oral intake, the absorption of phenolic compounds primarily occurs throughout the small intestine, although the exact nature of the bioactive uptake varies with molecular complexity of the bioactive. Simple phenols are easily absorbed, even starting in the stomach, due to their low molecular weight and high hydrophilicity (Palafox-Carlos, Fernando, & Gonzalez-Aguilar, 2011). Thus, gallic, caffeic, ferulic, coumaric, and chlorogenic acids have been detected in blood plasma as soon as 1–2 h following oral ingestion. By contrast, more complex polyphenols, such as glycosides, have difficulty diffusing across intestinal walls due to the bulkiness of the additional functional groups (Kamil, Chen, & Blumberg, 2015). Thus, polyphenols tend to exhibit low bioavailability, reported as low as 20% for isoflavones (Kamil et al., 2015). However, studies suggest glycones can be cleaved at the ester bond in the small intestine by enzymes such as 5-glucosidase to improve bioaccessibility. Nonetheless, absorption of aglycone polyphenols is still lower in the small intestine due to their higher molecular weight and lower solubility than the simple phenols (Palafox-Carlos et al., 2011).

In the colon, microflora can metabolize polyphenols and release metabolites into blood plasma, approximately seven to 8 h after oral ingestion of the polyphenols (Lafay & Gil-Izquierdo, 2008). Flavonoid metabolites continue to modify in the liver and kidneys. Recent studies have shown that secondary phenolic metabolites, especially glucuronidated, sulfated, or methylated flavonoid backbones, are detected in blood plasma and are indicators of phenolic bioaccessibility, besides being physiologically useful (Nagy et al., 2011; Palafox-Carlos et al., 2011). For instance, 3,4-dimethoxycinnamic acid was detected in human blood samples 60 min after coffee consumption (Nagy et al., 2011).

Despite the many benefits of phenolics, the incorporation of phenolic compounds into food matrices is challenging due to instability when exposed to common processing conditions, such as temperature, pH, oxygen, light, and enzymes (Massounga Bora et al., 2018). Furthermore, most polyphenols have poor solubility in water and are challenging to incorporate into aqueous food systems (Wei, Zhang, Memon, & Liang, 2017). Polyphenols also have poor oral bioavailability, due to low solubility in gastrointestinal fluids and insufficient residence times in the GI tract. Other issues are poor diffusivity across the lipid bilayer of intestinal epithelial cells, acidic pH conditions in the stomach, and poor absorption in the gut (Hu et al., 2017; Kamil et al., 2015; Sessa et al., 2014).

## 2.2 Carotenoids

Carotenoids are secondary metabolites of biosynthesis that produce yellow, orange, and red pigments in fruits and vegetables (Mutsokoti et al., 2017). While plants are the primary source of carotenoids, these bioactives may also be found in animal products such as egg yolk, shrimp, lobster, salmon, as well as some bacteria and fungi (Yonekura & Nagao, 2007). Carotenoids are classified as either carotenes or xanthophylls. Carotenes, such as lycopene, have a parent hydrocarbon chain based on the C40 isoprenoid backbone, while xanthophylls, such as lutein or zeaxanthin, have a functional group containing at least one oxygen atom (Saini et al., 2015; Yonekura & Nagao, 2007). The carbon-carbon conjugation in the backbone is responsible for both the emitted wavelengths towards the red-orange side of the visible electromagnetic spectrum as well as free radical scavenging properties (Yonekura & Nagao, 2007). Carotenoids have multiple physiological functions and health benefits, including hormone and immune regulation, gap-junction and intercellular communication, gastrointestinal function, improved immune system function, antioxidant activity, and anti-cancer and -cardiovascular disease properties (Mutsokoti et al., 2017; Saini et al., 2015). β-Carotene, specifically, is highly recognized for its provitamin-A activity. The role of accumulated lutein in the ocular region and of β-carotene as a vitamin A precursor has been associated with reduced risks of night blindness, age-related macular degeneration, and cataract formation (Donhowe, Flores, Kerr, Wicker, & Kong, 2014; Liu, Glahn, & Liu, 2004; Mutsokoti et al., 2017).

The bioaccessibility of carotenoids is largely dependent on food matrix and varies between chemical species (Mutsokoti et al., 2017; Palafox-Carlos et al., 2011). During the first stage of digestion, carotenoids entrapped in food matrices are released by mastication, digestive enzymes and acid in the stomach (Yonekura & Nagao, 2007). In the small intestine, carotenoids are solubilized by dietary lipids and their hydrolysis products. The solubilized lipids are then incorporated into mixed micelles formed with bile salts, phospholipids, dietary lipids and their hydrolysis products (Mutsokoti et al., 2017; Palafox-Carlos et al., 2011; Saini et al., 2015; Yonekura & Nagao, 2007). Thus, the accompaniment of carotenoids with dietary fatty acids plays a critical role in their bioaccessibility. The resulting micelles passively diffuse through epithelial cells and are secreted via chylomicrons to the lymphatic system.

Carotenoids have limited oral bioavailability due to their hydrophobicity, which impedes dissolution and absorption in the small intestine. Furthermore, the presence of unsaturated bonds makes these molecules vulnerable to oxidative degradation, which can be initiated by food processing conditions (e.g., heat, light, oxygen) as well as other constituents present in the food matrix. Another challenge is inadequate release of carotenoids from the plant matrix, such as protein and fibers (Saini et al., 2015). Consumption of dietary fiber with carotenoids also poses concern for reduced bioaccessibility of carotenoids due to entrapment of carotenoids preventing formation of mixed micelles in the intestinal phase (Palafox-Carlos et al., 2011).

## 2.3 Dietary fiber

According to the Nutrition Facts Label rule issued in 2016, the United States Food and Drug Administration (FDA) agency defines dietary fiber as "non-digestible soluble and insoluble carbohydrates (with three or more monomeric units), and lignin that are intrinsic and intact in plants; isolated or synthetic non-digestible carbohydrates (with three or more monomeric units) determined by FDA to have physiological effects that are beneficial to human health" (FDA, 2018). Dietary fiber is resistant to digestion because of their inability to be hydrolyzed by endogenous enzymes or any physicochemical means. They are thus excreted from the body, leaving remnants of health-beneficial compounds (Codex Alimentarius, 2009). Dietary fiber can be further divided into two categories—soluble dietary fiber (SDF) and insoluble dietary fiber (IDF)—each of which have their own nutritional benefits and nutritional obstacles.

IDF is classified as fiber that is not soluble throughout the digestive tract and acts as a bulking agent in foods; hemicellulose, cellulose, and lignin are included in IDF. They are not soluble in water, but still provide physiological health benefits, such as decreased transit time in the digestive tract, increased colon health and prevention of related diseases (i.e., colon cancer, diverticulosis, and constipation), and overall gastrointestinal health (Mehta, 2005).

SDF can solubilize in water; some are fermentable (i.e., inulin, resistant starches, and oligosaccharides) while others are not (i.e., psyllium) (McRorie, 2015). SDF forms a gel-like substances in the GIT. Fermentable SDF leads to an increase in bacterial mass, and results in increased fecal mass (Slavin, 2013). Some SDFs from whole grains have been found to reduce cardiovascular disease, cholesterol levels, Type II diabetes, and an improved postprandial glucose response (Ye, Chacko, Chou, Kugizaki, & Liu, 2012). On the other hand, soluble dietary gel-forming fibers show promise in controlling satiety. When soluble fiber comes in contact with chyme, it creates a more viscous solution, thereby reducing both the water-binding capacity and activity of digestive enzymes, resulting in delayed absorption (Gibb, McRorie, Russell, Hasselblad, & D'Alessio, 2015).

Dietary fibers have characteristics of an entrapping matrix that could reduce the bioavailability of macronutrients. Furthermore, dietary fibers can prevent the lipophilic antioxidants from being passively transported across the lumen of the small intestine because of the absence of a micellar phase. Dietary fiber also impacts the bioaccessibility of polyphenols through non-covalent binding through a weak van der Waals attraction between phenolic hydroxide group and the polar groups of the dietary fiber. This bond must be broken via digestive enzymes in the upper gut to promote absorption in the small intestine (Palafox-Carlos et al., 2011). Although dietary fiber is not absorbed by the body, recent studies have shown that they are promising in its application as an encapsulating agent for control release of bioactives through the GIT, which will be discussed later in the chapter.

## 2.4 Plant sterols

Plant sterols, also known as phytosterols, found in plants have a similar structure to cholesterol commonly found in animal-derived foodstuffs and in cellular membranes. Saturated or hydrogenated phytosterols are called stanols. They improve the low-density lipoprotein (LDL) cholesterol levels in humans via displacing cholesterol molecules from mixed micelles during the intestinal phase of digestion, thus partially preventing the absorption of cholesterol and ultimately lowering cholesterol levels in systemic circulation (Malinowski & Gehret, 2010; Schonfeld, 2010). Phytosterols are digested and absorbed primarily in the small intestine due to hydrolysis of their lipophilic constituents by lipase and micellarized via bile salts. Phytosterols vary in molecular structure that affects the solubility and thus absorption (Alemany et al., 2013). Sterols compete with each other for space in mixed micelles. The rigid structure of phytosterols, due to their long hydrophobic side-chain, compromises their ability to fit into mixed micelles (Alemany et al., 2013; Matsuoka, Kajimoto, Horiuchi, Honda, & Endo, 2010). The main concern with phytosterols is when they are oxidized either under endogenous conditions or from consuming oxidized sterols in foodstuffs. Oxidized phytosterols may pose adverse health effects, including cell apoptosis, necrosis, inflammation, carcinogenesis, gallstone formation, and immunosuppression (Lütjohann, 2004).

## 2.5 Bioactive lipids

Bioactive lipids have beneficial effects including altering circulating lipid levels. Fatty acids are comprised of a chain of carbon atoms with a carboxyl group (-COOH) located at the terminal end of the molecule. Essential fatty acids fall into the category of bioactives because they are not synthesized by the body but essential in fulfilling important functions related to wellbeing. Some essential polyunsaturated fatty acids (PUFA) have anti-inflammatory, -carcinogenic, -diabetic, -artherosclerotic, and -nociceptive effects in the body (Viladomiu, Hontecillas, & Bassaganya-Riera, 2016). More specifically, conjugated linoleic acid (CLA) has been shown to inhibit cell proliferation and promote apoptosis of cancer cells (Hernandez & Kamal-Eldin, 2013). PUFAs include ω-3 and ω-6 fatty acids that are precursors to other bioactive lipids, including eicosanoids, sphingolipids, and endocannabinoids (eCBs) (Chiurchiù, Leuti, & Maccarrone, 2018). ω-3 and ω-6 fatty acids contain double bonds on the third and sixth carbon from the methyl group ($CH_3$), respectively. Docosahexaenoic acid (DHA) and eicosapentaenoic acid (EPA) are the most common ω-6 fatty acids and are found in marine animals (e.g., salmon, tuna). EPA and DHA have been found to reduce risks of coronary heart disease and hypertension, thus reducing the morbidity and mortality rates (Cardoso, Bandarra, Lourenço, Afonso, & Nunes, 2010).

Digestion of bioactive lipids begins in the stomach where mechanical emulsification occurs. Without this step, the pancreatic lipase in the small intestine would be unable to hydrolyze lipids into fatty acids and thus absorption would not occur (Hernandez & Kamal-Eldin, 2013; Wang, Liu, Portincasa, & Wang, 2013). Bile salts then interact with the hydrophobic fatty acids liberated from glycerol backbone and form mixed micelles. These hydrophilic mixed micelles are readily absorbed into the intestinal lumen. Short chain fatty acids (<12 carbon atoms) and glycerol are able to cross the enterocyte and enter the portal vein via passive diffusion. On the other hand, long chain fatty acids must be formed into mixed micelles in order to be transported to the apical membrane of enterocytes for absorption. Once medium and long-chain fatty acids have crossed the enterocyte, they are formed into chylomicrons in the lymphatic system in order to aid in absorption (Hernandez & Kamal-Eldin, 2013; Wang et al., 2013). Even though fatty acids are hydrophobic, these mechanisms allow for the body to absorb such nutrients for utilization. However, some challenges are faced with bioactive lipids and their bioavailability.

PUFAs are largely water-insoluble and susceptible to oxidation, which limit their bioavailability (Heo, Kim, Pan, & Kim, 2016). Also, oxidized lipids contained in chylomicrons can result in adverse effects on the artery walls, constituting a link between postprandial lipaemia and atherogenesis (Kanner & Lapidot, 2001). Nanotechnology can offer a potential solution to prevent the oxidation PUFAs and increase their bioaccessibility and bioavailability.

## 2.6 Bioactive peptides

Bioactive peptides are typically peptides that contain anywhere from 3 to 20 amino acid chains that are derived from protein sources in their native form. These native proteins transform into their bioactive form via proteolysis either through endogenous enzymes, *in vitro* enzymatic hydrolysis, or bacterial fermentation (Hartmann & Meisel, 2007; Kitts & Weiler, 2003). The bioactive peptides are beneficial to the body in many ways, including antihypertensive properties, antioxidant activity, and immunomodulation (Escudero, Mora, & Toldrá, 2014; Reyes-Diaz, Gonzalez-Codova, Hernandez-Mendoza, & Vallejo-Cordoba, 2016; Rui, Lujuan, Qingquan, Guang-Hong, & Wan-Gang, 2016).

Peptides are versatile due to the variability in side chains that can react in many different environments. Peptides that contain a histidine residue function in scavenging free radicals and have antioxidant activity (Chen, Muramoto, Yamauchi, Fujimoto, & Nokihara, 1998). Hydrophobic amino acids, such as proline and leucine, have shown to enhance the anti-oxidant capacity of a His-His dipeptide when located at the N-terminus because of their capability to attach to the hydrophobic components on the cell membrane (e.g., polyunsaturated fatty acids) (Chen, Muramoto, Yamauchi, & Nokihara, 1996). Metal ions can be chelated by amino acids that contain an electron-dense aromatic ring (e.g., phenylalanine, tyrosine, and tryptophan residues) (Chen et al., 1996).

The acidic environment of the stomach and the action of the proteases impacts the bioavailability of bioactive peptides. The physiological function of bioactive peptides may be compromised when exposed to the peptidases contained at the brush border in the small intestine. By the time it reaches systemic circulation, bioactive functionality of the peptides may be inactivated (Kompella & Lee, 2001; Toldrá, Reig, Aristoy, & Mora, 2018). Peptides with a high molecular weight may not penetrate the tight junctions or transport themselves across the lumen wall in the small intestine (Renukuntla et al., 2013).

## 2.7 Micronutrients

Micronutrients are compounds that are only required in minimal amounts which aid in the growth, development, and maintenance of the body. The most common micronutrients are vitamins and minerals (e.g., iron, zinc, vitamins A, D, E, and K). Micronutrient absorption occurs in the small intestine, along with other compounds previously mentioned. Absorption in the small intestine may be impeded due to reactivity, limited solubility or lack of stability of the micronutrients (Li, Hu, Du, Xiao, & McClements, 2011). For example, the anti-nutritional factor, phytate (or phytic acid) inhibits the absorption of iron and zinc absorption, possibly attributing to worldwide prevalence in deficiencies for both minerals (Pullakhandam, Nair, Pamini, & Punjal, 2011). Calcium is another micronutrient that has low bioavailability due to its poor solubility and low permeability to the membrane, as well as calcium phosphate deposition (Gao, Dong, Wang, Li, & Chen, 2018). Vitamins that are of particular concern of absorption and bioavailability are the fat-soluble vitamins, i.e., A, D, E, and K. For example, vitamin $D_3$ (cholecalciferol) undergoes degradation when exposed to light, oxygen, and heat. Moreover, its poor solubility in water decreases absorption and bio-accessibility. By increasing the bioaccessibility of these micronutrients through different nanotechnological delivery vehicles, human well-being can be improved.

## 3. Bioaccessibility and bioavailability evaluation models

Various models have been developed to assess and quantify the bioaccessibility and bioavailability of bioactives and nutraceuticals. This section will provide a brief description of the common models/methods used. Table 1 highlights the main advantages and disadvantages of each model.

### 3.1 *In vitro* models

*In vitro* digestion models are commonly used for determining the bioaccessibility of bioactive compounds. These methods are inexpensive, rapid, and pose no ethical concerns pertaining to clinical trials. *In vitro* models simulate the GIT condition by adjusting the ionic strength and pH, as well as addition of enzymes, bile salts and even fermentation reactions to simulate colon conditions. The incorporation of bioactives into micelles, with bile salts as surfactants, has been closely correlated with *in vivo* studies (Donhowe et al., 2014). The conditions of each step of a typical *in vitro* digestion model are depicted in Fig. 3.

**Table 1** Advantages and disadvantages of the different models to evaluate bioavailability.

| Model | Advantages | Disadvantages |
|---|---|---|
| *In vitro* model | Inexpensive, rapid, and poses no ethical concerns | Can only evaluate bioaccessibility and not bioavailability |
| Cell models | Rapid and pose no ethical concerns. Evaluation of actual epithelial cells permeation | Does not incorporate GIT conditions. Can only assess absorption |
| Animal models | Less expensive than clinical studies. Possibility to harvest organs. Possibility to conduct toxicology studies | Not always possible to correlate results in animal models with human trials. Labor-intensity. Higher expense than In Vitro or cell studies. Ethical concerns |
| Clinical trials | Most accurate method | Labor-intensity. Highest expense. Ethical concerns, and time consumption. Not suitable for toxicology studies |

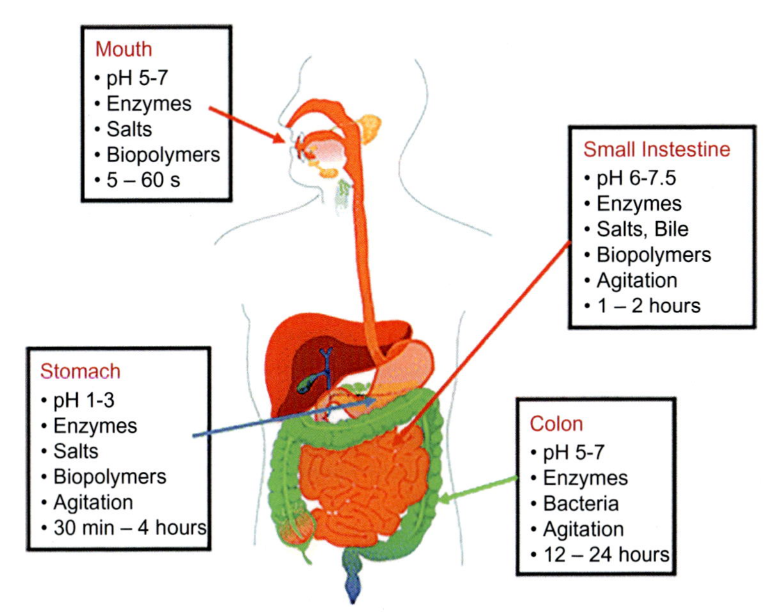


**Fig. 3** Conditions of each step of an in vitro digestion model. *Reprinted with permission from McClements, D. J., & Li, Y. (2010). Review of in vitro digestion models for rapid screening of emulsion-based systems.* Food & Function, 1*(1), 32–59. https://doi.org/10.1039/C0FO001. Copyright (2010) Royal Society of Chemistry.*

## 3.2 Cell models

A common cell model to assess bioavailability is Caco-2 cells isolated from the human colorectal adenocarcinoma. These cells differentiate spontaneously after 14–21 days of incubation in culture medium. They polarize on semi-permeable membranes to form a continuous monolayer with the expression of tight junctions, transports, microvilli, and enzymes (Feng & Betti, 2017; Schulz, 2011; Sessa et al., 2014). Caco-2 cells are typically cultured in Dulbecco's modified Eagle's medium supplemented with other enrichment solutions, depending on the nature of the analyte (Ding, Wang, Yu, Zhang, & Liu, 2016; Feng & Betti, 2017). The cell monolayer divides the apical and basolateral sides of absorption. With Hank's balanced salt solution (HBSS) as the carrier fluid, the analyte is introduced onto the apical side, and the absorbed moiety is collected on the basolateral side at the desired time intervals. Percent of analyte that is absorbed through the Caco-2 cells is determined by analytical analyses.

### 3.3 Animal studies

Animal studies are commonly performed during the more developed stages of research, allowing researchers to study the effects of the bioactives in a living organism. Determining bioavailability using animal models can be done through different techniques (Weis & LaVelle, 1991). One technique determines the area under the blood concentration versus time curve (AUC). Another approach is performing a mass balance, comparing the mass of bioactive consumed and the mass of the bioactive excreted in urine and feces. Finally, tissues and organs can be harvested at specific periods of time to compare the concentration of bioactives and its metabolites with the concentration ingested. Common animals used in nutrition and pharmacological research include mice, rats, chicks, pigs and non-human primates. Aside from non-human primates, pigs would be the species that best resembles humans (Baker, 2008).

### 3.4 Clinical studies

Clinical studies occur during the final stage of research because they are labor-intensive, costly, and time consuming, in addition to ethical concerns. Therefore, clinical studies are less common compared to the other alternative models discussed above (Reboul et al., 2006). Human clinical trials are essential because *in vitro* models and animal studies cannot fully replicate the complexity of how the human digestive system functions. Typically, bioavailability is measured following the AUC methods and employing a crossover study to reduce variability by individual specific factors (FDA, 2002). In crossover studies, each individual receives all treatments, only at different times (Stoney & Johnson, 2018).

## 4. Nanotechnology used to increase bioavailability/bioaccessibility

Nutraceuticals and bioactives exhibit different hydrophilic and hydrophobic properties. Hydrophilic compounds are soluble in water and capable of being absorbed in the digestive tract across the lumen and into systemic circulation. However, hydrophobic bioactives tend to have low bioaccessibility and bioavailability due to their poor solubility, low permeability, bulky side chains, reactivity, and so on (Table 2). Examples of hydrophobic

**Table 2** Comparisons of the main hydrophobic bioactive compounds covered in this chapter, including their sources, benefits to human health and challenges related to their bioaccessibility/bioavailability.

| Bioactive compound | Source(s) | Benefit(s) | Challenge(s) to bioavailability/ bioaccessibility |
|---|---|---|---|
| *Polyphenols* | | | |
| Curcuminoids | Turmeric | Anti-carcinogenic<br>Anti-cardiovascular disease<br>Anti-inflammatory<br>Free radical scavenging | Low chemical stability<br>Low water solubility |
| Flavonoids | Fruits, vegetables and green tea | | Low chemical stability<br>Low water solubility<br>Bulky size (glycosides) |
| Stilbenes<br>*Resveratrol* | Grape skin | | Low chemical stability<br>Low water solubility |
| *Carotenoids* | | | |
| β-Carotene | Carrots and other yellow/orange fruits and vegetables | Free radical scavenging<br>Provitamin A<br>Reduced risk of macular degeneration and night blindness<br>Hormone and immune regulation | Prone to oxidation<br>Low water solubility |
| Lutein | Marigolds, corn and egg yolks | Free radical scavenging<br>Reduced risk of macular degeneration and night blindness<br>Hormone and immune regulation | |
| Lycopene | Tomato | Free radical scavenging<br>Hormone regulation<br>Immune regulation | |

*Continued*

**Table 2** Comparisons of the main hydrophobic bioactive compounds covered in this chapter, including their sources, benefits to human health and challenges related to their bioaccessibility/bioavailability.—cont'd

| Bioactive compound | Source(s) | Benefit(s) | Challenge(s) to bioavailability/ bioaccessibility |
|---|---|---|---|
| *Fatty acids* | | | |
| ALA | Flaxseed, heart-healthy oils, walnuts, dairy, algae | Anti-diabetic<br>Anti-carcinogenic<br>Anti-atherosclerotic | Prone to oxidation<br>Low water solubility |
| CLA | Dairy and meat products | Anti-inflammatory<br>Anti-atherosclerotic<br>Anti-diabetic<br>Anti-carcinogenic | |
| DHA/EPA | Fish, aquatic sources | Anti-diabetic<br>Anti-carcinogenic<br>Anti-atherosclerotic<br>Anti-nociceptive | |
| *Fat soluble vitamins* | | | |
| Vitamins A, D, E (tocopherol), K | *Vitamin A*: Carrots, green leafy vegetables, sweet potato<br>*Vitamin D*: Animal sources<br>*Vitamin E*: Oils, nuts, seeds, green leafy vegetables<br>*Vitamin K*: Green leafy vegetables | *Vitamin A*: Vision health, anti-cancer,<br>*Vitamin D*: Aid in calcium absorption<br>*Vitamin E*: Antioxidant, anti-inflammatory<br>*Vitamin K*: Heart health, bone health | Low water solubility.<br>Some are prone to photooxidation<br>Thermal degradation |
| *Dietary fibers* | | | |
| Soluble | Oat bran, barley, nuts, psyllium, beans, lentils | Reduced risk of cardiovascular disease<br>Reduced cholesterol<br>Control of satiety and postprandial glucose response | Adverse effects on bioavailability of other nutrients<br>Can be used to create control release delivery systems |
| Insoluble | Wheat bran, vegetables, whole grains | Decreased digestion transit time<br>Increased colon health and related diseases | Adverse effects on bioavailability of other nutrients |

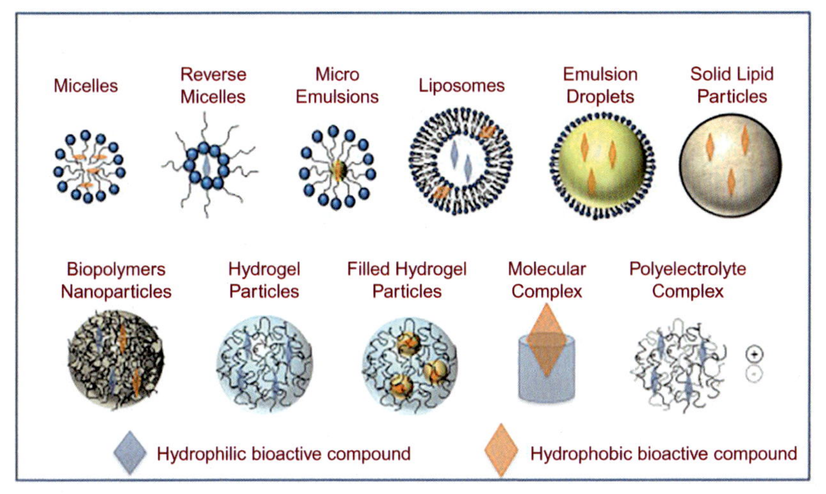


**Fig. 4** Depiction of the different delivery systems. *Reprinted with permission from* Joye, I. J., Davidov-Pardo, G., & McClements, D. J., (2014). Nanotechnology for increased micronutrient bioavailability. Trends in Food Science & Technology, 40 (2), 168–182. https://doi.org/10.1016/j.tifs.2014.08.006. *Copyright (2014) Elsevier.*

compounds include carotenoids, polyphenols, tocopherols, fat-soluble vitamins, phytosterols, and fatty acids. The remainder of the chapter focuses on exploring important nanometric delivery systems (Fig. 4) that aid in increasing bioaccessibility and bioavailability of hydrophobic bioactives (Table 3).

## 4.1 Surfactant based delivery systems

### *4.1.1 Liposomes*

Liposomes are spherical vesicles composed of one or more layers of phospholipid bilayers with the hydrophobic tails facing each other and an aqueous core. Due to their unique nature, liposomes are capable of loading lipophilic, amphiphilic, and hydrophilic components either embedded in the phospholipid bilayer, on the liposome-continuous phase interface, or in the aqueous core, respectively (Gonçalves, Martins, Duarte, Vicente, & Pinheiro, 2018; McClements, 2015). Typical phospholipids are sourced from eggs, soy, milk, or sunflower lecithins, while mixtures of compounds are typically found in commercial lecithins. Liposomes are formed when phospholipid bilayers spontaneously enclose into spherical vesicles in an aqueous solutions. For instance, in solvent evaporation/rehydration, a phospholipid is dissolved in an organic solvent, such as ethanol, and then solvent

**Table 3** A comparison of the different delivery systems in food nanotechnology and their impact on human bioavailability/bioaccessibility.

| Delivery system | Characteristics | Bioavailability/bioaccessibility |
|---|---|---|
| Liposomes | Phospholipid bilayers with aqueous core | Can deliver multiple bioactives (hydrophobic and hydrophilic)<br>Depends on permeability of membrane that can be enhanced by coating the phospholipid by-layer |
| Microemulsions/ swollen micelles | Thermodynamically stable, contain surfactant and carrier oil | Increase due to increased surface area<br>Increase solubility in intestines<br>Passive permeation |
| Nanoemulsions | Thermodynamically unstable, kinetically stable | Increase due to increased surface area<br>Increase solubility in intestines<br>Possibility to incorporate different size fatty acids to modulate micellization |
| Polymer nanoparticles | Dense protein and polysaccharide matrix | Increase with excipient oil droplets<br>Increase due to higher water solubility<br>Depends on bioactive-polymer interactions and susceptibility to digestive enzymes |
| Polyelectrolyte complexes | Protein and polysaccharide matrix stabilized by ionic attractions | Highly pH-dependent<br>Can increase due to higher water solubility and chemical stability |
| Filled hydrogels | Nanoemulsions trapped within porous biopolymer complex | Low in small intestine<br>High in colon<br>Increase with decreasing diameter<br>Depends on digestive enzymes susceptibility |
| Molecular inclusion | Host molecule (β-cyclodextrins) with hydrophobic interior | Increase due to higher solubility, biocompatibility with intestinal walls<br>Can increase permeability of epithelial cells |

is evaporated leaving a thin film of phospholipid. Subsequent addition of an aqueous solution to the film causes spontaneous formation of bilayer or multi-bilayer liposomes. A similar technique for liposome formation is solvent displacement, in which phospholipids are dissolved in an organic solvent and then injected into an aqueous solution. Other methods include surfactant displacement, homogenization, sonication, or microfluidization (Gonçalves et al., 2018; McClements, 2015).

Liposomes are advantageous in many aspects due to their biocompatibility, biodegradability, low toxicity, and they are relatively small in size ($r < 100$ nm) (Gonçalves et al., 2018; Tan et al., 2014). However, they are sensitive to processing and gastrointestinal conditions, such as light, high temperatures, oxygen, and pH. Also, depending on the nature of the system, liposomes can have a low encapsulation efficiency and short release times (Gonçalves et al., 2018). Poor release characteristics are attributable to excess flexibility of the phospholipid membrane which is susceptible to swelling and increased penetrability, leading to low bioaccessibility in the intestinal phase (Tan et al., 2014). The incorporation of bioactive into the mixed micelle phase also depends on liposome loading characteristics and encapsulation efficiency of the compound. For instance, more polar carotenoids inserted closer to the membrane outer surface can be more readily transferred to the mixed micelle phase in the small intestine (Tan et al., 2014).

Instability and penetrability of liposome membranes can be addressed by modification of the membrane's surface with a polymeric layer (Tan, Feng, Zhang, Xia, & Xia, 2016). Phospholipids can be partially replaced with amphiphilic polymers to form hybrid liposomes. Peng et al. (2017) prepared hybrid phospholipid-amphiphilic chitosan liposomes by a solvent evaporation-rehydration method. The hybrid liposomes had a more sustained *in vitro* release of curcumin compared to the pure phospholipid liposomes due to amphiphilic chitosan decreasing membrane permeability. Prematurely released curcumin could not be dissolved in simulated intestinal fluids and was not available for uptake by the epithelial cells. The hybrid liposomes also had a higher cellular uptake measured using Caco-2 cell model, due to the increased positive charge on the membrane surface that allowed higher adsorption to epithelial cells, and thus leading to a higher rate of endocytosis.

### 4.1.2 Micelles and microemulsions

Micelles are colloidal particles that have a hydrophobic core and a hydrophilic surface. Micelles are spontaneously formed when surfactants are introduced into a polar hydrophilic solution (i.e., water or aqueous solutions). The hydrophilic head of a surfactant molecule is oriented to interact with the hydrophilic aqueous solution while the hydrophobic tail orients itself into the core. Hydrophobic bioactives will be contained within the core of the micelle. The hydrophilic components become less concentrated as the concentration of the hydrophobic core materials increases (Rangel-Yagui, Pessoa, & Tavares, 2005). Micelles and microemulsions are similar in composition, except that micelles include only surfactant, whereas microemulsions include both surfactant and carrier lipid. The addition of carrier lipid to micelles can cause them to swell. In this case, microemulsions and swollen micelles can be interchangeable (McClements, 2015).

Microemulsions are thermodynamically stable colloidal systems that are optically isotropic and can be either transparent or slightly opalescent. They consist of a surfactant, a co-surfactant, oil and water, with typical micro droplets of $d \ll 100\,nm$ (Hu et al., 2012; McClements, 2012; Rai & Pandey, 2014). Their thermodynamic stability and small droplet size allows for resistance to sedimentation, flocculation, and coalescence, making them more stable. Microemulsions exhibit Brownian diffusion which prevents sedimentation or creaming, while steric hindrance prevents any coalescence or flocculation from occurring (Zhu et al., 2015). Microemulsions can spontaneously form when a ratio of surfactants, oil, and water are combined. However, the creation of a microemulsion system may require the input of activation energy, such as thermal treatment, sonication, or homogenization (McClements, 2015).

Microemulsions, just as nanoemulsions and micelles, may form an oil-in-water (O/W) emulsion in which the oil phase is dispersed in water (Fig. 5). However, they can also be formed into water-in-oil (W/O) emulsions, in which case water droplets are dispersed in an oil phase. W/O microemulsions are disadvantageous due to the use of organic solvents to create the continuous phase which can pose a threat to the environment (Rai & Pandey, 2014). A recent study focused on the use of water-in-ionic liquid (W/IL) microemulsions to address the concern of W/O microemulsion systems. This type of technology has water droplets in an ionic liquid (IL) continuous phase that are identical of a microemulsion. The utilization of nonionic or zwitterionic surfactant and W/IL microemulsions has grown in popularity due to their widespread use in chromatography, tendency to solubilize membrane proteins, and safe environmental conditions (Rai & Pandey, 2014). Ionic surfactants are still being utilized in microemulsion delivery systems. Ionic surfactants can either be cationic or anionic, depending on the application. Water-soluble core materials have been known to have an increased oral bioavailability when an ionic surfactant is utilized (Lawrence, 1996).

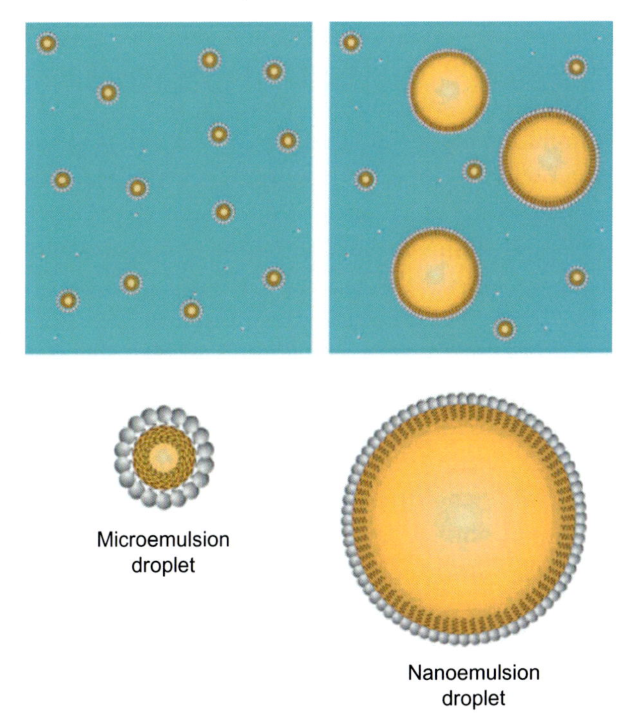


**Fig. 5** Differences in microemulsions and nanoemulsion droplets in an O/W matrix. *Reprinted with permission from McClements, D. J. (2012). Nanoemulsions versus microemulsions: Terminology, differences, and similarities.* Soft Matter, 8, *1719–1729. https://doi.org/10.1039/c2sm06903b. Copyright (2015) Royal Society of Chemistry.*

Microemulsions have a large surface area and low interfacial tension, allowing for increased bioavailability upon digestion (Fanun, 2012). Zhu et al. (2015) studied the bioavailability of capsaicin-loaded microemulsions made with Cremophor EL and medium triacylglyceride oil (MCT) and found that there was a reduction in capsaicin release in small intestine conditions, leading to larger releases in the colon, which is the most efficient absorption site for capsaicin. Also, *in vivo* studies with the capsaicin-loaded microemulsions and free capsaicin revealed that the capsaicin-loaded microemulsions had significantly increased oral bioavailability. The increased bioavailability was caused by the surfactant/co-surfactant which acted as absorption enhancers, as well as keeping the capsaicin solubilized in the GIT (Ting, Jiang, Ho, & Huang, 2014; Zhu et al., 2015).

Many studies have focused on increasing the bioavailability of bioactive compounds or nutraceuticals in the body. Other studies examined strategies to decrease absorption of certain dietary components. For example, Ostlund, Spilburg, and Stenson (1999) conducted a human clinical trial looking at efficacy of a phytosterol, sitostanol, on reducing the absorption of dietary cholesterol (Ostlund et al., 1999). They incorporated hexadeuterated cholesterol tracer in a standard meal and measured the test subjects plasma tracer concentration 4 and 5 days later. Sitostanol was formulated either as a powder or loaded into a lecithin micelle. They reported that sitostanol in lecithin micelles reduced cholesterol absorption by 37%, while the unencapsulated phytosterol by 11%. The reduced bioavailability for unencapsulated sitostanol was attributed to the crystallization of the phytosterol, which sequestered the hydrophilic hydroxyl groups inside the matrix, preventing its solubilization in fluids for further digestion and absorption (Ostlund et al., 1999). This finding highlighted the importance of proper formulation of phytosterols to achieve cholesterol lowering effect. Conceivably, naturally occurring phytosterol-phospholipid complexes may be useful for cholesterol lowering in the body.

## 4.2 Nanoemulsion delivery systems

Nanoemulsions have dispersed spherical droplets with $d < 100$ nm that typically have a protein or surfactant shell and lipophilic core (McClements, 2017a, 2017b). They are convenient carriers for lipophilic materials dispersed into a bulk aqueous phase (Davidov-Pardo & McClements, 2014; Joye & McClements, 2013). The primary difference between microemulsions and nanoemulsions is their thermodynamic stability. According to McClements (2012), microemulsions are colloidal dispersion that are thermodynamically stable because the dispersion of droplets have a lower free energy than the separate oil and water phases. On the other hand, for nanoemulsions, the free energy of the separated oil and water phases is lower than the dispersed conditions, and hence thermodynamically unstable (Fig. 4) (Abbasi & Radi, 2016; McClements, 2012). Just as in microemulsions, due to their small size and Brownian motion, nanoemulsions can exhibit high kinetic stability against creaming, sedimentation, and coalescence, as well as require less surfactant than other colloidal dispersions (Singh et al., 2017).

Nanoemulsions may be formed using high or low energy methods, described in detail by Singh et al. (2017) and McClements (2015). High energy methods include microfluidization, piston gap homogenizer and ultrasonication. Microfluidization uses shear, turbulence, and cavitation to promote size reduction. A positive displacement pump operating at 500–50,000 psi forces the emulsion through microchannels in multiple passes to achieve nanoparticles. A piston gap homogenizer uses a rotor and stator to create size reduction through high shear, stress, and grinding. Finally, ultrasonication uses sonic waves to produce cavitation and shock waves, which disrupt emulsion droplets to cause size reduction.

On the other hand, low energy methods involve spontaneous emulsification and phase inversion, which tend to be more energy efficient. Nanoemulsions can form spontaneously if an organic phase containing the bioactive, which is then added to an aqueous phase. Phase inversion utilizes a mixture of oil, water and surfactant that is prepared at the phase inversion temperature, at which the affinities of the surfactant for water are oil are balanced, i.e., the mixture forms a bicontinuous phase. Rapid heating results in the formation of a water-in-oil emulsion, while rapid cooling traps the oil droplets, forming an oil-in-water nanoemulsion.

Bioactives dispersed in nanoemulsions may diffuse through transcellular pathways in the intestinal epithelium (Singh et al., 2017). Alternatively, the bioactive may be solubilized and incorporated into mixed micelles. Larger surface areas carries a distinct advantage in the bioavailability of lipophilic bioactives. For instance, nanoemulsions of lycopene fabricated with Tween 20 had increased bioavailability with decreased droplet diameter (Ha et al., 2015). The increased droplet surface area increases exposure to lipases, bile salts, and phospholipids, which helps to facilitate lycopene solubilization in mixed micelles as well as increase the number of mixed micelles (Ha et al., 2015; Sotomayor-Gerding et al., 2016). Carotenoid bioavailability can be further improved by adding dietary triglycerides to the oil phase, resulting in metabolites which form mixed micelles large enough to solubilize carotenoids in the small intestine (Liu, Bi, Xiao, & McClements, 2016). Similarly, emulsions of astaxanthin or lycopene with Tween 20 and linseed oil exhibited 93% *in vitro* bioaccessibility, largely due to the presence of long chain triglycerides in the linseed oil, particularly linolenic acid (Sotomayor-Gerding et al., 2016).

Heo et al. (2016) investigated nanoemulsions of CLA in free fatty acid and triglyceride forms, using lecithin as a surfactant. The triglyceride nanoemulsion had smaller droplet size (70–120 nm) than the free fatty acid emulsion (230–260 nm). The triglyceride nanoemulsion had higher thermal stability than the free fatty acid counterpart. *In vitro* bioavailability analysis using Caco-2 cells showed that nano-emulsification increased the cellular uptake of CLA in both forms, increasing the bioavailability of essential fatty acids (Heo et al., 2016). In a randomized, cross-over clinical trial with

10 healthy volunteers, Raatz, Redmon, Wimmergren, Donadio, and Bibus (2009) demonstrated that emulsified fish oil supplements had a higher absorption of ω-3 fatty acids when compared to the intake of bulk fish oil, The reason behind the increased ω-3 fatty acids absorption is caused by the enhanced action of pancreatic lipase due to the increase in surface area of the nanoemulsion droplets.

Emulsifiers composed of a ternary complex of chlorogenic acid, lactoferrin, and dextran increased the oxidative stability and bioavailability of β-carotene as compared with binary complexes (Liu, Ma, McClements, & Gao, 2016). This enhanced effect of the ternary complexes was attributable to the formation of small droplets that remained stable during the *in vitro* gastric digestion, while creating an increased surface area for lipid digestion and carotenoid solubilization. In another study, it was observed that 1% curcumin nanoemulsion in a 10% Tween 20 solution resulted in higher curcumin bioaccessibility for nanoemulsions with smaller particle size (1–100 nm) as compared to those with particles of larger than 100 nm (Wang et al. 2008). High pressure homogenization produced small droplets, with synergistic effect with the lipid profile that enhanced the absorption of curcumin into the small intestine by the mechanism of bile salts creating micelles with free fatty acids (Wang et al., 2008).

Besides droplet size, bioavailability also depends on interfacial structure. Nanoemulsions using whey protein-dextran conjugates as surfactants improved the oxidative stability of β-carotene, but too thick of an interfacial layer could prevent lipases from accessing the lipophilic interior, impeding the release of the bioactive and thus reducing its bioaccessibility (Fan, Yi, Zhang, Wen, & Zhao, 2017). However, optimum protein-polysaccharide conjugates can provide physical stability to the emulsion and chemical stability to the bioactive without negatively impacting bioaccessibility (Gumus, Davidov-Pardo, & McClements, 2016). Another study compared the use of (i) lecithin, (ii) lecithin-chitosan, and (iii) lecithin-chitosan-pectin as emulsifiers to prepare corn oil-in-water emulsions and found that, in the presence of bile extract, the nanoemulsion containing lecithin-chitosan showed lower release of fatty acids as compared with the other two systems, possibly due to the formation of thicker cationic layer around the droplets that reduced the lipase access to the lipids (McClements, 2010).

## 4.3 Biopolymer-based delivery systems

### *4.3.1 Biopolymer nanoparticles*

Common techniques for size reduction of hydrophobic bioactive crystals are supercritical carbon dioxide, physical comminution (e.g., milling, grinding), and spray drying. For example, changing the pressure at which supercritical carbon dioxide is used allows for optimizing particle size, solubility properties, stability, and encapsulation efficiencies (Fahim et al., 2014; Vandana, Prasanna Raju, Harini Chowdary, Sushma, & Vijay Kumar, 2014). Micronization is a common technique for size reduction of hydrophobic bioactives to improve solubility and bioavailability, which is commonly followed by with a biopolymer to improve the delivery of the core materials for food applications. Biopolymer nanoparticles are typically assembled from food grade proteins and polysaccharides, which can take the form of polyelectrolyte complexes, hydrogel particles, filled hydrogel particles, and so on (Joye & McClements, 2013; McClements, 2017a, 2017b).

Biopolymer nanoparticles can be fabricated using antisolvent precipitation. For the antisolvent precipitation method, biopolymer and bioactive are dissolved in an ideal solvent, and then the solvent is rapidly changed to non-ideal condition, resulting in spontaneous formation of particles (Joye & McClements, 2013; McClements, 2015). To promote the particle formation, salts, acids, or alkali can be added to the ideal solvent, or the ideal solvent can be injected into an antisolvent. For instance, zein and resveratrol can be dissolved in ethanol and then rapidly injected into water as an antisolvent to form zein nanoparticles encapsulating resveratrol (Huang et al., 2017). Nanoparticle exteriors made by antisolvent precipitation are hydrophobic and prone to aggregation, necessitating an outer shell containing a surfactant or protein-polysaccharide complex (McClements, 2015; Davidov-Pardo, Perez-Ciordia, Marin-Arroyo, & McClements, 2015). Also, polyphenols entrapped in biopolymers tend to have low bioaccessibility, unless mixed with lipid droplets to promote solubilization of the bioactive in the micelle phase of the small intestine (Liu et al., 2018). To illustrate, curcumin and resveratrol loaded into a core-shell nanoparticle had higher *in vitro* bioavailability when co-encapsulated with an excipient nanoemulsion of lipid droplets (Liu et al., 2018). The lipid droplets and their hydrolysis products (from lipase action) promoted the formation of mixed micelles in the small intestinal fluids and helped to solubilize the polyphenols. This was demonstrated by higher volume of NaOH required to titrate intestinal fluids during small intestine phase digestion, attributed to the release of free fatty acids and monoacylglycerols from the triaclyglycerols in the lipid droplets.

When properly fabricated, nanoparticles have potential to increase bioaccessibility of otherwise poorly absorbed chemical species. For instance, amorphous curcumin loaded into zein nanoparticles demonstrated increased *in vitro* bioaccessibility due to higher water solubility of the amorphous form compared to the free crystalline curcumin. In addition, encapsulated curcumin has higher resistance to gastric fluids (Yao, Chen, Song, McClements, & Hu, 2018). Similarly, resveratrol bound in biopolymer nanoparticles or biopolymer complexes had improved *in vitro* bioaccessibility compared to free resveratrol that tended to crystalize and precipitate (Davidov-Pardo, Perez-Ciordia, Marin-Arroyo, & McClements, 2015). Nanoparticle behavior in the GIT is affected by the material properties of the biopolymer. For instance, lutein encapsulated in low-molecular-weight-chitosan (LMWC)-based nanoparticles had a higher percent micellization in *in vitro* gastric and intestinal digestion compared to purified lutein, as well as higher concentration in the plasma, liver, and eyes of mice 8 h after oral administration (Arunkumar, Harish Prashanth, & Baskaran, 2013). The higher *in vitro* absorption could be due to the LMWC widening paracellular gaps in the intestinal epithelium cells and thus facilitating uptake of lutein mixed micelles. The release of bioactives from nanocapsules also depends on bioactive-polymer interactions; stronger hydrophobic interactions lead to a more sustained released and a longer residence time in blood plasma.

### 4.3.2 Polyelectrolyte complexes

Polyelectrolyte complexes (PEC) are nanoparticles or microparticles (60 nm–10 μm) formed from the electrostatic interaction of biopolymers, typically proteins and/or polysaccharides, with small bioactive molecules (McClements, 2015; Tan, Selig, Lee, & Abbaspourrad, 2018). Association strength between the biopolymers and bioactives is governed by pH-dependent changes in the ionic composition of the species. The bioactive can be either directly bound to the polymer or can be trapped within an assembly of different biopolymers, such as the hydrophobic pockets and three-dimensional structure (e.g., helix) (McClements, 2015).

The formation of PEC involves spontaneous rearrangement of counter ion-polymer pairs into polymer-polymer pairs, which is relatively easy, cost-effective, and does not require organic solvents (Bourganis, Karamanidou, Kammona, & Kiparissides, 2017; Siyawamwaya et al., 2015). PEC can be formed through extrusion, layer-by-layer self-assembly, or solution combination (Siyawamwaya et al., 2015). For extrusion, low-melting-point polymers are melted while entrapping the bioactive. In layer-by-layer assembly, biopolymers are deposited in uniform layers on a solid substrate that provides ionic stability, such as calcium carbonate (Tan et al., 2018).

Typically, oppositely charged biopolymers are dissolved in ideal solutions, with the bioactive included in one solution. The solutions are then combined to form multi-layer complexes (Siyawamwaya et al., 2015). Finally, the pH is adjusted to precipitate the complexes, and the PEC are removed from the pellet.

Since PEC are fundamentally governed by electrostatic interactions, effective delivery of bioactives to the GIT can be challenging. For instance, vitamin $D_3$ has been encapsulated in a PEC consisting of oppositely charged β-lactoglobulin and egg lysozyme (Diarrassouba et al., 2015a). However, in simulated gastrointestinal conditions, the presence of pancreatin caused a rapid vitamin $D_3$ release, which is undesirable from a bioavailability standpoint. The rapid release was possibly due to protein digestion by trypsin and chymotrypsin in the microsphere, thus disrupting the binding of vitamin $D_3$. Also, the complex was disrupted due to charge alterations under gastric pH conditions. Despite these challenges, vitamin $D_3$ encapsulated in the PEC was more bioavailable in the blood than the free vitamin (*in vivo* studies in rat), likely due to the increased solubility within the complex. To increase the stability against gastric pH and proteolytic enzymes, the same researchers encapsulated vitamin $D_3$ within β-lactoglobulin/$D_3$ complex by acidification (Diarrassouba et al., 2015b). In the presence of pancreatin, the increase in ionic strength of the simulated intestinal fluids decreased the erosion rate of the PEC matrix, leading to higher vitamin $D_3$ recovery. Similar to the β-lactoglobulin-lysozyme complex, *in vivo* bioavailability of vitamin $D_3$ in rat blood serum increased when the vitamin was encapsulated in coagulum due to higher stability and biocompatibility. Yan, Qiu, Wang, and Wu (2017) reported a similar observation on rapid *in vitro* release of curcumin from lactoferrin-pectin polyelectrolyte complexes, with higher release rates corresponding to more acidic pH. The rapid release was likely due to the pH-dependent weakening of electrostatic interactions between the lactoferrin and pectin, specifically the protonation of carboxyl groups on the latter, leading to rupturing and leakage of the bioactive.

PEC may also provide a suitable vehicle for fat-soluble nutraceuticals in reduced-fat or fat-free products. In a study to determine the relative impacts of fat and reassembled casein micelles (rCM) on the bioaccessibility of vitamin $D_3$, 117 healthy adults were fed yogurt formulations: (1) 3% milk fat yogurt with vitamin D3 in the fat phase; (2) skim milk yogurt with vitamin D3 delivered in rCM; (3) 3% milk fat yogurt with vitamin D3 delivered in rCM; and (4) skim milk yogurt without vitamin D3 (placebo) (Cohen et al., 2017). Blood samples were tested for 25-hydroxyvitamin D

[25(OH)D] levels at 1, 7, and 14 days after initial ingestion. The researchers determined that 25(OH)D levels were not significantly different between the groups administered with vitamin $D_3$ rCM in 3% fat milk yogurts and in skim milk yogurts, suggesting that the absence of fat did not prevent the absorption of the lipophilic micronutrient. They speculated that besides fatty acids, peptides and amino acids could trigger the release of bile salts and lipases from the pancreas. The study indicates that rCM, or similar protein-based PEC, may be used in nonfat or low-fat foods for the enrichment of vitamin D. Overall, PEC are simple, convenient delivery systems for the increased bioavailability of bioactives, but their high sensitivity to pH may complicate their effective end use applications in food.

### 4.3.3 Filled hydrogels

Filled hydrogel particles consist of nanoemulsions trapped within hydrogel beads, which are also called "Trojan-horse" nanoparticles (McClements, 2017a, 2017b). They are considered as an $O/W_1/W_2$ emulsion, since oil droplets are emulsified within an aqueous network of biopolymers, which are suspended in a bulk aqueous phase. Filled hydrogels can be prepared by forming an O/W emulsion with microfluidization, and then mixing the emulsion with an alginate solution, followed by injecting this into a calcium chloride solution. Alternatively, an O/W emulsion can be prepared by high pressure homogenization, followed by a secondary homogenization step to produce an oil-in-water in oil emulsion. A thermal treatment above the denature temperature of the biopolymer, such as whey protein, creates a gelled aqueous phase in the dispersed particles. Subsequent washing with organic solvent isolates the dispersed particles, which can then be incorporated into a bulk aqueous phase. However, this method is time consuming, energy-intensive, and requires organic solvents (Sung, Xiao, Decker, & McClements, 2015). The particle release characteristics and size of nanoparticles can be modulated by controlling gelling agent and biopolymer concentrations and extrusion conditions (Pool et al., 2012). The release of the hydrophobic bioactive from the lipid phase depends on the pH and ionic strength of the solution, as well as the pore size in the polymer matrix.

Filled hydrogels have low *in vitro* bioavailability in the small intestine compared to nanoemulsions, due to reduced digestion of the lipid phase, leading to fewer mixed micelles (Zhang, Zhang, & McClements, 2016; Zhang, Zhang, Zou et al., 2016). Hydrogels protect oil droplets and control their bioaccessibility in the upper GI tract, while allowing the bioactives to be released in the colon (McClements, 2017a, 2017b). By carefully selecting biopolymers, such as dietary fiber or proteins, encapsulated bioactives have high stability during storage and can retain their integrity throughout the upper GI tract (Zhang, Zhang, Chen, Tong, & McClements, 2015). The challenge of filled hydrogels as delivery systems is tailoring release characteristics and location of the bioactive, by controlling digestibility of lipid droplets in the GI tract (Sung, Xiao, Decker, & McClements, 2015; Zhang, Zhang, Chen, Tong, & McClements, 2015). Lipid digestion can be controlled by increasing particle diameter or increasing concentrations of gel components, such as calcium cation and alginate (Li et al., 2011). Just as with nanoemulsions, increasing the diameter of the lipid phase reduces lipase absorption to the droplet surface and thus decreases release of lipid digestion products. Also, an increase in concentration of gelling agents creates a tighter gel network within the beads, limiting diffusion of lipase through the network to reach the lipid phase as well as other lipid hydrolysis products to the small intestinal fluids (Li et al., 2011; Zhang, Zhang, & McClements, 2016; Zhang, Zhang, Zou et al., 2016).

By altering the composition and structure of the biopolymer phase, the rate of pancreatic lipase digestion of the lipid phase can be controlled. Overall, lipid droplet digestion can be controlled by manipulating bead size, composition, and lipid type. For instance, β-carotene entrapped in rice-starch hydrogel beads had poor stability due to α-amylase digestion in the saliva phase, causing premature release of the lipid droplets and their aggregation in the stomach (Park, Mun, & Kim, 2018). However, when a dietary fiber, such as xanthan gum, was combined with rice starch to resist digestion in the oral phase, *in vitro* bioavailability in the upper GI tract was reduced. This was attributed to detrimental interactions between the xanthan gum and micelle components. Another study investigated β-carotene entrapped in calcium-alginate beads, which had over 87% stability throughout the small intestinal phase (Zhang, Zhang, & McClements, 2016). The β-carotene remained entrapped in lipid droplets, and limited release of free fatty acids and monoacylglycerols to form mixed micelles was observed. However, the hydrogel could be useful for delivery of bioactives to the colon. Similarly, curcumin had higher *in vivo* bioavailability in free lipid droplets than in polymer beads of carrageenan (25.5%) and alginate (12.5%) in the small intestine (Zhang, Zhang, & McClements, 2016; Zhang, Zhang, Zou et al., 2016). To different extents, both biopolymer matrices inhibited lipase access to the lipid droplets.

Liposomes are able to be entrapped by larger molecules in order to enhance their functionality, release characteristics, and stability. As mentioned previously, bioactive peptides are susceptible to degradation due to endogenous enzymes and the acidic gastric environment. Liposome-in-alginate beads also encapsulate oyster hydrolysate increasing its bioavailability and antihypertensive activity (Xie, Lee, Choung, Kang, & Choi, 2016). The alginate beads withstand enzyme and acid degradation during the gastric phase which allows for an increased bioavailability in the intestine phase. As well, the delayed swelling time increased bioavailability of the antihypertensive properties of the bioactive peptide that is encapsulated (Xie et al., 2016).

Due to the resistance of dietary fiber networks to digestion, filled hydrogel beads tend to negatively impact bioavailability in the small intestine. However, such delivery systems could be useful to deliver an intact bioactive to the colon, where the microflora could better digest the biopolymer network. Also, filled hydrogels properties can be tailored to slow digestion of dietary lipids, controlling satiety. Hydrogel beads allow flexibility to tailor release kinetics for a specific system. By changing particle composition, the effectiveness of the delivery system can be optimized.

### *4.3.4 Molecular inclusion*

Molecular inclusion consists of a "host" molecule that entraps a "guest" bioactive molecule with non-covalent interactions (Joye & McClements, 2013). One of the most common host molecule is cyclodextrins (also called Schardinger dextrins). Cyclodextrins are macrocyclic oligosaccharides composed of α-1,4-linked glucopyranose units (Raza et al., 2017; Wadhwa, Kumar, Chhabra, Mahant, & Rao, 2017). They are formed through enzymatic action on starch by cyclodextrin-transferases produced by *Bacillus macerans* (Wei et al., 2017). They have hydrophobic cavities, which can accommodate a variety of small, hydrophobic guest molecules, and hydrophilic outer surfaces (Shulman et al., 2011). While α-, β-, and γ-cyclodextrins, composed of 6, 7, and 8 glucopyranose rings, respectively, can be produced, β-cyclodextrin is the most common class for molecular inclusion due to its cost effectiveness (Raza et al., 2017; Wei et al., 2017). β-Cyclodextrins can be derivatized, such as with the addition of a functional group (e.g., hydroxypropyl-β-cyclodextrin) to improve water solubility and *in vivo* metabolism (Raza et al., 2017). Essential oils have been encapsulated into cyclodextrin complexes to mask objectionable flavors and odors, control release, and improve solubility, stability, and bioavailability (Wadhwa et al., 2017). β-Caryophyllene (βCP), a hydrophobic sesquiterpene found in essential oils of plants, had increased oral bioavailability in rats when complexed with β-cyclodextrin (Liu et al., 2013).

The higher availability of complexed βCP than the free βCP was attributed to the enhanced water solubility and high biocompatibility of β-cyclodextrin, as well as rapid *in vivo* release. Also, 2-hydroxypropyl-β-cyclodextrin (HPβCD) allowed 11-fold enhancement of naringenin, a bitter flavanone found in citrus pomace, across a Caco-2 cell monolayer compared with the free naringenin (Shulman et al., 2011). The HPβCD complex also increased naringenin oral bioavailability in rats, as revealed by urine analysis, as well as liver, kidney, and bowel sample histology. The increased bioavailability could be due to enhanced solubility of the complex, higher stability of the bioactive, alteration of membrane properties, and close proximity of the complex to the intestinal walls.

## 5. Conclusions

The unique characteristics of nanosized delivery systems (e.g., higher stability, higher surface area, etc.) have led to the development of a wide variety of nanoparticles for enhancement of bioaccessibility and/or bioavailability for a broad arrange of hydrophobic bioactives. As discussed in this chapter, by understanding the types of interaction that govern formation of colloidal systems and the conditions in the GIT during digestion, one can tailor a controlled release delivery system that protects and slowly delivers bioactive compound to a specific site of absorption. Nevertheless, there are still areas to be explored in relation to the impact of incorporating the delivery systems in complex food matrices, as well as the toxicology of such systems. Further studies should focus on assessing the toxicology and environmental impact of the nanosized delivery systems, the effect of food matrix on the bioavailability of the nanoencapsulated bioactives, as well as the impact of the delivery systems on the food matrices. Studies related to scale up of the manufacturing process of nano-delivery systems will be important for industrial applications.

## References

Abbasi, S., & Radi, M. (2016). Food grade microemulsion systems: Canola oil/lecithin: n-propanol/water. *Food Chemistry*, *194*(C), 972–979. https://doi.org/10.1016/j.foodchem.2015.08.078.

Alemany, L., Cilla, A., Garcia-Llatas, G., Rodriguez-Estrada, M. T., Cardenia, V., & Alegría, A. (2013). Effect of simulated gastrointestinal digestion on plant sterols and their oxides in enriched beverages. *Food Research International*, *52*(1), 1–7. https://doi.org/10.1016/j.foodres.2013.02.024.

Arunkumar, R., Harish Prashanth, K. V., & Baskaran, V. (2013). Promising interaction between nanoencapsulated lutein with low molecular weight chitosan: Characterization and bioavailability of lutein in vitro and in vivo. *Food Chemistry*, *141*(1), 327–337. https://doi.org/10.1016/j.foodchem.2013.02.108.

Baker, D. H. (2008). Animal models in nutrition research. *The Journal of Nutrition*, *138*(2), 391–396. https://doi.org/10.1093/jn/138.2.391.

Bourganis, V., Karamanidou, T., Kammona, O., & Kiparissides, C. (2017). Polyelectrolyte complexes as prospective carriers for the oral delivery of protein therapeutics. *European Journal of Pharmaceutics and Biopharmaceutics*, *111*, 44–60. https://doi.org/10.1016/j.ejpb.2016.11.005.

Cardoso, C., Bandarra, N., Lourenço, H., Afonso, C., & Nunes, M. (2010). Methylmercury risks and EPA + DHA benefits associated with seafood consumption in Europe. *Risk Analysis*, *30*(5), 827–840. https://doi.org/10.1111/j.1539-6924.2010.01409.x.

Chen, H. M., Muramoto, K., Yamauchi, F., Fujimoto, K., & Nokihara, K. (1998). Antioxidant properties of histidine-containing peptides designed from peptide fragments found in the digests of a soybean protein. *Journal of Agricultural and Food Chemistry*, *46*, 49–53.

Chen, H. M., Muramoto, K., Yamauchi, F., & Nokihara, K. (1996). Antioxidant activity of designed peptides based on the antioxidant peptide isolated from digests of a soybean protein. *Journal of Agricultural and Food Chemistry*, *44*, 2619–2623.

Chiurchiù, V., Leuti, A., & Maccarrone, M. (2018). Bioactive lipids and chronic inflammation: Managing the fire within. *Frontiers in Immunology*, *9*, 38. https://doi.org/10.3389/fimmu.2018.00038.

Codex Alimentarius. (2009). *Session of the codex committee on nutrition and foods for specific dietary uses*. Report of the 31St. Accessed on: 20 Aug 2018.

Cohen, Y., Ish-Shalom, S., Segal, E., Nudelman, O., Shpigelman, A., & Livney, Y. D. (2017). The bioavailability of vitamin D3, a model hydrophobic nutraceutical, in casein micelles, as model protein nanoparticles: Human clinical trial results. *Journal of Functional Foods*, *30*, 321–325. https://doi.org/10.1016/j.jff.2017.01.019.

Davidov-Pardo, G., & McClements, D. J. (2014). Resveratrol encapsulation: Designing delivery systems to overcome solubility, stability and bioavailability issues. *Trends in Food Science & Technology*, *38*, 88–103. https://doi.org/10.1016/j.tifs.2014.05.003.

Davidov-Pardo, G., Perez-Ciordia, S., Marin-Arroyo, M. R., & McClements, D. J. (2015). improving resveratrol bioaccessibility using biopolymer nanoparticles and complexes: Impact of protein-carbohydrate maillard conjugation. *Journal of Agricultural and Food Chemistry*, *63*, 3915–3923. https://doi.org/10.1021/acs.jafc.5b00777.

Diarrassouba, F., Garrait, G., Remondetto, G., Alvarez, P., Beyssac, E., & Subirade, M. (2015a). Food protein-based microspheres for increased uptake of vitamin D3. *Food Chemistry*, *173*(C), 1066–1072. https://doi.org/10.1016/j.foodchem.2014.10.112.

Diarrassouba, F., Garrait, G., Remondetto, G., Alvarez, P., Beyssac, E., & Subirade, M. (2015b). Improved bioavailability of vitamin D3 using a β-lactoglobulin-based coagulum. *Food Chemistry*, *172*, 361–367. https://doi.org/10.1016/j.foodchem.2014.09.054.

Ding, L., Wang, L., Yu, Z., Zhang, T., & Liu, J. (2016). Digestion and absorption of an egg white ACE-inhibitory peptide in human intestinal Caco-2 cell monolayers. *International Journal of Food Sciences and Nutrition*, *67*(2), 111–116. https://doi.org/10.3109/09637486.2016.1144722.

Donhowe, E., Flores, F., Kerr, W., Wicker, L., & Kong, F. (2014). Characterization and in vitro bioavailability of beta-carotene: Effects of microencapsulation method and food matrix. *LWT-Food Science and Technology*, *57*(1), 42–48. https://doi.org/10.1016/j.lwt.2013.12.037.

Escudero, E., Mora, L., & Toldrá, F. (2014). Stability of ACE inhibitory ham peptides against heat treatment and in vitro digestion. *Journal of Food Chemistry*, *161*, 305–311. https://doi.org/10.1016/j.foodchem.2014.03.117.

Fahim, T. K., Zaidul, I. S. M., Abu Bakar, M. R., Salim, U. M., Awang, M. B., Sahena, F., et al. (2014). Particle formation and micronization using non-conventional techniques-review. *Chemical Engineering & Processing: Process Intensification*, *86*, 47–52. https://doi.org/10.1016/j.cep.2014.10.009.

Fan, Y., Yi, J., Zhang, Y., Wen, Z., & Zhao, L. (2017). Physicochemical stability and in vitro bioaccessibility of β-carotene nanoemulsions stabilized with whey protein-dextran conjugates. *Food Hydrocolloids*, *63*, 256–264. https://doi.org/10.1016/j.foodhyd.2016.09.008.

Fanun, M. (2012). Microemulsions as delivery systems. *Current Opinion in Colloid & Interface Science*, *17*(5), 306–313. https://doi.org/10.1016/j.cocis.2012.06.001.

Food & Drug Administration. (2002). *Guidance for industry bioavailability and bioequivalence studies for orally administered drug products—General considerations*. U.S. Department of Health and Human Services.

Food & Drug Administration. (2018). *Guidance for industry: Scientific evaluation of the evidence on the beneficial physiological effects of isolated or synthetic non-digestible carbohydrates submitted as a citizen petition (21 CFR 10.30)*. MD: College Park. FDA.

Feng, M., & Betti, M. (2017). Transepithelial transport efficiency of bovine collagen hydrolysates in a human Caco-2 cell line model. *Food Chemistry*, *224*, 242–250. https://doi.org/10.1016/j.foodchem.2016.12.044.

Gao, A., Dong, S., Wang, X., Li, S., & Chen, Y. (2018). (Preparation, characterization and) calcium release evaluation in vitro of casein phosphopeptides-soluble dietary fibers copolymers as calcium delivery system. *Food Chemistry*, *245*, 262–269. https://doi.org/10.1016/j.foodchem.2017.10.036.

Gibb, R. D., McRorie, J. W., Russell, D., Hasselblad, V., & D'Alessio, D. (2015). Psyllium fiber improves glycemic control proportional to loss of glycemic control: A meta-analysis of data in euglycemic subjects, patients at risk of type 2 diabetes mellitus, and patients being treated for type 2 diabetes mellitus. *American Journal of Clinical Nutrition*, *102*(6), 1604–1614. https://doi.org/10.3945/ajcn.115.106989.

Gonçalves, R. F. S., Martins, J. T., Duarte, C. M. M., Vicente, A. A., & Pinheiro, A. C. (2018). Advances in nutraceutical delivery systems: From formulation design for bioavailability enhancement to efficacy and safety evaluation. *Trends in Food Science & Technology*, *78*, 270–291. https://doi.org/10.1016/j.tifs.2018.06.011.

Gumus, C. E., Davidov-Pardo, G., & McClements, D. J. (2016). Lutein-enriched emulsion-based delivery systems: Impact of maillard conjugation on physicochemical stability and gastrointestinal fate. *Food Hydrocolloids*, *60*, 38–49. https://doi.org/10.1016/j.foodhyd.2016.03.021.

Ha, T. V. A., Kim, S., Choi, Y., Kwak, H.-S., Lee, S. J., Wen, J., et al. (2015). Antioxidant activity and bioaccessibility of size-different nanoemulsions for lycopene-enriched tomato extract. *Food Chemistry*, *178*(C), 115–121. https://doi.org/10.1016/j.foodchem.2015.01.048.

Hartmann, R., & Meisel, H. (2007). Food-derived peptides with biological activity: From research to food applications. *Current Opinion in Biotechnology*, *18*(2), 163–169. https://doi.org/10.1016/j.copbio.2007.01.013.

Heo, W., Kim, J., Pan, J., & Kim, Y. (2016). Lecithin-based nano-emulsification improves the bioavailability of conjugated linoleic acid. *Journal of Agricultural and Food Chemistry*, *64*(6), 1355–1360. https://doi.org/10.1021/acs.jafc.5b05397.

Hernandez, E. M., & Kamal-Eldin, A. (2013). *Processing and nutrition of fats and oils*. John Wiley & Sons, Ltd: Chichester, UK.

Hu, L., Jia, Y., Niu, F., Jia, Z., Yang, X., & Jiao, K. (2012). Preparation and enhancement of oral bioavailability of curcumin using microemulsions vehicle. *Journal of Agricultural and Food Chemistry*, *60*(29), 7137–7141. https://doi.org/10.1021/jf204078t.

Hu, B., Liu, X., Zhang, C., & Zeng, X. (2017). Food macromolecule based nanodelivery systems for enhancing the bioavailability of polyphenols. *Journal of Food and Drug Analysis*, *25*(1), 3–15. https://doi.org/10.1016/j.jfda.2016.11.004.

Huang, X., Dai, Y., Cai, J., Zhong, N., Xiao, H., McClements, D. J., et al. (2017). Resveratrol encapsulation in core-shell biopolymer nanoparticles: Impact on antioxidant and anticancer activities. *Food Hydrocolloids*, *64*, 157–165. https://doi.org/10.1016/j.foodhyd.2016.10.029.

Joye, I. J., & McClements, D. J. (2013). Production of nanoparticles by anti-solvent precipitation for use in food systems. *Trends in Food Science & Technology*, *34*(2), 109–123. https://doi.org/10.1016/j.tifs.2013.10.002.

Kamil, A., Chen, C.-Y. O., & Blumberg, J. B. (2015). The application of nanoencapsulation to enhance the bioavailability and distribution of polyphenols. In C. M. Sabliov, H. Chen, & R. Y. Yada (Eds.), *Nanotechnology and functional foods: Effective delivery of bioactive ingredients*. New York: John Wiley & Sons, Ltd.

Kanner, J., & Lapidot, T. (2001). The stomach as a bioreactor: Dietary lipid peroxidation in the. Gastric fluid and the effects of plant-derived antioxidants. *Free Radical Biology and Medicine*, *31*(11), 1388–1395. https://doi.org/10.1016/S0891-5849(01)00718-3.

Kitts, D. D., & Weiler, K. (2003). Bioactive proteins and peptides from food sources. Applications of bioprocesses used in isolation and recovery. *Current Pharmaceutical Design*, *9*(16), 1309–1323. https://doi.org/10.2174/1381612033454883.

Kompella, U. B., & Lee, V. H. L. (2001). Delivery systems for penetration enhancement of peptide and protein drugs: Design considerations. *Advanced Drug Delivery Reviews*, *46*(1), 211–245. https://doi.org/10.1016/S0169-409X(00)00137-X.

Krishnaswamy, K., Orsat, V., & Thangavel, K. (2012). Synthesis and characterization of. nano-encapsulated catechin by molecular inclusion with beta-cyclodextrin. *Journal of Food Engineering*, *111*(2), 255–264. https://doi.org/10.1016/j.jfoodeng.2012.02.024.

Lafay, S., & Gil-Izquierdo, A. (2008). Bioavailability of phenolic acids. *Fundamentals and Perspectives of Natural Products Research*, 7(2), 301–311. https://doi.org/10.1007/s11101-007-9077-x.

Lall, R. K., Syed, D. N., Adhami, V. M., Khan, M. I., & Mukhtar, H. (2015). Dietary polyphenols in prevention and treatment of prostate cancer. *International Journal of Molecular Sciences*, *16*(2), 3350–3376. https://doi.org/10.3390/ijms16023350.

Lawrence, M. J. (1996). Microemulsions as drug delivery vehicles. *Current Opinion in Colloid & Interface Science*, *1*(6), 826–832. https://doi.org/10.1016/S1359-0294(96)80087-2.

Li, Y., Hu, M., Du, Y., Xiao, H., & McClements, D. J. (2011). Control of lipase digestibility of. emulsified lipids by encapsulation within calcium alginate beads. *Food Hydrocolloids*, *25*(1), 122–130. https://doi.org/10.1016/j.foodhyd.2010.06.003.

Liu, C. S., Glahn, R., & Liu, R. H. (2004). Assessment of carotenoid bioavailability of whole. Foods using a Caco-2 cell culture model coupled with an in vitro digestion. *Journal of Agricultural and Food Chemistry*, *52*(13), 4330–4337. https://doi.org/10.1021/jf040028k.

Liu, F., Ma, C., McClements, D. J., & Gao, Y. (2016). Development of. polyphenol-protein-polysaccharide ternary complexes as emulsifiers for nutraceutical emulsions: Impact on formation, stability, and bioaccessibility of β-carotene emulsions. *Food Hydrocolloids*, *61*, 578–588. https://doi.org/10.1016/j.foodhyd.2016.05.031.

Liu, F., Ma, D., Luo, X., Zhang, Z., He, L., Gao, Y., et al. (2018). Fabrication. and characterization of protein-phenolic conjugate nanoparticles for co-delivery of curcumin and resveratrol. *Food Hydrocolloids*, *79*, 450–461. https://doi.org/10.1016/j.foodhyd.2018.01.017.

Liu, H., Yang, G., Tang, Y., Cao, D., Qi, T., Qi, Y., et al. (2013). Physicochemical characterization and pharmacokinetics evaluation of β-caryophyllene/β-cyclodextrin inclusion complex. *International Journal of Pharmaceutics*, *450*(1–2), 304–310. https://doi.org/10.1016/j.ijpharm.2013.04.013.

Liu, X., Bi, J., Xiao, H., & McClements, D. J. (2016). Enhancement of nutraceutical bioavailability using excipient nanoemulsions: Role of lipid digestion products on bioaccessibility of carotenoids and phenolics from mangoes. *Journal of Food Science*, *81*(3), N754–N761. https://doi.org/10.1111/1750-3841.13227.

Lütjohann, D. (2004). Sterol autoxidation: From phytosterols to oxyphytosterols. *The British Journal of Nutrition*, *91*(1), 3–4. https://doi.org/10.1079/BJN20031048.

Malinowski, M. J., & Gehret, M. M. (2010). Phytosterols for dyslipidemia. *American Journal of Health-System Pharmacy*, *67*(14), 1165–1173. https://doi.org/10.2146/ajhp090427.

Massounga Bora, A. F., Ma, S., Li, X., & Liu, L. (2018). Application of microencapsulation for. The safe delivery of green tea polyphenols in food systems: Review and recent advances. *Food Research International*, *105*, 241–249. https://doi.org/10.1016/j.foodres.2017.11.047.

Matsuoka, K., Kajimoto, E., Horiuchi, M., Honda, C., & Endo, K. (2010). Competitive solubilization of cholesterol and six species of sterol/stanol in bile salt micelles. *Chemistry and Physics of Lipids*, *163*(4), 397–402. https://doi.org/10.1016/j.chemphyslip.2010.03.006.

McClements, D. J. (2010). Design of nano-laminated coatings to control bioavailability of lipophilic food components. *Journal of Food Science*, *75*(1), R30–R42.

McClements, D. J. (2012). Nanoemulsions versus microemulsions: Terminology, differences, and similarities. *Soft Matter*, *8*, 1719–1729. https://doi.org/10.1039/c2sm06903b.

McClements, D. J. (2015). *Nanoparticle- and microparticle-based delivery systems: Encapsulation, protection and release of active compounds*. New York: CRC Press Taylor and Francis Group.

McClements, D. J. (2017a). Recent progress in hydrogel delivery systems for improving nutraceutical bioavailability. *Food Hydrocolloids*, *68*, 238–245. https://doi.org/10.1016/j.foodhyd.2016.05.037.

McClements, D. J. (2017b). The future of food colloids: Next-generation nanoparticle delivery systems. *Current Opinion in Colloid & Interface Science*, *28*, 7–14. https://doi.org/10.1016/j.cocis.2016.12.002.

McRorie, J. W. (2015). Evidence-based approach to fiber supplements and clinically meaningful health benefits, part 1: What to look for and how to recommend an effective fiber therapy. *Nutrition Today*, *50*(2), 82–89. http://doi.org.proxy.library.cpp.edu/10.1097/NT.0000000000000082.

Mehta, R. (2005). Dietary fiber benefits. *Cereal Foods World*, *50*(2), 66–71.

Mutsokoti, L., Panozzo, A., Pallares Pallares, A., Jaiswal, S., Van Loey, A., Grauwet, T., et al. (2017). Carotenoid bioaccessibility and the relation to lipid digestion: A kinetic study. *Food Chemistry*, *232*, 124–134. https://doi.org/10.1016/j.foodchem.2017.04.001.

Nagy, K., Redeuil, K., Williamson, G., Rezzi, S., Dionisi, F., Longet, K., et al. (2011). First identification of dimethoxycinnamic acids in human plasma after coffee intake by liquid chromatography–mass spectrometry. *Journal of Chromatography A*, *1218*(3), 491–497. https://doi.org/10.1016/j.chroma.2010.11.076.

Ostlund, R., Jr., Spilburg, C., & Stenson, W. (1999). Sitostanol administered in lecithin micelles potently reduces cholesterol absorption in humans. *The American Journal of Clinical Nutrition*, *70*(5), 826–831.

Palafox-Carlos, H., Fernando, A.-Z. J., & Gonzalez-Aguilar, G. A. (2011). The role of dietary fiber in the bioaccessibility and bioavailability of fruit and vegetable antioxidants. *Journal of Food Science*, *76*(1), R6–R15. https://doi.org/10.1111/j.1750-3841.2010.01957.x.

Park, S., Mun, S., & Kim, Y.-R. (2018). Effect of xanthan gum on lipid digestion and bioaccessibility of β-carotene-loaded rice starch-based filled hydrogels. *Food Research International*, *105*, 440–445. https://doi.org/10.1016/j.foodres.2017.11.039.

Peng, S., Zou, L., Liu, W., Li, Z., Liu, W., Hu, X., et al. (2017). Hybrid liposomes composed of amphiphilic chitosan and phospholipid: Preparation, stability and bioavailability as a carrier for curcumin. *Carbohydrate Polymers*, *156*, 322–332. https://doi.org/10.1016/j.carbpol.2016.09.060.

Pool, H., Quintanar, D., Figueroa, J. D., Mano, C. M., Bechara, E., Godinez, L., et al. (2012). Antioxidant effects of quercetin and catechin encapsulated into PLGA nanoparticles. *Journal of Nanomaterials*. https://doi.org/10.1155/2012/145380.

Pérez-Abril, M., Lucas-Abellán, C., Castillo-Sánchez, J., Pérez-Sánchez, H., Cerón-Carrasco, J. P., Fortea, I., et al. (2017). Systematic investigation and molecular modeling of complexation between several groups of flavonoids and HP-β-cyclodextrins. *Journal of Functional Foods*, *36*, 122–131. https://doi.org/10.1016/j.jff.2017.06.052.

Pullakhandam, R., Nair, K., Pamini, H., & Punjal, R. (2011). Bioavailability of iron and zinc from multiple micronutrient fortified beverage premixes in Caco-2 cell model. *Journal of Food Science*, *76*(2), H38–H42. https://doi.org/10.1111/j.1750-3841.2010.01993.x.

Raatz, S. K., Redmon, J. B., Wimmergren, N., Donadio, J. V., & Bibus, D. M. (2009). Enhanced absorption of omega-3 fatty acids from emulsified compared with encapsulated fish oil. *Journal of the American Dietetic Association*, *109*(6), 1076–1081. https://doi.org/10.1016/j.jada.2009.03.006.

Rai, R., & Pandey, S. (2014). Evidence of water-in-ionic liquid microemulsion formation by nonionic surfactant Brij-35. *Langmuir*, *30*(34), 10156–10160. https://doi.org/10.1021/la502174a.

Schonfeld, G. (2010). Plant sterols in atherosclerosis prevention. *The American Journal of Clinical Nutrition*, *92*(1), 3–4. https://doi.org/10.3945/ajcn.2010.29828.

Rangel-Yagui, C. O., Pessoa, A., & Tavares, L. C. (2005). Micellar solubilization of drugs. *Journal of Pharmacy & Pharmaceutical Sciences*, *8*, 147–163.

Raza, A., Sun, H., Bano, S., Zhao, Y., Xu, X., & Tang, J. (2017). Preparation, characterization, and in vitro anti-inflammatory evaluation of novel water soluble kamebakaurin/hydroxypropyl-β-cyclodextrin inclusion complex. *Journal of Molecular Structure*, *1130*(C), 319–326. https://doi.org/10.1016/j.molstruc.2016.10.059.

Reboul, E., Richelle, M., Perrot, E., Desmoulins-Malezet, C., Pirisi, V., & Borel, P. (2006). Bioaccessibility of carotenoids and vitamin E from their main dietary sources. *Journal of Agricultural and Food Chemistry*, *54*(23), 8749–8755. https://doi.org/10.1021/jf061818s.

Renukuntla, J., Vadlapudi, A. D., Patel, A., Boddu, S. H. S., & Mitra, A. K. (2013). Approaches for enhancing oral bioavailability of peptides and proteins. *International Journal of Pharmaceutics*, *447*(1–2), 75–93. https://doi.org/10.1016/j.ijpharm.2013.02.030.

Reyes-Diaz, A., Gonzalez-Codova, A., Hernandez-Mendoza, A., & Vallejo-Cordoba, B. (2016). Immuno-modulating peptides obtained from milk proteins. *Interciencia*, *41*(2), 84–91.

Rui, L., Lujuan, X., Qingquan, F., Guang-Hong, Z., & Wan-Gang, Z. (2016). A review of antioxidant peptides derived from meat muscle and by-products. *Antioxidants*, *5*(3), 32. https://doi.org/10.3390/antiox5030032.

Saini, R. K., Nile, S. H., & Park, S. W. (2015). Carotenoids from fruits and vegetables: Chemistry, analysis, occurrence, bioavailability and biological activities. *Food Research International*, *76*(Pt 3), 735–750. https://doi.org/10.1016/j.foodres.2015.07.047.

Schulz, M. A. (2011). *CACO-2 cells and their uses*. Hauppauge: Nova Science. Publishers, Incorporated.

Sessa, M., Balestrieri, M. L., Ferrari, G., Servillo, L., Castaldo, D., D'onofrio, N., et al. (2014). Bioavailability of encapsulated resveratrol into nanoemulsion-based delivery systems. *Food Chemistry*, *147*, 42–50. https://doi.org/10.1016/j.foodchem.2013.09.088.

Shulman, M., Cohen, M., Soto-Gutierrez, A., Yagi, H., Wang, H., Goldwasser, J., et al. (2011). Enhancement of naringenin bioavailability by complexation with hydroxypropoyl-β-cyclodextrin (enhanced naringenin bioavailability). *PLoS One*, *6*(4), e18033. https://doi.org/10.1371/journal.pone.0018033.

Singh, Y., Meher, J. G., Raval, K., Khan, F. A., Chaurasia, M., Jain, N. K., et al. (2017). Nanoemulsion: Concepts, development and applications in drug delivery. *Journal of Controlled Release*, *252*, 28–49. https://doi.org/10.1016/j.jconrel.2017.03.008.

Siyawamwaya, M., Choonara, Y. E., Bijukumar, D., Kumar, P., Du Toit, L. C., & Pillay, V. (2015). A review: Overview of novel polyelectrolyte complexes as prospective drug bioavailability enhancers. *International Journal of Polymeric Materials and Polymeric Biomaterials*, *64*(18), 955–968. https://doi.org/10.1080/00914037.2015.1038816.

Slavin, J. (2013). Fiber and prebiotics: Mechanisms and health benefits. In *Vol. 5. Nutrients* (pp. 1417–1435).

Sotomayor-Gerding, D., Oomah, B. D., Acevedo, F., Morales, E., Bustamante, M., Shene, C., et al. (2016). High carotenoid bioaccessibility through linseed oil nanoemulsions with enhanced physical and oxidative stability. *Food Chemistry*, *199*(C), 463–470. https://doi.org/10.1016/j.foodchem.2015.12.004.

Stoney, C. M., & Johnson, L. L. (2018). Design of clinical trials and studies. In J. I. Gallin, F. P. Ognibene, & L. L. Johnson (Eds.), *Principles and practice of clinical research* (4th ed., pp. 249–268). Boston: Academic Press. Chapter 18.

Sung, M.-R., Xiao, H., Decker, E. A., & McClements, D. J. (2015). Fabrication, characterization and properties of filled hydrogel particles formed by the emulsion-template method. *Journal of Food Engineering*, *155*, 16–21. https://doi.org/10.1016/j.jfoodeng.2015.01.007.

Tan, C., Feng, B., Zhang, X., Xia, W., & Xia, S. (2016). Biopolymer-coated liposomes by electrostatic adsorption of chitosan (chitosomes) as novel delivery systems for carotenoids. *Food Hydrocolloids*, *52*, 774–784. https://doi.org/10.1016/j.foodhyd.2015.08.016.

Tan, C., Selig, M. J., Lee, M. C., & Abbaspourrad, A. (2018). Encapsulation of copigmented anthocyanins within polysaccharide microcapsules built upon removable CaCO3 templates. *Food Hydrocolloids*, *84*, 200–209. https://doi.org/10.1016/j.foodhyd.2018.05.036.

Tan, C., Zhang, Y., Abbas, S., Feng, B., Zhang, X., & Xia, S. (2014). Modulation of the carotenoid bioaccessibility through liposomal encapsulation. *Colloids and Surfaces B: Biointerfaces*, *123*, 692–700. https://doi.org/10.1016/j.colsurfb.2014.10.011.

Ting, Y., Jiang, Y., Ho, C.-T., & Huang, Q. (2014). Common delivery systems for enhancing in. vivo bioavailability and biological efficacy of nutraceuticals. *Journal of Functional Foods*, 7(1), 112–128. https://doi.org/10.1016/j.jff.2013.12.010.

Toldrá, F., Reig, M., Aristoy, M. C., & Mora, L. (2018). Generation of bioactive peptides during food processing. *Journal of Food Chemistry*, *267*, 395–404. https://doi.org/10.1016/j.foodchem.2017.06.119.

Vandana, K. R., Prasanna Raju, Y., Harini Chowdary, V., Sushma, M., & Vijay Kumar, N. (2014). An overview on in situ micronization technique—An emerging novel concept in advanced drug delivery. *Saudi Pharmaceutical Journal*, *22*(4), 283–289. https://doi.org/10.1016/j.jsps.2013.05.004.

Viladomiu, M., Hontecillas, R., & Bassaganya-Riera, J. (2016). Modulation of inflammation and immunity by dietary conjugated linoleic acid. *European Journal of Pharmacology*, *785*, 87–95. https://doi.org/10.1016/j.ejphar.2015.03.095.

Wadhwa, G., Kumar, S., Chhabra, L., Mahant, S., & Rao, R. (2017). Essential oil–cyclodextrin complexes: An updated review. *Journal of Inclusion Phenomena and Macrocyclic Chemistry*, *89*(1), 39–58. https://doi.org/10.1007/s10847-017-0744-2.

Wang, X., Jiang, Y., Wang, Y.-W., Huang, M.-T., Ho, C.-T., & Huang, Q. (2008). Enhancing anti-inflammation activity of curcumin through O/W nanoemulsions. *Food Chemistry*, *108*(2), 419–424. https://doi.org/10.1016/j.foodchem.2007.10.086.

Wang, T. Y., Liu, M., Portincasa, P., & Wang, D. Q. H. (2013). New insights into the molecular mechanism of intestinal fatty acid absorption. *European Journal of Clinical Investigation*, *43*, 1203–1223.

Wei, Y., Zhang, J., Memon, A. H., & Liang, H. (2017). Molecular model and in vitro antioxidant activity of a water-soluble and stable phloretin/hydroxypropyl-β-cyclodextrin inclusion complex. *Journal of Molecular Liquids*, *236*, 68–75. https://doi.org/10.1016/j.molliq.2017.03.098.

Weis, C. P., & LaVelle, J. M. (1991). Characteristics to consider when choosing an animal model for the study of lead bioavailability. *Chemical Speciation & Bioavailability*, *3*(3–4), 113–119. https://doi.org/10.1080/09542299.1991.11083162.

Xie, C. L., Lee, S. S., Choung, S. Y., Kang, S. S., & Choi, Y. J. (2016). Preparation and optimisation of liposome-in-alginate beads containing oyster hydrolysate for sustained release. *International Journal of Food Science and Technology*, *51*(10), 2209–2216. https://doi.org/10.1111/ijfs.13207.

Yan, J., Qiu, W., Wang, Y., & Wu, J. (2017). Biocompatible polyelectrolyte complex nanoparticles from lactoferrin and pectin as potential vehicles for antioxidative curcumin. *Journal of Agricultural and Food Chemistry*, *65*(28), 5720–5730. https://doi.org/10.1021/acs.jafc.7b01848.

Yao, K., Chen, W., Song, F., McClements, D. J., & Hu, K. (2018). Tailoring zein nanoparticle functionality using biopolymer coatings: Impact on curcumin bioaccessibility and antioxidant capacity under simulated gastrointestinal conditions. *Food Hydrocolloids*, *79*, 262–272. https://doi.org/10.1016/j.foodhyd.2017.12.029.

Ye, E., Chacko, S. A., Chou, E., Kugizaki, M., & Liu, S. M. (2012). Greater whole-grain intake is associated with lower risk of type 2 diabetes, cardiovascular disease, and weight gain. *The Journal of Nutrition*, *142*(7), 1304–1313. https://doi.org/10.3945/jn.111.155325.

Yonekura, L., & Nagao, A. (2007). Intestinal absorption of dietary carotenoids. *Molecular Nutrition & Food Research*, *51*(1), 107–115. https://doi.org/10.1002/mnfr.200600145.

Zhang, Z., Zhang, R., Chen, L., Tong, Q., & McClements, D. J. (2015). Designing hydrogel particles for controlled or targeted release of lipophilic bioactive agents in the gastrointestinal tract. *European Polymer Journal*, *72*(C), 698–716. https://doi.org/10.1016/j.eurpolymj.2015.01.013.

Zhang, Z., Zhang, R., & McClements, D. J. (2016). Encapsulation of β-carotene in alginate-based hydrogel beads: Impact on physicochemical stability and bioaccessibility. *Food Hydrocolloids*, *61*(C), 1–10. https://doi.org/10.1016/j.foodhyd.2016.04.036.

Zhang, Z., Zhang, R., Zou, L., Chen, L., Ahmed, Y., Al Bishri, W., et al. (2016). Encapsulation of curcumin in polysaccharide-based hydrogel beads: Impact of bead type on lipid digestion and curcumin bioaccessibility. *Food Hydrocolloids*, *58*(C), 160–170. https://doi.org/10.1016/j.foodhyd.2016.02.036.

Zhu, Y., Zhang, J., Zheng, Q., Wang, M., Deng, W., Li, Q., et al. (2015). In vitro and in vivo evaluation of capsaicin-loaded microemulsion for enhanced oral bioavailability. *Journal of the Science of Food and Agriculture*, *95*(13), 2678–2685. https://doi.org/10.1002/jsfa.7002.

## Further reading

Gregoriadis, G. (1990). Immunological adjuvants: A role for liposomes. *Immunology Today*, *11*(3), 89–97. https://doi.org/10.1016/0167-5699(90)90034-7.

Liu, M., Zhang, J., Zhu, X., Shan, W., Li, L., Zhong, J. J., et al. (2016). Efficient mucus permeation and tight junction opening by dissociable "mucus-inert". Agent coated trimethyl chitosan nanoparticles for oral insulin delivery. *Journal of Controlled Release*, *222*, 67–77.

Mukhija, K., Singhal, K., Angmo, S., Yadav, K., Yadav, H., Sandhir, R., et al. (2016). Potential of alginate encapsulated ferric saccharate microemulsions to ameliorate iron deficiency in mice. *Biological Trace Element Research*, *172*(1), 179–192. https://doi.org/10.1007/s12011-015-0564-4.

Ozturk, B., Argin, S., Ozilgen, M., & McClements, D. J. (2015). Nanoemulsion delivery systems for oil-soluble vitamins: Influence of carrier oil type on lipid digestion and vitamin D3 bioaccessibility. *Food Chemistry*, *187*(C), 499–506. https://doi.org/10.1016/j.foodchem.2015.04.065.

Xiao, J., Li, C., & Huang, Q. K. (2015). Nanoparticles-stabilized pickering emulsions as oral delivery vehicles: Physicochemical stability and in vitro digestion profile. *Journal of Agricultural and Food Chemistry*, *63*, 10,263–10,270.

Xiao, Y., Chen, X., Yang, L., Zhu, X., Zou, L., Meng, F., et al. (2013). Preparation and oral bioavailability study of curcuminoid-loaded microemulsion. *Journal of Agricultural and Food Chemistry*, *61*(15), 3654–3660. https://doi.org/10.1021/jf400002x.

# Interaction of nanoclay-reinforced packaging nanocomposites with food simulants and compost environments

**Yining Xia[a], Maria Rubino[b,*], Rafael Auras[b,*]**

[a]Institute of Quality Standard and Testing Technology for Agro-Products, Chinese Academy of Agricultural Sciences, Beijing, China
[b]School of Packaging, Michigan State University, East Lansing, MI, United States
*Corresponding authors: e-mail address: mariar@msu.edu; aurasraf@msu.edu

## Contents



## Abstract

The production of engineered nanomaterials (ENMs) has increased exponentially over the last few decades. ENMs, made from use of engineered nanoparticles (ENPs), have been applied to the food, agriculture, pharmaceutical, and automobile industries. Of particular interest are their applications in packaging nanocomposites for consumer and non-consumer goods. ENPs in nanocomposites are of interest as a packaging material because they reduce the amount of polymer needed, while improving the physical properties. However, the transformation of ENPs in nanocomposite production, their fate, and their toxicity remain unknown while in contact with the package content or after the end of life. The objectives of this chapter are (a) to provide an overview of the main nanoclays used in packaging; (b) to categorize the main polymeric packaging nanocomposites; (c) to provide an overview of the fate and mass transport of

*Advances in Food and Nutrition Research*, Volume 88
ISSN 1043-4526
https://doi.org/10.1016/bs.afnr.2019.02.001

ENPs, especially nanoclays; (d) to describe the mass transfer of nanoclays in food simulants and in compost environments; and (e) to identify current and future research needs.

# 1. Introduction

Engineered nanomaterials (ENMs) contain components with at least one dimension between 1 and 100 nm. By 2020, nanotechnology products are expected to achieve a $3 trillion (U.S.) market value employing 6 million workers (National Science and Technology Council, 2016; Roco, Mirkin, & Hersam, 2010). The use and production of engineered nanoparticles (ENPs) have likewise expanded for consumer and non-consumer applications in the food, agriculture, pharmaceutical, automobile, and packaging industries (Duncan, 2011).

The incorporation of ENPs, such as nanoclays, graphene, and carbon nanotubes, can enhance the material properties of many polymers. This trend in interest and increased rate of publications (Fig. 1) illustrates the sustained R&D interests to exploit nanoclay and nanoparticles for nanocomposite applications. Nanoclays, such as montmorillonite (MMT),

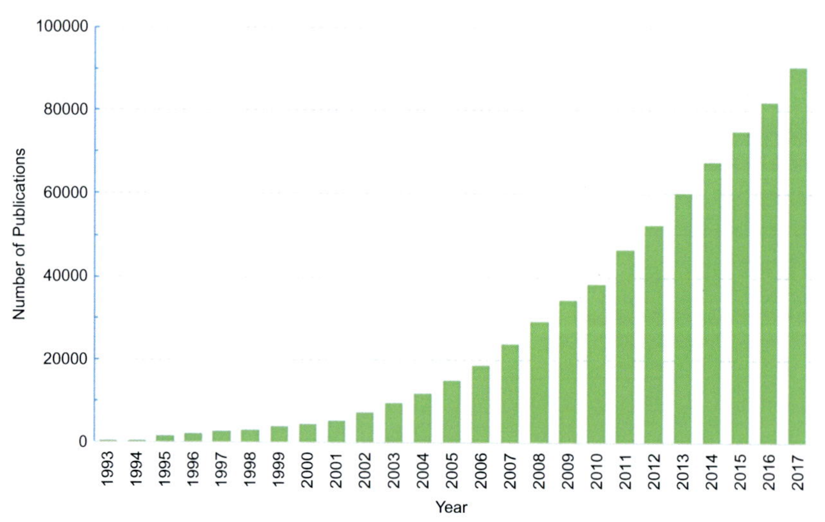


**Fig. 1** Number of peer-reviewed publications about nanocomposites and nanoparticles as related to packaging between 1993 and 2017. *From Web of Science® Core Collection search results with keywords "nanocomposite," AND "packaging," AND "nanoclay," or "nanoparticles."*

Laponite® and halloysite, are ENPs widely used in consumer goods because of their low cost, significant enhancement of composite performance, high stability, and ease of processing (Duncan, 2011). The addition of ENPs, especially nanoclays, into polymers is useful for enhancing the mechanical/barrier properties and reducing the weight of the polymer-clay nanocomposite material (Azeredo, 2009; Castro-Aguirre, Auras, Selke, Rubino, & Marsh, 2018; Duncan, 2011; Jordan, Jacob, Tannenbaum, Sharaf, & Jasiuk, 2005; Mohanty, Vivekanandhan, Pin, & Misra, 2018; Ray & Okamoto, 2003).

The broad use of ENPs in nano-enabled consumer products is causing concern regarding their environmental impact and potential risk associated with chronic human exposure (Diaz et al., 2013; Xia, Rubino, & Auras, 2014; Xia, Uysal Unalan, Rubino, & Auras, 2017). Little is known about the fate and transport behaviors of ENPs when ENMs come in contact with food products or after disposal. Hence, the mobility and partition of these ENPs between the matrix and different environments need to be assessed.

## 2. Types of clay nanoparticles in packaging

Basic characteristics of select nanoclays are in Table 1. The most commercially significant and well-researched nanoclays are MMT, Laponite®, and halloysite. Specifically, MMT is extensively used due to its low cost.

### 2.1 Montmorillonite (MMT)

MMT is a naturally-occurring, layered smectite comprised of several tens of stacked nanolayers (tactoids) (~1 nm thick and 1 nm interlayer spacing) (Fig. 2A). Each platelet is made up of an aluminum-oxygen-hydroxyl octahedral sheet sandwiched between two silicon-oxygen tetrahedral sheets (Jayaraman & Kumar, 2006). van der Waals interactions, between platelts, make them difficult to separate from each other. The interlayer galleries usually contain sodium cations, which can be exchanged with surfactants, such as long-chain alkyl ammonium ions, to promote the interaction between the gallery faces and polymer chains. The modification of MMT causes swelling of clay galleries, thereby improving polymer chain intercalation and dispersion of the clay in the polymer matrix. The silicate layers of MMT can be delaminated through modification and compounding with a polymer to form a nanocomposite (Fig. 2B, exfoliated structure), resulting in improved tensile properties.

**Table 1** Examples of commercially available nanoclays.

| Name | Shape | Dimensions (nm) | SSA ($m^2$/g) | Commercial producers | References |
|---|---|---|---|---|---|
| Montmorillonite$^{N}$ | Sheet/ platelet | $t=1$<br>$w=70$–150 | 750 | Southern Clay Products, Nanocor | Nikolaidis, Achilias, and Karayannidis (2011) |
| Laponite$^{®S}$ (synthetic smectite) | Disk | $t=1$<br>$d=25$ | 900 | Rockwood Additives Ltd | Loiseau and Tassin (2006) |
| Halloysite$^{N}$ | Hollow nanotube | $d=30$–50<br>$l=300$–1200 | 60–70 | Transmit Technology Group, LLC | Hedicke-Höchstötter, Lim, and Altstädt (2009) |
| Kaolinite | Asymmetric layers | $t=1$<br>$w=300$ | 170 | China-Kaolin Company | Jia, Li, Zhang, Cheng, and Zhang (2008) and Quincoces, Basaldella, De Vargas, and González (2004) |
| Bentonite (sodium montmorillonite)$^{N}$ | Platelet/ flake | $t=1$<br>$w \times l=800 \times 800$ | 700–800 | S&B Industrial Minerals | Industrial_Minerals_GmbH (2017) |
| Hectorite$^{N,S}$ | Elongated platelet | $t=1$<br>$w \times l=800 \times 80$ | | Elementis Specialties | Elementis_Specialist_Inc (2011) |
| Saponite$^{S}$ | Platelet | | 100–500 | Kunimine Industries | Wu and Liao (2001) |
| Talc$^{N}$ | Platelet | $PS=1500$ | 13.4 | Luzenac America | Flaris, 2005 and Reignier, Tatibouët, and Gendron (2006) |

*N*, natural origin; *S*, synthetic; *t*, thickness; *w*, width; *d*, diameter; *l*, length; *PS*, particle size; *SSA*, surface area.

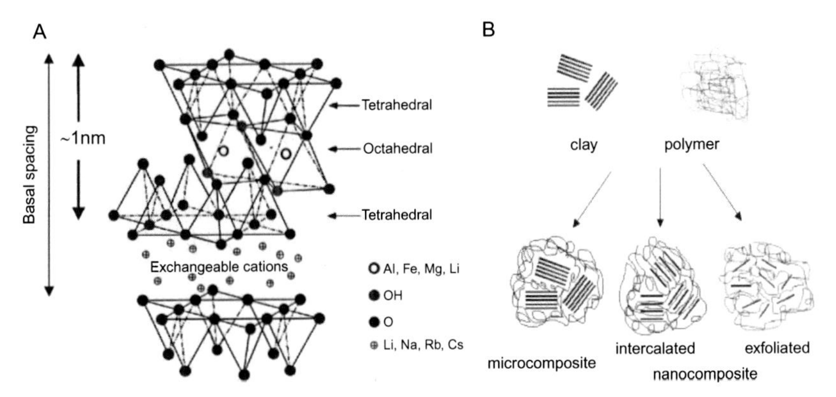


**Fig. 2** (A) Structure of MMT and (B) possible nanostructures obtained after mixing of polymer and nanoparticles. *Adapted from Ray, S. S., & Okamoto, M. (2003). Polymer/layered silicate nanocomposites: A review from preparation to processing.* Progress in Materials Science, 28, *1539–1641.*

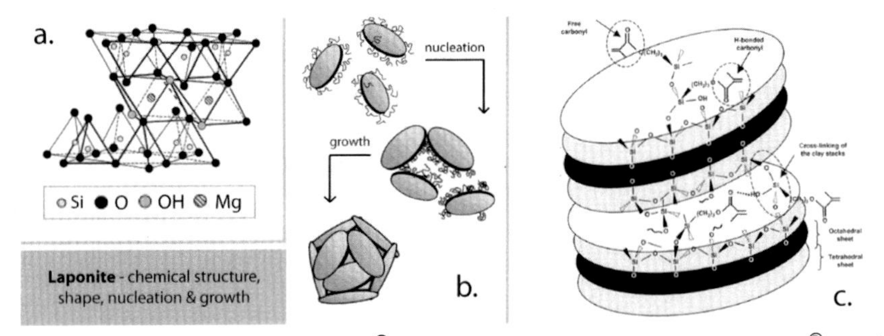


**Fig. 3** (A) Structures of Laponite® (LP); (B) typical shapes of Laponite®; and (C) monofunctional and trifunctional of Laponite® with a coupling agent. *Adapted from Herrera, N. N., Letoffe, J.-M., Reymond, J.-P., & Bourgeat-Lami, E. (2005). Silylation of laponite clay particles with monofunctional and trifunctional vinyl alkoxysilanes.* Journal of Materials Chemistry, 15*(8), 863–871. https://doi.org/10.1039/B415618H.*

## 2.2 Laponite® (LP)

LP is a synthetic smectite (primarily hectorite in origin) made up of discoid platelets that are ~1 nm in thickness. The basic unit is a hydrous magnesium silicate platelet (diameter: 25–30 nm) (Herrera, Letoffe, Reymond, & Bourgeat-Lami, 2005). The cations in LP are coordinated by 20 oxide ions and 4 hydroxyl groups (Fig. 3A). The edges of the LP disks have –OH groups (Fig. 3B). Small positive charges are located primarily at the broken edges of the crystal. The LP platelets form flocculating aggregates (Fig. 3B) that display stack-like structure in a monofunctional silane molecule (Fig. 3C). The clay platelets could be homogeneously distributed within a polymer to form a nanocomposite (Bourgeat-Lami & Lang, 2000; Herrera et al., 2005).

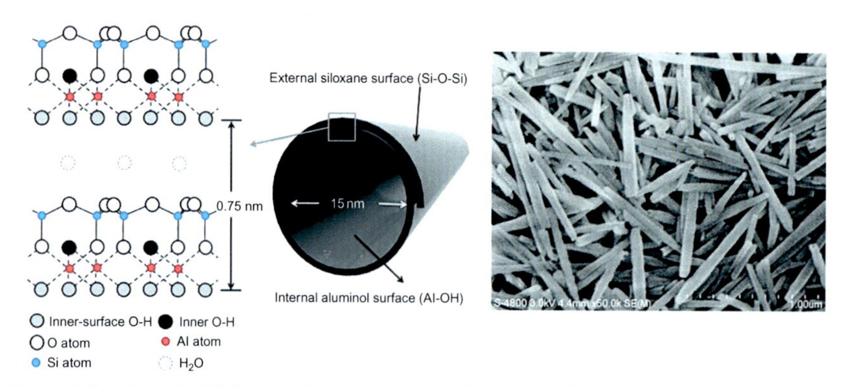


**Fig. 4** (A) Halloysite (HY) crystalline structure (*left*); (B) tubular morphology (*middle*); and (C) transmission electron microscopy image of HY nanotubes (*right*). *Reproduced from Yah, W. O., Takahara, A., & Lvov, Y. M. (2012). Selective modification of halloysite lumen with octadecylphosphonic acid: New inorganic tubular micelle.* Journal of the American Chemical Society, 134*(3), 1853–1859. https://doi.org/10.1021/ja210258y.*

## 2.3 Halloysite (HY)

HY is a clay found in various tropical and subtropical regions (Hashemifard, Ismail, & Matsuura, 2011). The structure and chemical composition of HY is similar to kaolinite, but the unit layers in halloysite are separated by a monolayer of water molecules. Halloysite is most commonly used as a fully hydrated mineral, with a basal (d001) spacing of 10 Å (Fig. 4). HY clay has up to 12.8%wt. of $Fe_2O_3$, in which $Fe^{3+}$ could be substituted by $Al^{3+}$ (Joussein et al., 2005).

# 3. Polymeric nanocomposites

Polymer nanocomposites produced with nanoclays show enhanced material properties compared with pure polymer or conventional composites (e.g., micrometer reinforcement) and have been a subject of intensive research and commercialization (Choudalakis & Gotsis, 2009; Ray & Okamoto, 2003). For example, in packaging applications, monolayer nanocomposite can replace complex multilayer structures, which may improve the potential for recycling and lower their environmental footprint due to material reduction.

Many polymers have been studied for the manufacture of nanocomposites, including polyamides (nylon), polypropylene (PP), low-density polyethylene (LDPE), polystyrene (PS), thermoplastic polyurethane, ethylene vinyl alcohol (EVOH), poly(ethylene terephthalate) (PET), and so on

(Calcagno, Mariani, Teixeira, & Mauler, 2007). PP- and LDPE-clay nanocomposites have shown improved barrier properties to water vapor and oxygen, due to the incorporation of the nanofillers (Choudalakis & Gotsis, 2009; Pereira de Abreu, Paseiro Losada, Angulo, & Cruz, 2007). As a result, similar barrier properties are achieved with a thinner structure compared to unreinforced polymer, while reducing cost. For novel bio-based plastics, such as poly(lactic acid) (PLA) and thermoplastic starch, nanotechnology is expanding the range of applications for these polymers by overcoming their performance limitations (e.g., low barrier to moisture, low heat deflection temperature) (Lagaron & Lopez-Rubio, 2011; Ray & Okamoto, 2003).

In addition to modifying the clay, compatibilizers are often incorporated during the preparation of nanocomposites to enhance the dispersion of the filler particles. For example, polyolefins, due to their non-polar chemical structure, are modified with maleic anhydride (MA) to form a polymeric compatibilizer—maleated polypropylene (MA-*g*-PP). Similar compatibilizers are used to promote interactions between polymers with the organoclays (Table 2). The coupling and interaction between the clay and the polymer matrix can be further enhanced by silane treatment of the clays to obtain an exfoliated structure (Ray & Okamoto, 2003). For polyolefins, the combined clay modification and the use of compatibilizers are key factors that contribute to the production of well-dispersed exfoliated fillers in the nanocomposites (Fig. 2B).

## 4. Fate and transport of engineered nanoparticles

ENPs may enter biological systems through either direct or indirect contact with the atmospheric, aquatic, and/or terrestrial environment (Monica & Cremonini, 2009), as well as food and medical contact (Fig. 5). For example, ENPs can be released into the air from manufacturers, into soil and water from landfills where consumer goods are buried, and from human-made nanomaterials (Fig. 6). In food packaging, one possible route could be a package made of a polymer nanocomposite that is in direct contact with a food product, consumption of which can lead to direct human exposure. Another route could be *via* consumer waste buried in landfills where ENPs could migrate into leachate, runoff, or water streams that may reach plants or human beings. The mass transfer of ENP through the matrix material will be critical for determining the exposure dose and safety of related ENMs.

**Table 2** Examples of commercially exfoliated nanocomposite systems.

| Polymer | Clay type | Surfactant | Compatibilizer | References |
|---|---|---|---|---|
| PET | Cloisite® 15A | Dimethyl, dehydrogenated tallow, quaternary ammonium | None | Calcagno et al. (2007) |
| PP | Nanomer® I.44P | Quaternary ammonium bis(hydrogenated tallow alkyl)dimethyl, chlorides | PP-g-MA | Diaz et al. (2013) |
| HDPE | Cloisite® 20 A | Dimethyl, dehydrogenated tallow, quaternary ammonium | HDPE-g-MA | Lee, Jung, Hong, Rhee, and Advani (2005) |
| LDPE | Nanomer® I.30P | Octadecylamine | PE-g-MA | Morawiec et al. (2005) |
| Nylon | Nanomer® I.34TCN | Methyl dihydroxyethyl hydrogenated tallow ammonium | None | Shen, Phang, Chen, Liu, and Zeng (2004) |
| PLA | Cloisite® 30B | Methyl, tallow, bis-2-hydroxyethyl, quaternary ammonium | None | Ray and Okamoto (2003) |

Cloisite® and Nanomer® are trade names of nanoclays supplied by Southern Clay Products and Nanocor, respectively.
Adapted from Xia, Y. (2014). *Release of nanoclay and surfactant from polymer-clay nanocomposite systems*. Michigan State University.

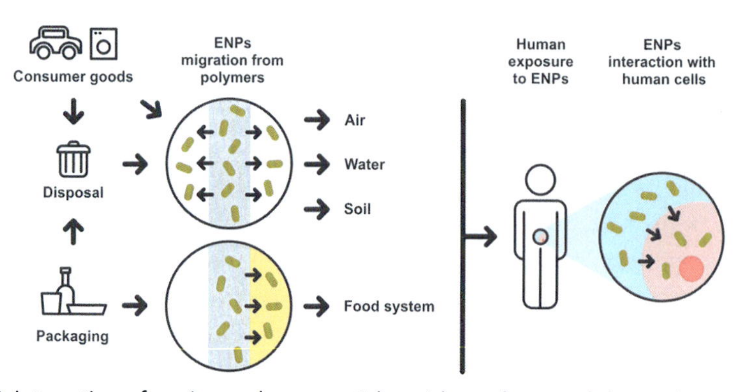


**Fig. 5** Interaction of engineered nanoparticles with products and the environment.

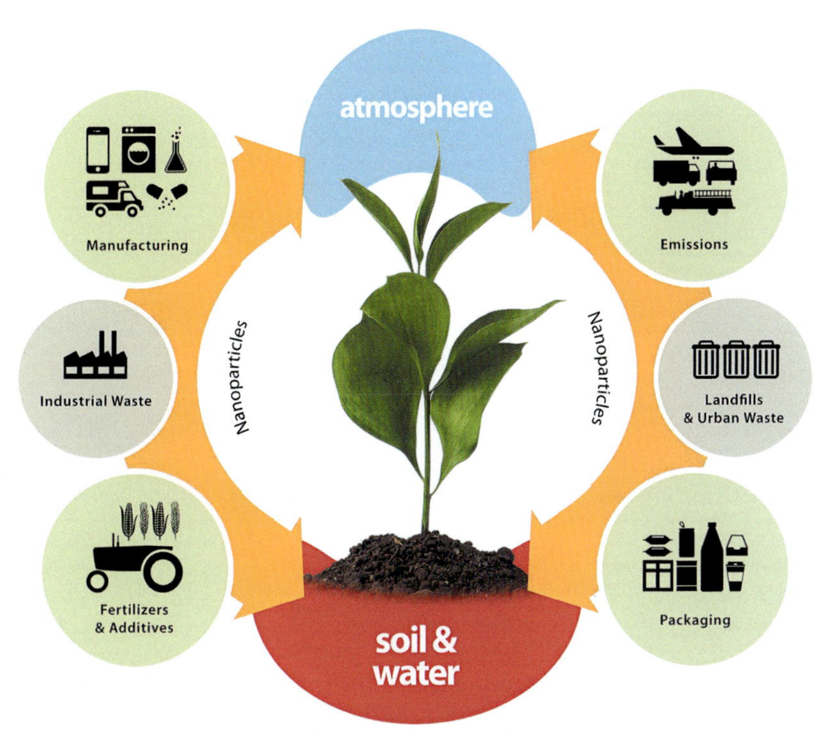


**Fig. 6** Atmospheric, aquatic, and terrestrial routes of exposure to nanoparticles.

Due to their potential toxicity, public concerns regarding ENPs safety have increased over the past decade (Maurer-Jones, Gunsolus, Murphy, & Haynes, 2013). Characteristics of ENPs, such as nano dimension, large surface binding capacity, enhanced reactivity, and ability to translocate away from the site of entrance within the biological system, could result in adverse effects in biological systems. For example, due to their inherent high binding capacity, ENPs tend to act as scavengers of essential biomolecules, such as proteins and hormones, through a process known as the corona effect (Röcker, Pötzl, Zhang, Parak, & Nienhaus, 2009). Metal oxide ENPs may induce cytotoxicity by adsorbing serum protein and calcium ions, as demonstrated in cultured cells (Horie et al., 2009). ENPs could attach and internalize into plants, affecting their germination and growth (Khot et al., 2012).

Several groups have addressed the importance of characterizing and investigating the biological consequences of nanoparticles (Haase, Tentschert, & Luch, 2012; Hristozov & Malsch, 2009; Seaton, Tran, Aitken, & Donaldson, 2009). Prior to an assessment of safety or toxicity,

**Table 3** Physicochemical characteristics of ENPs and their possible impact on biological systems.

| Composition | Inherent toxicity of the nanomaterial |
|---|---|
| Size | Uptake, translocation, elimination |
| Shape | Uptake, clearance |
| Surface modification | Uptake, interaction with biomolecules |
| Charge | Uptake, interaction with biomolecules |
| Conductibility | Interference with transport processes or signaling |
| Surface area | Interaction with and binding of biomolecules |
| Solubility | Release of potentially toxic ions, translocation, elimination |
| Strength/biopersistance | Clearance |

Adapted from Haase, A., Tentschert, J., & Luch, A. (2012). Nanomaterials: A challenge for toxicological risk assessment? In: A. Luch (Ed.), *Molecular, clinical and environmental toxicology: Vol. 3: Environmental toxicology* (pp. 219–250). Basel: Springer Basel. https://doi.org/10.1007/978-3-7643-8340-4_8.

however, the physicochemical characteristics of the ENP need to be defined. The relationship between the physicochemical characteristics of ENPs and some toxicity and environmental consequences are provided in Table 3.

Until recently, it has been assumed that the use of nanoparticles in consumer and non-consumer goods and food contact materials is unlikely to cause significant exposure to consumers and the environment. Since nanoparticles have an effective radius larger than 1 nm and have theoretically calculated diffusion coefficients $\approx 10^{-19}$ $m^2$ $s^{-1}$, they are assumed to be permanently entrapped in the polymer matrix (Šimon, Chaudhry, & Bakoš, 2008). Šimon et al. (2008) concluded that the migration of nanoparticles, such as nanosilver and surface-modified MMT embedded in a different polymeric matrix, will not be detectable and quantifiable. However, diffusion may not be the only physical phenomenon underlying the migration of nanoparticles. Migration could take place once the materials are either in the landfill or even during recycling, which then could become the sources for environmental contamination.

Non-diffusive mechanisms for mass transfer of nanoparticles have been investigated in the context of flame-retardant polymer nanocomposites (Hao, Lewin, Wilkie, & Wang, 2006; Lewin & Tang, 2008; Tang & Lewin, 2008). Annealing maleated PP-clay nanocomposites at temperatures near the melting point of the polymer matrix and below the temperatures required for combustion of the polymer or the degradation temperature of the surfactant present in the interlayer galleries resulted in lowering the surface energy of the nanoparticles well below the surface energy of the matrix, thus leading to surface aggregation of the nanoparticles. At higher temperatures, decomposition reactions of the surfactant and compatibilizer lead to gas generation, which can carry the nanoparticles by convection. As a result, the clay particles tend to migrate to the surface, through a multi-step mechanism: (1) diffusion of oxygen into the melt; (2) oxidation of the PP matrix molecules producing polar groups; (3) intercalation of the polar molecules into the clay gallery; (4) exfoliation of the plates; and (5) migration of the exfoliated plates (Hao et al., 2006). The migration of clay to the surface of the composite samples was reported to increase with an increase in the percentage of MA grafted onto the PP. Decomposition of PP-*g*-MA led to $CO_2$ emission and then the gas flow through the polymer carrying the particles to the surface. Furthermore, the rate of migration of clay is dependent on the mobility of the polymeric chains, $M_n$ of the polymer, aspect ratio and orientation of the platelets, and the molecular structure of the polymer (Hao et al., 2006; Lewin, 2006).

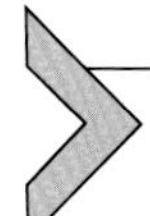

# 5. Mass transfer of nanoparticles in packaging

## 5.1 Theoretical background

In food packaging systems, the driving force for migration is the chemical potential (i.e., concentration) difference between the polymer and the food in contact with the polymer. Migration of a substance occurs from the high concentration side (polymer) to the low concentration side (food), so as to balance the chemical potential difference between the two sides (Sonchaeng et al., 2018). However, the chemical potential difference is not the only prerequisite for the migration of a substance. Polymers usually consist of crystalline and amorphous domains. The crystalline domain is considered non-permeable since the polymer chains are tightly packed. On the other hand, migration takes place in the amorphous region since the polymer chains are more randomly and loosely packed, creating "space" for the movement of substances in the polymer. The term "space" here has two meanings. First, the space can be referred to free volumes not occupied by the polymer molecules, which vary in size and shape and are mutually connected. Second, the space can be considered as the temporary channels formed by the movement of polymer segments due to chain relaxation, so small free volume cavities have the probability to conjoin a large cavity to accommodate the migrating substance. The substance can migrate in the temporary channels, moving from one cavity to another, and eventually leaving the surface of the polymer.

The migration behavior of small molecules (e.g., additives, monomers, and oligomers) from polymers has been well investigated. These substances are small in size (around 1 nm), which is comparable to or smaller than the size of free volume in the polymer. Accordingly, their movement in the polymer matrix is relatively easy. On the other hand, the movements of the relatively larger nanoparticles, with size ranging from tens to hundreds of nm, in the polymer matrix are more restricted.

Since nanoparticles are different from small molecules, the mathematical models (i.e., Fick's second law of diffusion and its numerical solutions) used to describe the migration of small molecules may not be optimal to describe the migration process for nanoparticles. By far, the most frequently used mathematical model to describe the movement (or diffusion) of nanoparticles in polymers is the Stokes–Einstein (SE) equation (Zwanzig & Harrison, 1985):

$$D = \frac{k_B T}{6\pi\eta R} \quad (1)$$

where the diffusion of nanoparticles (as indicated by the diffusion coefficient $D$) depends on the particle size (in term of radius $R$), polymer viscosity ($\eta$), and temperature ($T$). Šimon et al. (2008) modeled the diffusion of nanoparticles in polymers using the SE equation, it was found that the calculated $D$ values for nanoparticles were extremely small, which were several orders of magnitude lower than those for small molecules. This finding suggests that the migration of nanoparticles from polymers is negligible within a short time scale (i.e., 1 year or less).

Although the SE equation is useful for predicting the diffusion of nanoparticles in polymers, the prediction may not be always correct. For nanoparticle diffusion in polymer melts, several studies (Grabowski & Mukhopadhyay, 2014; Tuteja, Duxbury, & Mackay, 2007) reported measured $D$ values of up to two orders of magnitude higher than those predicted from the SE equation. The underestimated diffusion of nanoparticles was attributed to the lower local viscosity of the polymer in the vicinity of the nanoparticles than that of the polymer bulk. A question arises as of how small the nanoparticles should be in order to arrest their migration? The answer could be revealed by the relationship between the particle size and various polymer characteristic length scales, including (descending order) coil radius ($R_g$), entanglement spacing ($d_t$), and correlation length ($\xi$, close to monomer size) (Brochard Wyart & de Gennes, 2000; Cai, Panyukov, & Rubinstein, 2011). The breakdown of the SE relation starts at $R < R_g$, when a higher $D$ value may be expected than the predicted one. When $R < \xi$, nanoparticles are not affected by the polymer chains and can move freely in the polymer matrix. When $\xi < R < d_t$, the movement of nanoparticles requires the rearrangement of polymer segments of scale comparable with the nanoparticle size—a scenario similar to the movement of small molecules in the polymer matrix. When $R > d_t$, the movement of nanoparticles requires the removal of entire polymer chains, which could happen in a polymer melt, but not for packaging polymers under typical end-use conditions.

In summary, size is an important factor that determines the migration of nanoparticles in the polymer matrix. It should be noted that since both $\xi$ ($<1$ nm) and $d_t$ (several nm) are very small in magnitude, the size of ENPs used in packaging should be in the range of $R > d_t$, making their migration unlikely to happen in the polymer matrix. However, it is worthwhile to further carry out experiments to better understand the mass transfer of ENPs in polymers.

## 5.2 Migration of nanoparticles

Over the past few decades, there have been debates on whether nanoparticles can migrate from a polymer matrix and release from its surface. Although experimental studies have been carried out to investigate the migration of nanoparticles under controlled laboratory conditions (Diaz et al., 2013), challenges remain due to the lack of tools and techniques for reliable detection of nanoparticles in different media. Currently, the detection of nanoparticles is often relied on the measurement of elements as markers of the nanoparticles. Advanced microscopies (i.e., scanning electron microscopy [SEM] and transmission electron microscopy [TEM]) are also applied to provide direct observation of nanoparticles present in food simulants. Based on the findings reported in the literature, nanoparticles migrate from polymers into food simulants (solvents) mainly *via* three different routes (Fig. 7). In the first route, nanoparticles are partially dissolved in the solvent, releasing ions into the solvent. Although the ion release is not equivalent to the nanoparticle migration, the released ions can represent a safety concern, especially if they are heavy metals. In the second route, the nanoparticles migrate from the polymer into the solvent, which can

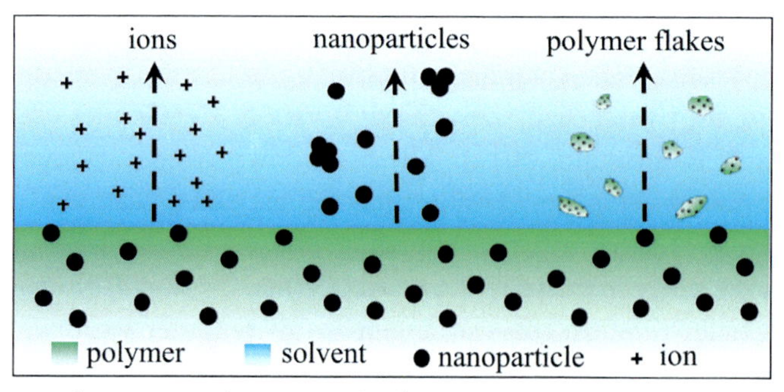


**Fig. 7** Routes of migration of nanoparticles from a polymer into a food simulant liquid.

be observed directly by SEM and TEM techniques, as individual particles or aggregates. In the third route, nanoparticles embedded in polymer may detach from the polymer surface as flakes that are vary in size and shape. Here, the nanoparticles in polymer flake have the same structure and morphology as those present in the bulk polymer. Conceivably, the migration of nanoparticles can also take place *via* a combination of different routes, which will complicate the mass transport analysis.

The migration of nanoparticles could be affected by many factors. For nanoclays (Table 4), the amount of migration varies with the type of polymer, type of food or food simulant, and the testing conditions. Whatever the migration route, the amounts of nanoparticles detected, based on the values reported in the literature, are always very small, at the levels of μg/L, μg/kg, or μg/dm$^2$ (Duncan & Pillai, 2015; Froggett, Clancy, Boverhof, & Canady, 2014; Pillai et al., 2016; Störmer, Bott, Kemmer, & Franz, 2017). However, a large amount of migration has been recorded for the surfactant used in modifying the nanoclays. The migration of the surfactant from polymer-clay nanocomposites can cause changes in the nanoclay structure, thereby affecting the mobility of the filler in the polymer (Xia et al., 2014).

On the basis of the published results, it is fair to deduct that the migration of nanoparticles is unlikely to take place from the bulk polymer. The small amount of nanoparticles found in the food simulant fluid is assumed to be originated from the nanocomposite surfaces, i.e., the migration of nanoparticles can be considered as a surface phenomenon. This hypothesis is in accordance with Simon's theory (Šimon et al., 2008) on the migration of nanoparticles, where the average travel distance ($r$) of nanoparticles is calculated by the equation:

$$r = \left(\frac{2k_B T t}{3\pi^2 \eta R}\right)^{1/2} \tag{2}$$

**Table 4** Migration of nanoclays from nanocomposite packaging materials.

| Polymer | wt% added | Food or food simulant | Conditions | Max. migration | References |
|---|---|---|---|---|---|
| PLA | 5 | 95% ethanol | 40°C, 10 d | ND | Schmidt et al. (2009) |
| LDPE | 2, 4, 6 | Surfactant solution | 70°C, 2 h 100°C, 15 min | ND | Bott and Franz (2018) |
| PE | — | 3% acetic acid, 10% ethanol | 40°C, 10 d 70°C, 2 h | 5.16 μg/dm$^2$ ($Al^{3+}$) | Echegoyen, Rodríguez, and Nerín (2016) |
| PET | 3 | 3% acetic acid | 45°C, 90 d | 0.34 mg/kg ($Al^{3+}$) 9.5 mg/kg ($Si^{2+}$) | Farhoodi, Mousavi, Sotudeh-Gharebagh, Emam-Djomeh, and Oromiehie (2013) |
| Starch | 4 | Lettuce | 40 C, 10 d | 0.19 mg/kg ($Si^{2+}$) | Avella et al. (2005) |
| Wheat gluten | 5 | Water, 3% acetic acid, 15% ethanol, olive oil | 40 C, 10 d | 1 mg/kg ($Al^{3+}$) 4.5 mg/kg ($Si^{2+}$) | Mauricio-Iglesias, Peyron, Guillard, and Gontard (2010) |
| PP & nylon | PP (3) nylon (5) | Ethanol | 70 C, 10 d | PP: 6 μg/dm$^2$ (clay) nylon: 3.5 μg/dm$^2$ (clay) | Xia et al. (2014) |

Consider LDPE, the equation predicted that there is a decreasing trend of travel distance of nanoparticles with increasing particle size (Fig. 8). For the ENPs commonly used in packaging, the travel distance (after 30 days) should be no more than 1000 nm (corresponding to a particle size of 20 nm). Since nanoclay particles are hundreds of nm in size, the travel distance (after 30 days) would further reduce to below 500 nm. These travel distance values are very small as compared to the typical thickness (e.g., 50–100 μm) of packaging materials. For other polymers with a viscosity higher than that of LDPE, the travel distance would be even smaller. Therefore, it can be assumed that only nanoparticles on the polymer surface can migrate into the food simulant.

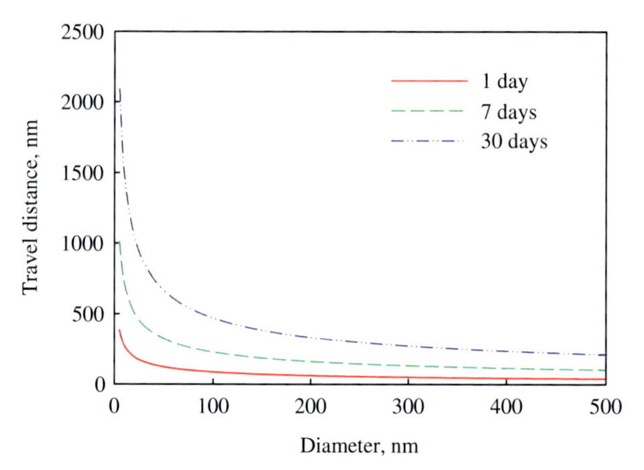


**Fig. 8** Travel distance of nanoparticles in LDPE as a function of particle size (diameter) after different days at 25°C (LDPE viscosity $6.6 \times 10^4$ Pa.s; Šimon et al., 2008).

In summary, experimental studies indicate that the migration of nanoparticles happens at the polymer surface but not in the bulk polymer. Such behavior is different from the migration of small molecules. For this reason, it may be more appropriate to use the word "release" instead of "migration" when describing the mass transfer of nanoparticles in nanocomposite systems. To better understand the mass transfer of nanoparticles, a careful investigation is needed to look at the interactions among the polymer, the nanoparticle, and the solvent (food or food simulant). Polymer nanoparticle interaction will affect the compatibility of nanoparticles with the polymer, and therefore the mobility of nanoparticles in the polymer. Solvent-polymer interaction may induce surface swelling to facilitate the release of nanoparticles from the polymer surface. In addition, the penetration depth of solvent could influence the dissolution of nanoparticles in the polymer and their subsequent release. Finally, solvent-nanoparticle interaction will dictate the solubility of nanoparticles in the solvent and therefore the amount of the filler released.

## 6. Effects of nanoclays in compost environments

Although nanoparticles are naturally occurring in the environment and they have been used for many centuries. There are many possible entry points of ENPs to the environment. Additionally, ENPs can undergo dissolution, agglomeration, sedimentation, and change of surface activities in the environment. Therefore, risk assessments of ENPs have been based on experimental ecosystem models. For packaging nanocomposite materials, after their useful life, ENMs can enter the environment *via* different levels of the waste management hierarchy, such as during composting, energy recovery, and landfill disposal.

Composting is increasingly being sought as a waste management technique for contaminated organic products (Kale et al., 2007; Kijchavengkul & Auras, 2008). There is an increasing use of ENMs for compostable packaging materials, which can pose a greater risk for ENPs to leach into the compost soil and ultimately absorb by plants. To date, most of the nanotoxicity studies have been conducted on germination of plants, plant cell cultures, and plants for human consumption. However, the effects of ENPs on compost microorganisms and plants are largely unknown. Since plant cell walls have pore sizes of around 5 nm or less, ENPs smaller than 5 nm could in theory penetrate and translocate in plant cells.

Composting is predominantly an aerobic process in which biodegradable materials, such as manure and leaves, are decomposed and transformed by microorganisms into a humus-like substance called compost, as well as $CO_2$, water, and minerals, through a controlled biological process. In a composting environment, compostable packages are biodegraded and can be finally assimilated to added soil, including any component ENPs. In industrial composting processes, mixed wastes, including compostable packaging materials, are combined in an optimal carbon: nitrogen ratio to advance through the biodegradation process. At the end of the process, compost is produced and sold for soil amendment/enrichment.

ENM-based packaging materials must be certified as compostable if they are meant to be disposed in the compost stream. Compostable packaging is certified by certifying bodies such as the Biodegradable Product Institute (BPI) in the United States or Vinçotte in Belgium. Certification processes for these compostable materials are based on specifications such as ASTM D6400-12 (Standard Specification for Labeling of Plastics Designed to be Aerobically Composted in Municipal or Industrial Facilities) or EU directive EN13432 (Packaging: Requirements For Packaging Recoverable Through Composting and Biodegradation) (Kijchavengkul & Auras, 2008). It is important to note that none of these standards or certifying bodies at the time of this review have procedures to test or to certify ENMs and/or ENPs.

Extensive research has been conducted to study the effect of nanoclays on the biodegradation of compostable polymers, such as poly (3-hydroxybutyrate-*co*-3-hydroxyvalerate), polycaprolactone, PLA, and thermoplastic starch (Castro-Aguirre et al., 2018). Reportedly, these nanocomposites biodegrade more rapidly than their pristine polymer. So, the incorporation of nanoclays into a biodegradable polymer matrix may be promising not only for enhancing the polymer performance but also for improving the biodegradation rate of these nanocomposites in composting and/or environmental conditions. However, the effects of ENPs on the compost and the compost environment have not yet been adequately addressed. Castro-Aguirre et al. (2018) conducted a study to evaluate the effect of organo-modified MMT (OMMT), HY nanotubes (HNT), and LP in compost environments, and demonstrated that the presence of HNT in the compost environment can inhibit the evolution of $CO_2$ (Fig. 9), indicating a possible obstruction effect by HY on the growth of microorganisms in the compost.

In the same composting study, the researchers also studied the effect of OMMT and the compatibilizer surfactant for OMMT (i.e., Tomamine™ Q-T-2 with 60–70% purity of a methyl tallow bis-2-hydroxyethyl quaternary ammonium chloride; QAC) (Castro-Aguirre et al., 2018). When OMMT or QAC was added to the compost, the $CO_2$ evolution rates were always significantly lower than the controls, indicating that there was an inhibition of the microbial activity when these materials were present (Fig. 10). However, the researchers did not observe the same effect when testing the respective nanocomposites. The exact mechanisms remain elusive; further studies are needed to determine whether the presence of nanoclays can affect the compost environments.

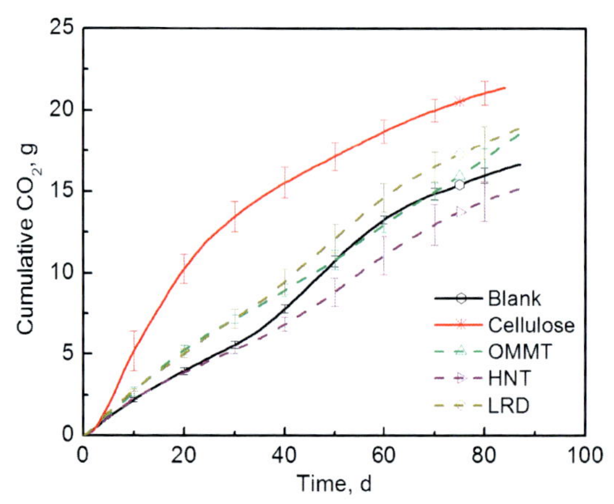


**Fig. 9** $CO_2$ evolution of three different nanoclays in a compost environment. *Reproduced from Castro-Aguirre, E., Auras, R., Selke, S., Rubino, M., & Marsh, T. (2018). Impact of nanoclays on the biodegradation of poly(lactic acid) nanocomposites.* Polymers, 10*(2), 202. https://doi.org/10.3390/polym10020202.*

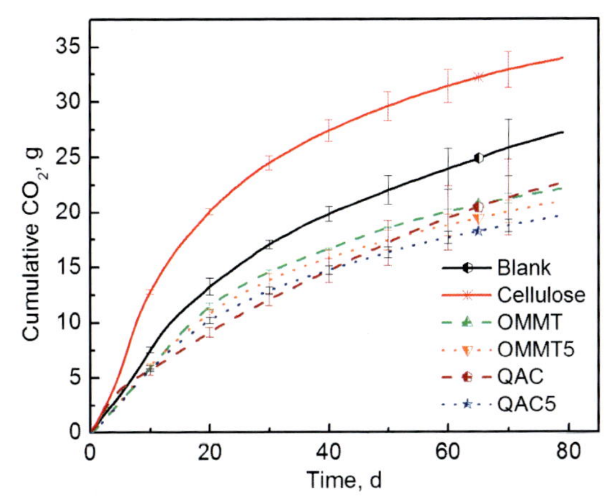


**Fig. 10** Cumulative $CO_2$ evolution vs time in days for bioreactors loaded with cellulose, OMMT, and QAC. QAC refers to a bioreactor loaded with 8g and QAC5 refers to a bioreactor loaded with the same amount that should be used for PLA films containing nanoclay. *Reproduced from Castro-Aguirre, E., Auras, R., Selke, S., Rubino, M., & Marsh, T. (2018). Impact of nanoclays on the biodegradation of poly(lactic acid) nanocomposites.* Polymers, 10*(2), 202. https://doi.org/10.3390/polym10020202.*

## 7. Final remarks

ENMs are increasingly being used in consumer and non-consumer goods, including packaging. In particular, nanoclays have been extensively explored as ENPs in nanocomposites to enhance their barrier and mechanical properties. However, these nanoparticles may contribute to safety issues when their size is below 5 nm, which can translocate in living tissues. Furthermore, to facilitate the dispersion into polymer matrices, many ENPs are modified with organic compounds, some of which have been shown to be toxic. To mitigate the potential nanotoxicity of ENPs to humans and the environment, research should be directed to better understand and assess the transport, fate, and effects of ENPs in food, aquatic systems, and soils. Although the focus of this overview is on nanoclays, there are several other ENPs used in consumer and non-consumer goods. For example, nanoscale metals are used for their antimicrobial activity (e.g., silver nanoparticles, zinc oxide) (Duncan, 2011; Egger, Lehmann, Height, Loessner, & Schuppler, 2009; Radheshkumar & Münstedt, 2006; Tang & Dong, 2009), UV resistance (e.g., titanium dioxide) (Shun-Xing, Feng-Ying, Wen-Lian, Ai-Qin, & Yu-Kun, 2006), and heat stability (e.g., titanium nitride) (Polyakova & Hübert, 2001). Cellulose-based nano-reinforcements are used to increase strength and reduce weight of polymers (Azeredo, 2009), and carbon nanotubes are used to enhance thermal, mechanical, and electrical properties of polymers (Bal & Samal, 2007; Lewin & Tang, 2008). Nanosensors are ENPs that respond to environmental changes (e.g., temperature, humidity, or oxygen exposure) (Anderson, Torres-Chavolla, Castro, & Alocilja, 2011; Yuk, Rose, & Alocilja, 2010). These nanotechnologies may be promising for active and intelligent packaging applications, which should also be further investigated in the future.

## References


Anderson, M. J., Torres-Chavolla, E., Castro, B. A., & Alocilja, E. C. (2011). One step alkaline synthesis of biocompatible gold nanoparticles using dextrin as capping agent. *Journal of Nanoparticle Research*, *13*(7), 2843–2851. https://doi.org/10.1007/s11051-010-0172-3.

Avella, M., De Vlieger, J. J., Errico, M. E., Fischer, S., Vacca, P., & Volpe, M. G. (2005). Biodegradable starch/clay nanocomposite films for food packaging applications. *Food Chemistry*, *93*(3), 467–474. https://doi.org/10.1016/J.FOODCHEM.2004.10.024.

Azeredo, H. M. C. d. (2009). Nanocomposites for food packaging applications. *Food Research International*, *42*(9), 1240–1253. https://doi.org/10.1016/J.FOODRES.2009.03.019.

Bal, S., & Samal, S. S. (2007). Carbon nanotube reinforced polymer composites—A state of the art. *Bulletin of Materials Science*, *30*(4), 379. https://doi.org/10.1007/s12034-007-0061-2.

Bott, J., & Franz, R. (2018). Investigation into the potential migration of nanoparticles from laponite-polymer nanocomposites. *Nanomaterials*, *8*. https://doi.org/10.3390/nano8090723.

Bourgeat-Lami, E., & Lang, J. (2000). Silica-polystyrene composite particles. *Macromolecular Symposia*, *151*, 377–385. https://doi.org/10.1002/1521-3900(200002)151:1<377::AID-MASY377>3.0.CO;2-H.

Brochard Wyart, F., & de Gennes, P. G. (2000). Viscosity at small scales in polymer melts. *The European Physical Journal E*, *1*(1), 93–97. https://doi.org/10.1007/s101890050011.

Cai, L.-H., Panyukov, S., & Rubinstein, M. (2011). Mobility of nonsticky nanoparticles in polymer liquids. *Macromolecules*, *44*(19), 7853–7863. https://doi.org/10.1021/ma201583q.

Calcagno, C. I. W., Mariani, C. M., Teixeira, S. R., & Mauler, R. S. (2007). The effect of organic modifier of the clay on morphology and crystallization properties of PET nanocomposites. *Polymer*, *48*(4), 966–974. https://doi.org/10.1016/J.POLYMER.2006.12.044.

Castro-Aguirre, E., Auras, R., Selke, S., Rubino, M., & Marsh, T. (2018). Impact of nanoclays on the biodegradation of poly(lactic acid) nanocomposites. *Polymers*, *10*(2), 202. https://doi.org/10.3390/polym10020202.

Choudalakis, G., & Gotsis, A. D. (2009). Permeability of polymer/clay nanocomposites: A review. *European Polymer Journal*, *45*(4), 967–984. https://doi.org/10.1016/J.EURPOLYMJ.2009.01.027.

Diaz, C. A., Xia, Y., Rubino, M., Auras, R., Jayaraman, K., & Hotchkiss, J. (2013). Fluorescent labeling and tracking of nanoclay. *Nanoscale*, *5*(1), 164–168. https://doi.org/10.1039/c2nr32978f.

Duncan, T. V. (2011). Applications of nanotechnology in food packaging and food safety: Barrier materials, antimicrobials and sensors. *Journal of Colloid and Interface Science*, *363*(1), 1–24.

Duncan, T. V., & Pillai, K. (2015). Release of engineered nanomaterials from polymer nanocomposites: Diffusion, dissolution, and desorption. *ACS Applied Materials & Interfaces*, 7(1), 2–19. https://doi.org/10.1021/am5062745.

Echegoyen, Y., Rodríguez, S., & Nerín, C. (2016). Nanoclay migration from food packaging materials. *Food Additives & Contaminants: Part A*, *33*(3), 530–539. https://doi.org/10.1080/19440049.2015.1136844.

Egger, S., Lehmann, R. P., Height, M. J., Loessner, M. J., & Schuppler, M. (2009). Antimicrobial properties of a novel silver-silica nanocomposite material. *Applied and Environmental Microbiology*, *75*(9), 2973 LP–2976.

Elementis_Specialist_Inc. (2011). *Comparison of smectite clays in underarm products*. Available at: https://www.elementisspecialties.com/esweb/webprodliterature.nsf/9edb597806ad101c852575fb004a6987/ee53701122eb5210852576400062 4605/$FILE/Comparison%20of%20Smectite%20Clays%20in%20Underarm%20Products.pdf. accessed 02/24/2019.

Farhoodi, M., Mousavi, S. M., Sotudeh-Gharebagh, R., Emam-Djomeh, Z., & Oromiehie, A. (2013). Migration of aluminum and silicon from PET/clay nanocomposite bottles into acidic food simulant. *Packaging Technology and Science*, *27*(2), 161–168. https://doi.org/10.1002/pts.2017.

Flaris, V. (2005). Talc. In M. Xanthos (Ed.), In functional fillers for plastics. Weinheim: Wiley-VCH Verlag GmbH &Co. KGaA. https://doi.org/10.1002/3527605096.ch12.

Froggett, S. J., Clancy, S. F., Boverhof, D. R., & Canady, R. A. (2014). A review and perspective of existing research on the release of nanomaterials from solid nanocomposites. *Particle and Fibre Toxicology*, *11*(1), 17. https://doi.org/10.1186/1743-8977-11-17.

Grabowski, C. A., & Mukhopadhyay, A. (2014). Size effect of nanoparticle diffusion in a polymer melt. *Macromolecules*, *47*(20), 7238–7242. https://doi.org/10.1021/ma501670u.

Haase, A., Tentschert, J., & Luch, A. (2012). Nanomaterials: A challenge for toxicological risk assessment? In A. Luch (Ed.), Molecular, clinical and environmental toxicology: Volume 3: Environmental toxicology (pp. 219–250). Basel: Springer Basel. https://doi.org/10.1007/978-3-7643-8340-4_8.

Hao, J., Lewin, M., Wilkie, C. A., & Wang, J. (2006). Additional evidence for the migration of clay upon heating of clay–polypropylene nanocomposites from X-ray photoelectron spectroscopy (XPS). *Polymer Degradation and Stability*, *91*(10), 2482–2485. https://doi.org/10.1016/J.POLYMDEGRADSTAB.2006.03.023.

Hashemifard, S. A., Ismail, A. F., & Matsuura, T. (2011). Mixed matrix membrane incorporated with large pore size halloysite nanotubes (HNT) as filler for gas separation: Experimental. *Journal of Colloid and Interface Science*, *359*(2), 359–370. https://doi.org/10.1016/J.JCIS.2011.03.077.

Hedicke-Höchstötter, K., Lim, G. T., & Altstädt, V. (2009). Novel polyamide nanocomposites based on silicate nanotubes of the mineral halloysite. *Composites Science and Technology*, *69*(3–4), 330–334. https://doi.org/10.1016/J.COMPSCITECH.2008.10.011.

Herrera, N. N., Letoffe, J.-M., Reymond, J.-P., & Bourgeat-Lami, E. (2005). Silylation of laponite clay particles with monofunctional and trifunctional vinyl alkoxysilanes. *Journal of Materials Chemistry*, *15*(8), 863–871. https://doi.org/10.1039/B415618H.

Horie, M., Nishio, K., Fujita, K., Endoh, S., Miyauchi, A., Saito, Y., et al. (2009). Protein adsorption of ultrafine metal oxide and its influence on cytotoxicity toward cultured cells. *Chemical Research in Toxicology*, *22*(3), 543–553. https://doi.org/10.1021/tx800289z.

Hristozov, D., & Malsch, I. (2009). Hazards and risks of engineered nanoparticles for the environment and human health. *Sustainability*, *1*(4), 1161–1194. https://doi.org/10.3390/su1041161.

Industrial_Minerals_GmbH. (2017). Altonit®-bentonite products for the paper industry. Available at: https://www.imerys-performance-additives.com/your-market/paper/altonitr. accessed date: 02/24/18.

Jayaraman, K., & Kumar, S. (2006). Chapter 4—Polypropylene layered silicate nanocomposites. In Y.-W. Mai & Z.-Z. Yu (Eds.), *Polymer Nanocomposites* (pp. 130–150). https://doi.org/10.1533/9781845691127.1.130.

Jia, X., Li, Y., Zhang, B., Cheng, Q., & Zhang, S. (2008). Preparation of poly(vinyl alcohol)/kaolinite nanocomposites via in situ polymerization. *Materials Research Bulletin*, *43*(3), 611–617. https://doi.org/10.1016/J.MATERRESBULL.2007.04.008.

Jordan, J., Jacob, K. I., Tannenbaum, R., Sharaf, M. A., & Jasiuk, I. (2005). Experimental trends in polymer nanocomposites—A review. *Materials Science and Engineering A*, *393*(1–2), 1–11. https://doi.org/10.1016/J.MSEA.2004.09.044.

Joussein, E., Petit, S., Churcham, J., Theng, B., Righi, D., & Delvaux, B. (2005). Halloysite clay minerals—A review. *Clay Minerals*, *40*(4), 383–426.

Kale, G., Kijchavengkul, T., Auras, R., Rubino, M., Selke, S. E., & Singh, S. P. (2007). Compostability of bioplastic packaging materials: An overview. *Macromolecular Bioscience*, *7*(3), 255–277. https://doi.org/10.1002/mabi.200600168.

Khot, M., Kamat, S., Zinjarde, S., Pant, A., Chopade, B., & RaviKumar, A. (2012). Single cell oil of oleaginous fungi from the tropical mangrove wetlands as a potential feedstock for biodiesel. *Microbial Cell Factories*, *11*(1), 71. https://doi.org/10.1186/1475-2859-11-71.

Kijchavengkul, T., & Auras, R. (2008). Compostability of polymers. *Polymer International*, *57*(6), 793–804. https://doi.org/10.1002/pi.2420.

Lagaron, J. M., & Lopez-Rubio, A. (2011). Nanotechnology for bioplastics: Opportunities, challenges and strategies. *Trends in Food Science & Technology*, *22*(11), 611–617. https://doi.org/10.1016/J.TIFS.2011.01.007.

Lee, J.-H., Jung, D., Hong, C.-E., Rhee, K. Y., & Advani, S. G. (2005). Properties of polyethylene-layered silicate nanocomposites prepared by melt intercalation with a PP-g-MA compatibilizer. *Composites Science and Technology*, *65*(13), 1996–2002. https://doi.org/10.1016/J.COMPSCITECH.2005.03.015.

Lewin, M. (2006). Reflections on migration of clay and structural changes in nanocomposites. *Polymers for Advanced Technologies*, *17*(9–10), 758–763. https://doi.org/10.1002/pat.762.

Lewin, M., & Tang, Y. (2008). Oxidation – migration cycle in polypropylene-based nanocomposites. *Macromolecules*, *41*(1), 13–17. https://doi.org/10.1021/ma702094e.

Loiseau, A., & Tassin, J.-F. (2006). Model nanocomposites based on laponite and poly(ethylene oxide): Preparation and rheology. *Macromolecules*, *39*(26), 9185–9191. https://doi.org/10.1021/ma061324w.

Maurer-Jones, M. A., Gunsolus, I. L., Murphy, C. J., & Haynes, C. L. (2013). Toxicity of engineered nanoparticles in the environment. *Analytical Chemistry*, *85*(6), 3036–3049. https://doi.org/10.1021/ac303636s.

Mauricio-Iglesias, M., Peyron, S., Guillard, V., & Gontard, N. (2010). Wheat gluten nanocomposite films as food-contact materials: Migration tests and impact of a novel food stabilization technology (high pressure). *Journal of Applied Polymer Science*, *116*(5), 2526–2535. https://doi.org/10.1002/app.31647.

Mohanty, A. K., Vivekanandhan, S., Pin, J.-M., & Misra, M. (2018). Composites from renewable and sustainable resources: Challenges and innovations. *Science*, *362*(6414), 536–542.

Monica, R. C., & Cremonini, R. (2009). Nanoparticles and higher plants. *Caryologia*, *62*(2), 161–165. https://doi.org/10.1080/00087114.2004.10589681.

Morawiec, J., Pawlak, A., Slouf, M., Galeski, A., Piorkowska, E., & Krasnikowa, N. (2005). Preparation and properties of compatibilized LDPE/organo-modified montmorillonite nanocomposites. *European Polymer Journal*, *41*(5), 1115–1122. https://doi.org/10.1016/J.EURPOLYMJ.2004.11.011.

National Science and Technology Council. (2016). *National nanotechnology initiative strategic plan*. National Science and Technology Council.

Nikolaidis, A. K., Achilias, D. S., & Karayannidis, G. P. (2011). Synthesis and characterization of PMMA/organomodified montmorillonite nanocomposites prepared by in situ bulk polymerization. *Industrial & Engineering Chemistry Research*, *50*(2), 571–579. https://doi.org/10.1021/ie100186a.

Pereira de Abreu, D. A., Paseiro Losada, P., Angulo, I., & Cruz, J. M. (2007). Development of new polyolefin films with nanoclays for application in food packaging. *European Polymer Journal*, *43*(6), 2229–2243. https://doi.org/10.1016/J.EURPOLYMJ.2007.01.021.

Pillai, K. V., Gray, P. J., Tien, C.-C., Bleher, R., Sung, L.-P., & Duncan, T. V. (2016). Environmental release of core–shell semiconductor nanocrystals from free-standing polymer nanocomposite films. *Environmental Science: Nano*, *3*(3), 657–669. https://doi.org/10.1039/C6EN00064A.

Polyakova, I. G., & Hübert, T. (2001). Thermal stability of TiN thin films investigated by DTG/DTA. *Surface and Coatings Technology*, *141*(1), 55–61. https://doi.org/10.1016/S0257-8972(01)01042-8.

Quincoces, C. E., Basaldella, E. I., De Vargas, S. P., & González, M. G. (2004). Ni/γ-Al2O3 catalyst from kaolinite for the dry reforming of methane. *Materials Letters*, *58*(3–4), 272–275. https://doi.org/10.1016/S0167-577X(03)00468-3.

Radheshkumar, C., & Münstedt, H. (2006). Antimicrobial polymers from polypropylene/silver composites—Ag+ release measured by anode stripping voltammetry. *Reactive and Functional Polymers*, *66*(7), 780–788. https://doi.org/10.1016/J.REACTFUNCTPOLYM.2005.11.005.

Ray, S. S., & Okamoto, M. (2003). Polymer/layered silicate nanocomposites: A review from preparation to processing. *Progress in Materials Science*, *28*, 1539–1641.

Reignier, J., Tatibouët, J., & Gendron, R. (2006). Batch foaming of poly(ε-caprolactone) using carbon dioxide: Impact of crystallization on cell nucleation as probed by ultrasonic measurements. *Polymer*, *47*(14), 5012–5024. https://doi.org/10.1016/J.POLYMER.2006.05.040.

Röcker, C., Pötzl, M., Zhang, F., Parak, W. J., & Nienhaus, G. U. (2009). A quantitative fluorescence study of protein monolayer formation on colloidal nanoparticles. *Nature Nanotechnology*, *4*, 577.

Roco, M. C., Mirkin, C. A., & Hersam, M. C. (2010). *Nanotechnology research directions for societal needs in 2020 retrospective and outlook*. National Science Foundation.

Schmidt, B., Petersen, J. H., Bender Koch, C., Plackett, D., Johansen, N. R., Katiyar, V., et al. (2009). Combining asymmetrical flow field-flow fractionation with light-scattering and inductively coupled plasma mass spectrometric detection for characterization of nanoclay used in biopolymer nanocomposites. *Food Additives & Contaminants: Part A*, *26*(12), 1619–1627. https://doi.org/10.1080/02652030903225740.

Seaton, A., Tran, L., Aitken, R., & Donaldson, K. (2009). Nanoparticles, human health hazard and regulation. *Journal of the Royal Society Interface*, 7, S119–S129.

Shen, L., Phang, I. Y., Chen, L., Liu, T., & Zeng, K. (2004). Nanoindentation and morphological studies on nylon 66 nanocomposites. I. Effect of clay loading. *Polymer*, *45*(10), 3341–3349. https://doi.org/10.1016/J.POLYMER.2004.03.036.

Shun-Xing, L., Feng-Ying, Z., Wen-Lian, C., Ai-Qin, H., & Yu-Kun, X. (2006). Surface modification of nanometer size TiO2 with salicylic acid for photocatalytic degradation of 4-nitrophenol. *Journal of Hazardous Materials*, *135*(1–3), 431–436. https://doi.org/10.1016/J.JHAZMAT.2005.12.010.

Šimon, P., Chaudhry, Q., & Bakoš, D. (2008). Migration of engineered nanoparticles from polymer packaging to food—A physicochemical view. *Journal of Food and Nutrition Research*, *47*, 105–113. https://doi.org/10.1016/j.envres.2014.11.022.

Sonchaeng, U., Iñiguez-Franco, F., Auras, R., Selke, S., Rubino, M., & Lim, L.-T. (2018). Poly(lactic acid) mass transfer properties. *Progress in Polymer Science*, *86*, 85–121. https://doi.org/10.1016/J.PROGPOLYMSCI.2018.06.008.

Störmer, A., Bott, J., Kemmer, D., & Franz, R. (2017). Critical review of the migration potential of nanoparticles in food contact plastics. *Trends in Food Science & Technology*, *63*, 39–50. https://doi.org/10.1016/J.TIFS.2017.01.011.

Tang, E., & Dong, S. (2009). Preparation of styrene polymer/ZnO nanocomposite latex via miniemulsion polymerization and its antibacterial property. *Colloid and Polymer Science*, *287*(9), 1025–1032. https://doi.org/10.1007/s00396-009-2057-5.

Tang, Y., & Lewin, M. (2008). New aspects of migration and flame retardancy in polymer nanocomposites. *Polymer Degradation and Stability*, *93*(11), 1986–1995. https://doi.org/10.1016/J.POLYMDEGRADSTAB.2008.02.021.

Tuteja, A., Duxbury, P. M., & Mackay, M. E. (2007). Multifunctional nanocomposites with reduced viscosity. *Macromolecules*, *40*(26), 9427–9434. https://doi.org/10.1021/ma071313i.

Wu, T.-M., & Liao, C.-S. (2001). Polymorphism in nylon 6/clay nanocomposites. *Macromolecular Chemistry and Physics*, *201*(18), 2820–2825. https://doi.org/10.1002/1521-3935(20001201)201:18<2820::AID-MACP2820>3.0.CO;2-4.

Xia, Y., Rubino, M., & Auras, R. (2014). Release of nanoclay and surfactant from polymer–clay nanocomposites into a food simulant. *Environmental Science & Technology*, *48*(23), 13617–13624. https://doi.org/10.1021/es502622c.

Xia, Y., Uysal Unalan, I., Rubino, M., & Auras, R. (2017). Carbon nanotube release from polymers into a food simulant. *Environmental Pollution*, *229*, 818–826. https://doi.org/10.1016/J.ENVPOL.2017.06.067.

Yuk, J. S., Rose, J., & Alocilja, E. C. (2010). Characterization of polyaniline-coated magnetic nanoparticles for application in a disposable membrane strip biosensor. *The European Physical Journal Applied Physics*, *50*(1), 11401. https://doi.org/10.1051/epjap/2010023.

Zwanzig, R., & Harrison, A. K. (1985). Modifications of the Stokes–Einstein formula. *The Journal of Chemical Physics*, *83*(11), 5861–5862. https://doi.org/10.1063/1.449616.

# Is "nano safe to eat or not"? A review of the state-of-the art in soft engineered nanoparticle (sENP) formulation and delivery in foods

**Xiaobo Liu[a,†], Boce Zhang[a,*,†], Ikjot Singh Sohal[b], Dhimiter Bello[a,*], Hongda Chen[c,*]**

[a]Department of Biomedical and Nutritional Sciences, University of Massachusetts, Lowell, MA, United States
[b]Purdue University, Center for Cancer Research, West Lafayette, IN, United States
[c]U.S. Department of Agriculture, National Institute of Food and Agriculture, Washington DC, United States
*Corresponding authors: e-mail address: boce_zhang@uml.edu; dhimiter_bello@uml.edu; hchen@nifa.usda.gov

## Contents



† These authors contributed equally to this work.

*Advances in Food and Nutrition Research*, Volume 88
ISSN 1043-4526
https://doi.org/10.1016/bs.afnr.2019.03.004

## Abstract

With superior physicochemical properties, soft engineered nanoparticles (sENP) (protein, carbohydrate, lipids and other biomaterials) are widely used in foods. The preparation, functionalities, applications, transformations in gastrointestinal (GI) tract, and effects on gut microbiota of sENP directly incorporated for ingestion are reviewed herein. At the time of this review, there is no notable report of safety concerns of these nanomaterials found in the literature. Meanwhile, various beneficial effects have been demonstrated for the application of sENP. To address public perception and safety concerns of nanoscale materials in food, methodologies for evaluation of physiological effects of nanomaterials are reviewed. The combination of these complementary methods will be useful for the establishment of a comprehensive risk assessment system.

## 1. Introduction

The definition of "nanoparticles," relevant to food applications, refers to materials which has at least one dimensions that can be measured in the nanoscale (i.e., between 1 and 100 nm) (National Science and Technology Council Committee on Technology, 2016). The term "nanotechnology" usually refers to the manufacturing and utilization of nanoscale materials. In 1959, the vision of nanoscale science was first proposed by Richard Feynman, the 1965 Nobel Prize Laureate in physics, in his talk at the American Physics Society annual meeting at Caltech. The idea of self-assembly process at the nanoscale (bottom-up) rendering new properties of matters was articulated by Eric Drexler of the Massachusetts Institute of Technology in 1986 (Hulla, Sahu, & Hayes, 2015). In the United States and across the globe, the concept of manipulating matter at its nanometer scale, inspired by Feynman's 1959 speech, has profoundly influenced the scientific community research priorities. Advances in nanotechnology from the late 1980s to early 1990s prompted the formation of the National Nanotechnology Initiative (NNI) by the White House in 1990 and is managed by the National Science and Technology Council (NSTC). The NNI was also affirmed through legislative efforts in 2003 (National Science and Technology Council (NSTC), 2014).

The small size of nanoparticles endows them with remarkable physicochemical properties for numerous novel applications. Nanoparticles can be naturally presented in food and consumed. For example, casein micelles in milk are between 50 and 400 nm as the complexes of three subgroups of casein molecules ($\alpha_s$-casein, $\beta$-casein, and $\kappa$-casein). The phosphoric acid binds calcium and magnesium to form bonds between and within molecules. Casein micelles are polymers comprised of hundreds and thousands of individual molecules and form a colloidal solution, which has a whitish-blue appearance in skim milk. It is difficult today to find a field of science and technology that has not been impacted by developments at the nanoscale.

Engineered food nanoparticles result from routine food processing, such as homogenization, grinding, and others (Fellows, 2016; Gupta, Eral, Hatton, & Doyle, 2016). In addition to nanoparticles used in food packaging materials, nanoparticles specifically designed for incorporation into foods can be divided into two general categories according to their chemical composition: inorganic nanoparticles (e.g., metals, metal oxides, etc.) and soft engineered nanoparticles (sENP) (e.g., proteins, polysaccharides, lipids, etc.).

Nanoparticles have been employed to address challenges related to food quality, shelf life, cost, safety, and nutritional benefits. Fig. 1 illustrates the functionalities and safety assessments of edible nanoparticles. The chemical composition and physical microstructure are important factors that determine how nanoparticles interact in the gastrointestinal tract and their subsequent physiological effects (McClements & Xiao, 2017). The physiological effects of sENP are defined by their fate and transformation in the GI tract, impact on gut microbiome, and other potential effects to the body. In addition to the chemical composition, other factors such as size distribution, surface charge, surface area, surface contact, dissolution rate, and bioavailability are all essential when applying sENP in processed food and nutraceuticals. Fig. 1 illustrates the application matrix of nanoparticles in food.

While there are widespread applications of nanotechnology in ingredient manufacturing, printing, electronics, nanocomposites, catalysis, and energy storage, the acknowledgment of nanotechnologies in the food industry is relatively slow (Sozer & Kokini, 2009). Reasons for the slow uptake in the food industry pertain to public perception, subsequent labeling and potential regulatory burdens (Sozer & Kokini, 2009). In the early 2000s (ca.2005), the rapid development of nanotechnology was considered as a beneficial advance for science and technology, while the first series of nanotoxicology publications raised some safety concerns and subsequently shifted the public perception of nanotechnology to a more conservative

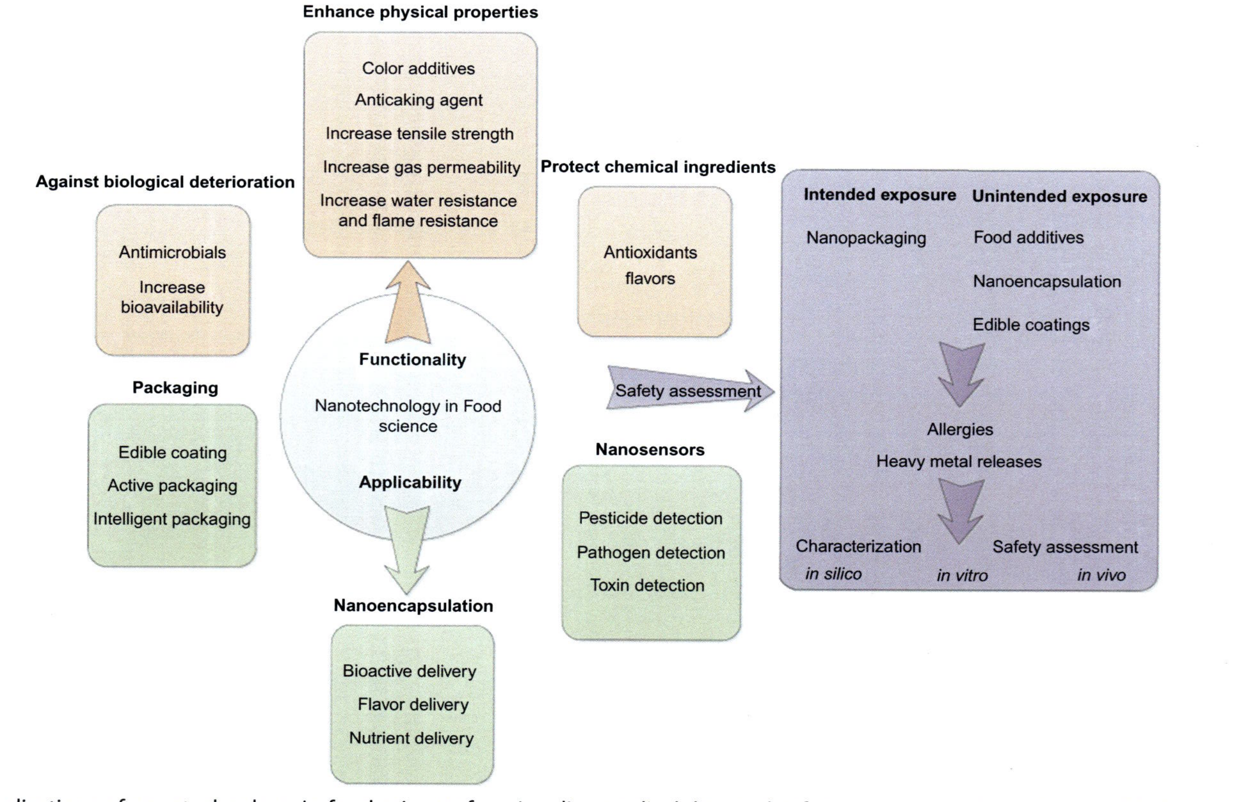


**Fig. 1** Applications of nanotechnology in food science: functionality, applicability, and safety assessments (He & Hwang, 2016).

attitude. The reluctant acceptance of employing nanotechnology in foods also reflected in the lack of transparency in the labeling of foods containing nano-ingredients.

Public perception of nanotechnologies is important in their consumer acceptance and commercialization. Statements such as "this product contains × nanoparticles" or "this product is manufactured using nanotechnologies" may raise public concerns, especially with sporadic reports about potential adverse effects of certain nanomaterials, especially inorganic engineered nanoparticles (Fröhlich & Fröhlich, 2016). However, such simplistic statements without reference to specific types of nanomaterials in question are not constructive in public engagement and may be misleading.

The methods and models developed to study acute and chronic physiological effects of nanomaterials are under development, especially for sENP in food. The chronic effects of nanoscale materials incorporated into products are especially difficult to characterize. Besides, it also demands high-level coordination and cooperation with risk assessment and regulatory actions. A series of four comprehensive reviews from the NanoRelease Food Additive project, which is an international joint effort from academia, industry, and government, address the gap in methods for assessment and evaluation of commercially used food nanomaterials (Alger, Momcilovic, Carlander, & Duncan, 2014; Noonan, Whelton, Carlander, & Duncan, 2014; Singh, Stephan, Westerhoff, Carlander, & Duncan, 2014; Yada et al., 2014). This series of reports analyzed the presence of engineered nanoparticles in food, summarized their methods of evaluation for engineered nanoparticles release from food matrices and uptake in GI tract, as well as methods to detect and quantify engineered nanoparticles in food. It should be noted that the engineered nanoparticles reviewed in these reports are primarily focused on metal or metal oxide nanoparticles. Some of the toxicity concerns and methods for nanoparticles detection, quantification, and evaluation of uptake may not apply to sENPs reviewed in this chapter.

The focus of this review is on the physiological effects on human health and wellbeing of sENP (e.g., proteins, polysaccharides, lipids, etc.) directly incorporated into foods. Fig. 2 presents a scheme for the potential fate of ingested sENP. Specifically, this review covers major types of sENP, with GRAS status (generally recognized as safe), from a chemical composition standpoint, including proteins, carbohydrates, lipids, and biopolymers. Biochemical transformations of sENP in GI tract and their physiological effects referred to as the fate of sENP in GI tract (such as uptake and

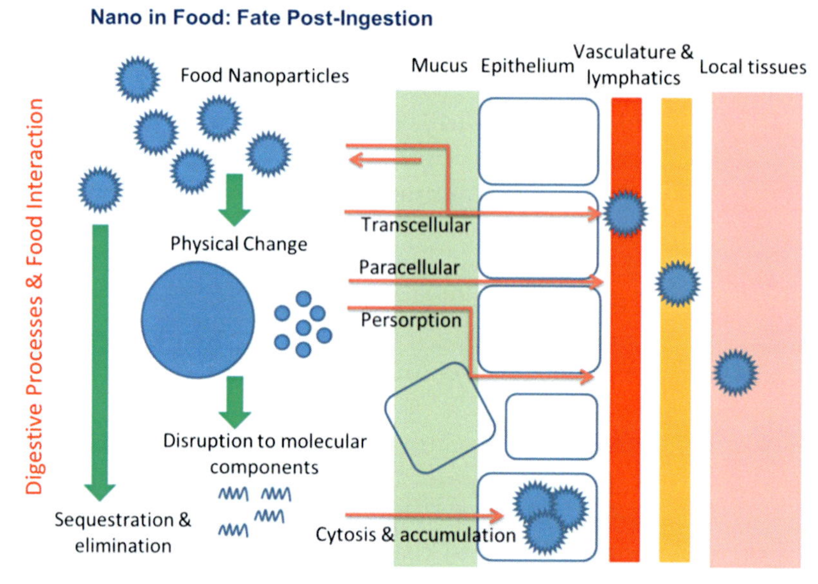


**Fig. 2** Potential fate of ingested sENP in the GI environment: sENP in the GI tract can undergo physiochemical and morphological changes within the lumen (cargo discharge, dissolution, disintegration, shrinking, re-agglomeration), and absorb *via* a number of potential routes (Yada et al., 2014).

dissolution) as well as their impact on the gut microbiome. The physiological characteristics of sENP reported in current literature involves release behaviors of protein nanoparticles in formulated media, such as buffered solutions or simulated GI solutions. The behavior of sENP in complex food matrices, as well as on the interaction of these particles with other food components and bioavailability are still underreported (Yada et al., 2014). This article summarizes existing methodologies for holistic assessment of overall physiological effects of sENP in food.

## 2. Soft engineered nanoparticles (sENP) in food

sENP in food refer to nanomaterials prepared from food ingredients, especially, lipids, proteins, carbohydrates, and other food-based biopolymers. sENP are widely found in prepared and processed foods. The physical state of sENP vary from liquid to semi-solid (such as hydrogels) or solid (crystalline or amorphous powder) based on their chemical composition and the matrix of the food, as well as processing and storage conditions. Spheres are the most commonly observed morphology for sENP, while non-spherical nanomaterials, such as nanofibers, also exist. Some nanoscale cellulosic fibers are explored as food packaging substrates and reinforcements to network scaffolds. sENP can go through drastically different fates (i.e., dissolution, aggregation, precipitation, and digestion) in the human GI (i.e., mouth, stomach, small intestines, and/or colon) tract due to unique physiochemical pathways based on the chemical and physical natures of the material. sENP in food are generally made of GRAS materials and no evidence of adverse health effects from consuming these materials has been reported to date. As a close analog of sENP in food, numerous drug delivery systems using food-based ingredients such as liposomes, nanocrystals, and emulsions, have been developed and approved by the US-FDA for pharmaceutical applications in the last two decades. As part of this evolving industry, 70% of all applications involving organic sENP, have been proven safe and effective, according to the Center for Drug Evaluation (D'Mello et al., 2017). In the following sections, different classes of sENP are discussed.

## 2.1 Proteins

Proteins are a vital building block and an energy-dense nutrient for human growth and function. The health significance, nutritional benefits, and versatile physicochemical functionality of proteins make them ideal candidates for development of edible protein-based nanomaterials. One of the main goals of protein sENP formulations is for oral delivery of bioactive protein-based medicines or nutraceuticals. Proteins have been used in delivery systems to replace polymeric carriers and to improve drug efficacy (Lohcharoenkal, Wang, Chen, & Rojanasakul, 2014). For example, albumin-bound paclitaxel (nab-paclitaxel, also known as Abraxane®) has been used to treat breast cancer (Hawkins, Soon-Shiong, & Desai, 2008). There have been no reported concerns on the safety of protein-based nanomaterials because the digestive pathways for protein sENP are not significantly different from the daily intake of bulk proteins from a variety of food sources. For nutraceutical delivery formulation, soy protein was used for the encapsulation of curcumin to achieve a very high (97.2%) encapsulation efficiency (Teng, Luo, & Wang, 2012). However, the study did not include further investigation on the health implications of the delivery system.

Nanoscale protein particulates are major constituents of many common food products and commodities. For instance, casein micelles are a natural component in milk and are utilized as a natural vehicle for water-insoluble or unstable nutrients through various processes (Livney, 2010). The primary benefits of using colloidal protein delivery systems include targeted delivery and enhanced nutrient uptake. Additionally, water-insoluble flavonoids or carotenoids, such as polymethoxyflavones (PMF) and β-carotene, also require colloidal delivery systems with advanced functionalities of controlled release. (Hans & Lowman, 2002; Weber, Coester, Kreuter, & Langer, 2000). For example, β-carotene is encapsulated in casein-graft-dextran nanoparticles using a green process that involves no hazardous chemicals. Whey protein isolate and sodium caseinate prepare a delivery system for β-carotene to provide better protection from degradation in the GI tract caused by drastic pH changes or digestive enzymes (Cornacchia & Roos, 2011; Pan, Yao, & Jiang, 2007). Aggregated protein molecules in protein sENP interact *via* various forces: weak van der Waals forces, hydrogen bonding, strong electrostatic attractions and covalent such as disulfide or isopeptide bond. The fabrication of protein nanoparticles involves controlled thermal denaturation, injection, molding and spray drying. These methods are common in preparation of other soft materials in food (Matalanis, Jones, & McClements, 2011). Applications of protein nanomaterials in food share many similarities to more common uses of proteins by the nutraceutical industry. For instance, β-carotene encapsulated by sodium caseinate, whey protein isolate and soybean protein isolate yielded nanostructures with diameters between 80 and 320 nm with improved antioxidant activities, as compared with free β-carotene (Yi, Lam, Yokoyama, Cheng, & Zhong, 2015). Zein/β-lactoglobulin colloidal systems provide targeted delivery of tangeretin, a flavonoid with anti-carcinogenic and anti-inflammatory activities (Chen, Zheng, McClements, & Xiao, 2014). Additional examples of nutrient delivery using protein nanoparticles include the encapsulation of ethyl hexanoate in whey protein nanoparticles (Giroux & Britten, 2011) and omega-3 fatty acids in casein nanoparticles (Zimet, Rosenberg, & Livney, 2011).

In general, pH, ionic strength, water activity, temperature, and presence of proteases are key factors influencing the stability of protein nanoparticles through the process of structural transformation and/or proteolytic hydrolysis. These key factors vary in different processes, such as thermal processing, chilling, freezing, drying and homogenization, and leads to protein nanostructures with potentially different stabilities and physiological fates. Unique physiochemical characteristics of proteins have significant impact on the stabilities and physiological fates of protein nanostructures, such as thermal denaturation temperatures, helix-coil transition temperatures, isoelectric points (pI), affinities and ligand stability constants ($K_{stab}$) of different ion-protein complexes, accessibility to cross-linking agents, and susceptibilities to proteolytic degradations (Matalanis et al., 2011; McClements, 2004).

Similarly, *in vitro* and *in vivo* studies investigating the fate and transport of protein nanoparticles in food, through the GI tract, are scarce, relating to the expectation that protein sENP behave similarly as bulk proteins. Another challenge is the selection of suitable, high throughput, cost-effective and predictive test systems. Fig. 3 illustrates a commonly used Caco-2 cell

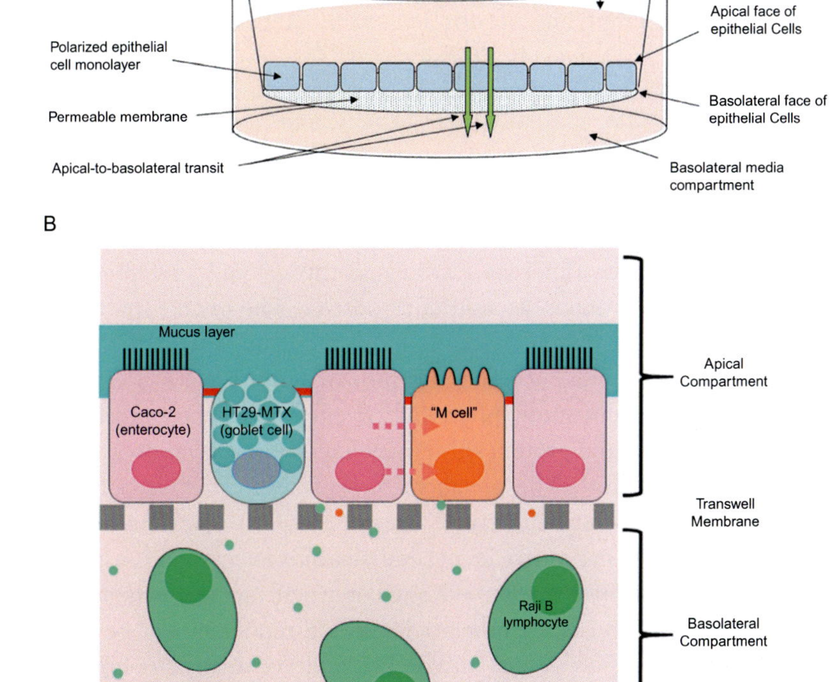


**Fig. 3** (A) Depiction of intestinal epithelial monolayer experiment (Caco-2 assay), which measures translocation of an analyte across a monolayer of cells with characteristics similar to the gut epithelium. Green arrows show possible unidirectional transit of the analyte across intestinal barriers (Alger et al., 2014). (B) A recently established triculture model that was more sensitive and representative to the GI environment (DeLoid et al., 2017).

*in vitro* model for sENP assessment and a recently established tri-culture model, which was more sensitive and representative to the real GI environment. One study demonstrated that protein nanoparticles, as a delivery system, showed very low immunogenicity and excellent biodegradability (Maghsoudi, Shojaosadati, & Vasheghani Farahani, 2008). Protein nanomaterials will undergo degradation in the GI due to the presence of different proteolytic enzymes and pH variations from the oral cavity to small intestines (Alger et al., 2014). Fig. 4 shows the interactions between proteins and proteolytic enzymes and surfactants in the human GI tract. Digestion of protein-based sENP in the GI, especially the small intestine, alter particle size and surface properties by inducing degradation or aggregation among sENP and other substances. Another important factor closely related with protein digestion, as well as other digestion of biomaterials, is gastric emptying. The gastric emptying rate for acid-stable emulsions is lower. Low gastric emptying rates may be a risk factor for several adverse health implications, including gastroparesis and obstructions (Marciani et al., 2008). Type II diabetic patients who administrated whey protein showed lower postprandial lipemia, which was attributed to the intake of whey protein (Mortensen et al., 2009). It is unclear at this time whether whey protein sENP have similar health benefits.

Gut microbiota has different catabolic pathways and preferences for carbohydrates and proteins as nutrient sources. The intake of protein nanoparticles and potentially with high protein and low carbohydrate diet have significant influences on the human gut microbiota (Portune et al., 2016). In a recent study, physiological effects of dietary proteins on the gut microbiota found protein sENP may have beneficial and/or deleterious effects on the host. The metabolites of protein catabolic pathways, which are yet to be fully elucidated, were believed to be important in this outcome. Protein nanoparticles in large quantities may significantly shift the nutrient-balance of the diet, although it is still unclear whether the amount and form of protein nanoparticles would render similar beneficial and/or deleterious effects on gut microbiota, and eventually the host itself (Portune et al., 2016).

Recent advancements in omics technologies may provide a simpler way of investigating physiological effects. For instance, genomics measures shifts in the microbiome populations, which represents fluctuation of microbiota in the gut in response to consumption of nanoparticles (Gokulan, Bekele, Drake, & Khare, 2018). Proteomics can be used to explore microbial catabolism and transformation of nanoparticles in the

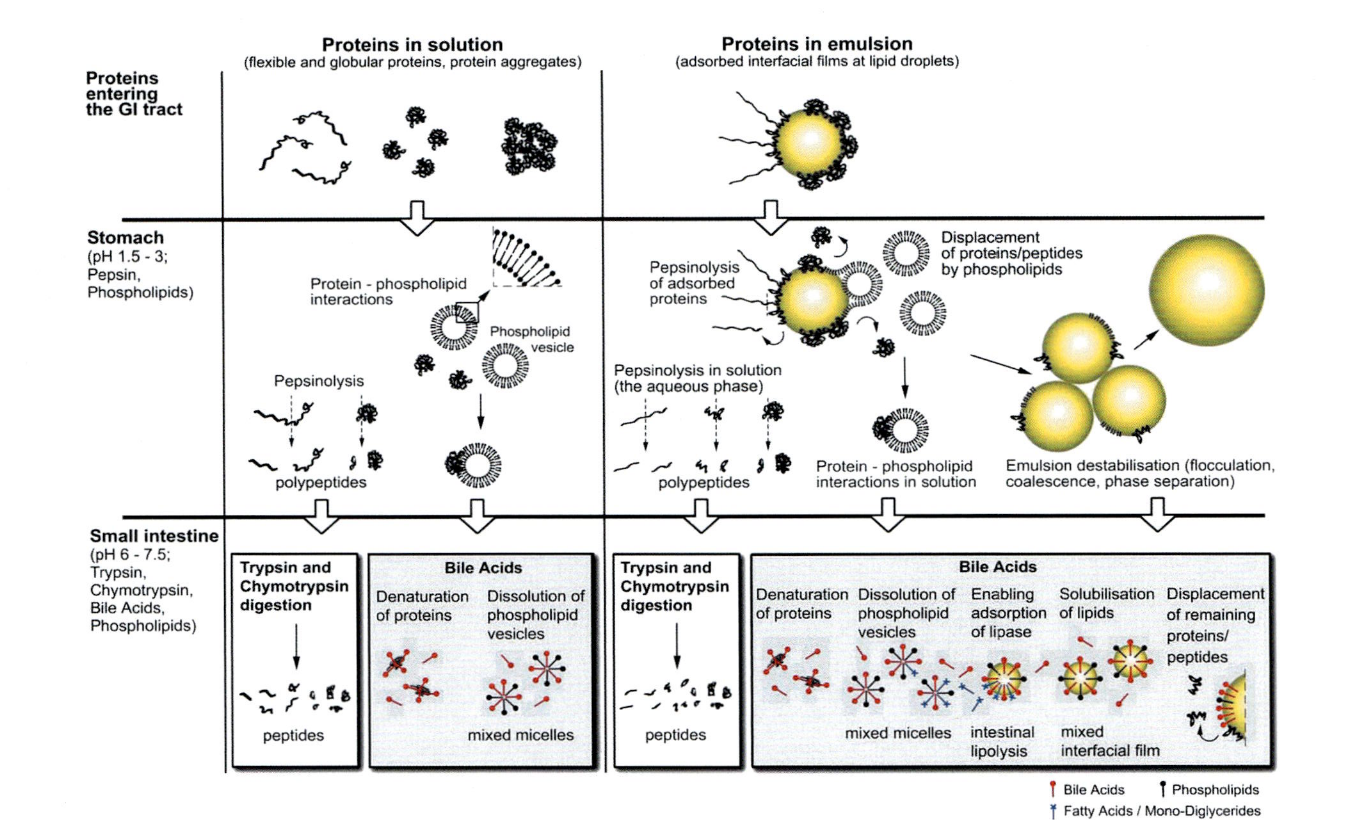


**Fig. 4** Schematic representation of the interactions of dietary proteins with proteolytic enzymes and physiological surfactants in the human GI tract. Only the principal interactions important for colloidal aspects of protein digestion have been shown. A comparison has been made between the proteins dissolved in aqueous solution (the left-hand side column) or adsorbed at the surface of droplets in the oil-in-water emulsion (the right-hand side column) (Mackie & Macierzanka, 2010).

gut (Pietroiusti, Magrini, & Campagnolo, 2016). Metabolomics approach was used for mechanistic study for health implication of nano-pesticides found in fresh produce (Zhao et al., 2016). In combination with other characterization techniques, it is possible to depict a clearer picture of the physiological fate of protein nanoparticles in the GI tract.

## 2.2 Carbohydrates

Carbohydrates are biomolecules comprised of carbon (C), hydrogen (H) and oxygen (O) atoms. Carbohydrate nanoparticles are made from digestible or indigestible polysaccharides. Common carbohydrate nanoparticles include, but not limited to, starch, cellulose, alginate, carrageenan, pectin, and xanthan (Myrick, Vendra, & Krishnan, 2014). These biomolecules/polymers are typically less than 100 nm and are up to 500 nm as sENP, or they can be transformed to nanomaterials using either top-down and bottom-up approaches, including breaking down of large natural carbohydrate molecules and growing polysaccharide molecules in microorganisms, *via* enzymes or cross-linking agents.

The applications of carbohydrate nanoparticles fall into two major categories. The first category involves carbohydrate nanoparticles as delivery systems, which in most cases are polysaccharides nanoparticles. Polysaccharide-based nanoparticles are advantageous as naturally occurring, hydrophilic and biocompatible materials (Mizrahy & Peer, 2012). Additionally, polysaccharide nanoparticles also enhance bioavailability and controlled-release properties of other bioactive components. Early studies on physiological effects on polysaccharide nanoparticles also originated in clinical and pharmaceutical applications. For example, β-cyclodextrin nanostructures were loaded with the antimalarial drug, artemisinin, to suppress the growth of *Plasmodium falciparum*, a parasite responsible for malaria (Yaméogo et al., 2012).

One of the most studied polysaccharides is chitosan; its nanoparticles are biocompatible delivery vehicles for peptide- and protein-based vaccines. Results show that chitosan nanoparticles travel faster to draining lymph nodes than chitosan microparticles, indicating an earlier interaction with naive T cells, and resulting in the desired immune response (Chua, Al Kobaisi, Zeng, Mainwaring, & Jackson, 2012). Due to these advantages, carbohydrate nanoparticles have been explored for nutrient and nutraceutical delivery systems. Insulin was encapsulated in starch nanoparticles with a permeability enhancer. *In vivo* experiments in rats reported no toxicity over a 6-h sustained release of insulin (Jain, Khar, Ahmed, & Diwan, 2008). Flax seed oil encapsulated in corn starch was used in bread dough formula to protect lipids from oxidation and thermal process contaminants during bread baking (Gökmen et al., 2011). V-amylose was used to encapsulate unsaturated fatty acids in order to provide a solution to stable delivery system for long chain fatty acids (Lesmes, Cohen, Shener, & Shimoni, 2009; Zabar, Lesmes, Katz, Shimoni, & Bianco-Peled, 2009). Tapioca starch formed complexes with flavors, mostly aliphatic alcohols and ketones alcohols with 6–10 carbons (Itthisoponkul, Mitchell, Taylor, & Farhat, 2007). The variance in human digestibility of different polysaccharides is a natural advantage of carbohydrate-based nanomaterials because it can be exploited to synthesize customizable delivery systems with different advanced functionalities. For instance, pectin—a linear anionic polysaccharide comprised of linearly connected galacturonic acid residues—has been used to increase the dissolution rate of encapsulants (Dutta & Sahu, 2012). As an FDA approved material for human therapy, poly(lactide-*co*-glycolide) (PLGA) based nanoparticles have been prepared and used for delivery of thymoquinone. Thymoquinone loaded nanoparticles showed an increased efficacy compared with free thymoquinone in inhibition of cancer cell growth (Ganea et al., 2010). Other applications of PLGA nanoparticles include the modification of PLGA nanoparticles, which enhanced antioxidant activity of encapsulated bioactive ingredients (Astete, Dolliver, Whaley, Khachatryan, & Sabliov, 2011), and controlled-release properties in the GI tract (Murugeshu, Astete, Leonardi, Morgan, & Sabliov, 2011). Preliminary assessment of modified PLGA nanoparticles showed no major safety concerns (Semete et al., 2010; Simon & Sabliov, 2013). Other examples of carbohydrates being used as nanoscale delivery systems include alginate and chitosan/alginate nanoparticles, which have been used to prepare encapsulations of lipids and enzymes (Liu, Chen, et al., 2012; Strasdat & Bunjes, 2013), and nanocomplexes formed using cyclodextrins and allyl isothiocyanate with improved antimicrobial activity compared to free allyl isothiocyanate (Piercey et al., 2012). These carbohydrate nanoparticles are usually fabricated by coacervation, spray drying, electrospinning and electrospray, supercritical fluids, emulsion-diffusion, and hydrogel methods (Fathi, Martín, & McClements, 2014).

The second category of carbohydrate nanomaterials in food involves emerging dietary fiber ENMs, the most notable being cellulose nanofibers and nanofibrils (CNF). These CNFs are produced using either mechanical methods (such as grinding and milling) or chemical methods. The CNFs can undergo additional chemical processing to modify their surfaces. The patent-pending Innofresh™ (CNF)/potassium sorbate (KSb) coating developed by Zhao et al. (Zhao, Simonsen, Cavender, Jung, & Fuchigami, 2018) with an optimum surfactant mixture ratio and glycerol concentration showed significant reduction of cherry fruit cracking in comparison with non-coated fruit in field trails in USA and Chile. The coatings showed no negative effects on fruit growth and quality (Jung, Deng, Simonsen, Bastías, & Zhao, 2016). More importantly, nano-cellulose showed no cellular and genetic adverse effects on the host. For example, one in-depth toxicity assessment using rainbow trout hepatocytes showed low toxicity and environmental risk (Kovacs et al., 2010). Pulmonary inflammation from inhaled and non-degradable CNFs remains a concern that needs further investigation for occupational safety of handlers (Lin & Dufresne, 2014; Shatkin & Kim, 2015), as well as a recent report about the reproductive toxicity of CNF due to pulmonary exposure (Farcas et al., 2016). However, there is no report of acute or chronic toxicities associated with oral uptake of CNF.

The fate (adsorption, distribution, metabolism, and excretion) and biokinetics of carbohydrate nanoparticles vary significantly based on their size, surface chemistry, and chemical composition. Some polysaccharide ENPs, a notable example being high amylose starch, can undergo rapid hydrolysis by amylases. As a result, oligosaccharides and glucose appear in the oral cavity and small intestines, while other types of starch, with higher amylopectin components, are structurally more resistant to chemical and enzymatic hydrolysis.

Other polysaccharides, especially dietary fibers, are not digestible in the upper GI tract but can be fermented by the microbiota in the lower GI tract. Diets high in dietary fibers have been shown to increase the circulating concentration of short-chain fatty acids (SCFAs), which provide protective effects in lungs again inflammation (Trompette et al., 2014). Dietary fibers originate from a diverse collection of plant polysaccharides and oligosaccharides and have varied physiochemical and structural properties, which are important in subsequent digestive processes by the gut microbiota (Hamaker & Tuncil, 2014). Yet, these structure-function-performance relationships remain understudied. Besides dietary fibers, guar gum is another emerging candidate material for colon-specific delivery and controlled release applications, due to its resistance to degradation by the gut microbiota (Prabaharan, 2011). Curcumin, a natural phenolic compound with antioxidant properties, was able to decrease the antimicrobial activity of ingested silver nanoparticles against gut microbiome (Khorshidi, Sarvi Moghanlou, Imani, & Behrouzi, 2018). Some concerns on the health implications, rather than safety, of carbohydrate-based nanoparticles have been raised in recent years. One potential issue of carbohydrate nanoparticle in the upper GI tract is the impact on insulin release in diabetic patients. In general, carbohydrate nanoparticles that are fully digested in the upper GI tract are unlikely to have health-related implications due to the simple fact that digestion produces sugars. However, there is no evidence suggesting digestible carbohydrate nanoparticles are of safety concerns when applied in food.

In contrast to digestible carbohydrate ENPs, indigestible carbohydrate nanoparticles are slightly more complex because of their absorption and ability to interact with the gut microbiota. These complex interactions in the GI yield potential adverse health effects. For instance, in a very recent study, over-processed and refined soluble fibers were found to have devastating effects by promoting the overgrowth of the gut microbiota or dysbiosis, and abnormal concentrations of fermented by-products could increase host's susceptibility to liver cancer by as much as 40% (Singh et al., 2018). Although there is overwhelming evidence on beneficial effects of consuming natural dietary fibers (Anderson et al., 2009; Lattimer & Haub, 2010), it appears that modifications in the chemical composition and structure due to refining processes, may be key factors that determine whether a carbohydrate nanoparticle/fiber is ultimately beneficial or deleterious to human health.

## 2.3 Lipids

Lipids are a group of hydrophobic biomolecules that play important life-sustaining roles in living organisms, including long-term energy storage, protection, insulation, and lubrication. Lipids are precursors for certain types of hormones and key constituents of cell membranes. The original use of lipids as an effective, safe delivery material originated from the pharmaceutical industry. Colloidal lipid emulsions and solid lipid nanoparticles have been used for delivery of insoluble drugs (Bunjes, 2010). In the food and beverage industry, lipid nanoparticles are used widely in commercial products (McClements, 2013). Colloidal oil-in-water (O/W) nanoemulsions are used in beverages, such as soft drinks, fortified waters, fruit juices, and dairy drinks (Piorkowski & McClements, 2014). Colloidal nutrient delivery systems are major applications of lipid nanoparticles (Li et al., 2009;

Qian, Decker, Xiao, & McClements, 2013; Tamjidi, Shahedi, Varshosaz, & Nasirpour, 2013), because of increased bioavailability of hydrophobic encapsulants.

Commonly encountered lipid nanoparticles include: liposomes, coacervates, O/W nanoemulsions, solid lipid nanoparticles (SLNs) and inclusion complexes. The particle size of these materials vary from a few nanometers to sub-micron level. Liposomes are nano- and/or micro-scale spherical particulates prepared from phospholipids and cholesterol (Beloglazova, Goryacheva, Shmelin, Kurbangaleev, & De Saeger, 2015). The recent application of liposomes is focused on their antimicrobial activities in dairy products, such as milk and cheese. (da Silva Malheiros, Daroit, & Brandelli, 2010; da Silva Malheiros, Daroit, da Silveira, & Brandelli, 2010; Malheiros Pda, Sant'Anna, Barbosa, Brandelli, & Franco, 2012). Coacervates are typically colloidal materials formed by electrostatic attractions of oppositely-charged macro-ions (e.g., protein, polysaccharide, and/or lipids) and hydrophobic interactions between hydrocarbon-based functional groups (Sagiri, Anis, & Pal, 2016). Quercetin loaded nanoparticles prepared from coacervation showed increased antimicrobial activities against common foodborne microorganisms (Dinesh Kumar, Verma, & Singh, 2016). Similarly, nisin-loaded chitosan/alginate coacervates are also used as food preservatives in tomato juice (Zohri et al., 2013).

To prepare O/W nanoemulsions, two approaches are commonly adopted: high-energy or low-energy approaches. Surface electrostatic properties of these nanoparticles are determined by the particles' exterior layer. High-energy approaches refer to using mechanical devices with extreme shear forces to create a balance between droplet disruption and droplet coalescence. Examples of this approach are high-pressure valve homogenizers, microfluidizers, and ultrasonic homogenizers. Low-energy approaches rely on the spontaneous formation of nanoscale oil droplets in the surfactant-oil-water mixtures by changing the formulation of the mixtures and temperature. SLNs are created by heating an O/W nanoemulsion prepared at a certain temperature to above the melting point of the lipid component, which is followed by gradual cooling of the O/W nanoemulsion to induce lipid crystallization (Mehnert & Mäder, 2001).

The major ingredients of lipid nanoparticles are neutral lipids, such as triacylglycerol (TAGs), diacylglycerols (DAGs), monoacylglycerols (MAGs), hydrocarbons, and terpenes or polar lipids, such as free fatty acids (FFAs), surfactants, and phospholipids. Similar to carbohydrates, most lipids undergo hydrolysis reactions catalyzed by lipases following ingestion (McClements & Rao, 2011; McClements & Xiao, 2012). Since lipid hydrolysis is rapid and often yields predictable products, there is no evidence to suggest safety concerns on digestible lipid nanoparticles.

Lipid nanoparticles can also go through a structural transformation in the GI tract, including changes in particle size (McClements, 2013). The chemical composition of lipid nanoparticles is also the decisive factor in structural stabilities of lipid nanoparticles in the GI tract. Some types of lipid nanoparticles are indigestible because the primary compositions of their oil phase consist of indigestible hydrocarbons (McClements & Xiao, 2012). Lemon oil is an example of indigestible hydrocarbons. The monoterpenes, sesquiterpenes and the corresponding oxygenates in this product cannot be hydrolyzed by lipase in GI tract (Rao, Decker, Xiao, & McClements, 2013). Based on "safe-by-design" principles, these indigestible hydrocarbons should be avoided to prevent potential adverse effects on human health (Kraegeloh, Suarez-Merino, Sluijters, & Micheletti, 2018).

When lipids are digested, they form free fatty acids and 2-monoglycerides as the first step of its catabolic pathway. The impact of free fatty acids on the microbiome has drawn increasing attention in recent studies. One study showed omega-3 supplementation alters the human gut microbiota. A recent study of 20 middle age individuals who consumed functional drinks supplemented with DHA (docosahexaenoic acid)/EPA (eicosapentaenoic acid) documented an increase of the *Bifidobacterium* and *Oscillospira* genera bacteria and a decrease of *Coprococcus* and *Faecalibacterium* genera bacteria. The study found only polyunsaturated fatty acids encapsulated in the emulsion and stabilized in drinks were associated with prolonged suppression of pro-inflammatory arachidonic acid in the blood, which represented a significant improvement from the conventional omega-3 capsule supplements (Watson et al., 2018). Despite this promising result, it is unclear what the emulsion droplet size distribution was in the study, and whether nano-encapsulated DHA/EPA would provide similar health benefits. Depending on the source and origin, omega-3 polyunsaturated fatty acids (PUFAs) can lower the population of gut bacteria *Bacteroidetes* or *Firmicutes*, affecting the ratio of *Firmicutes/Bacteroidetes, which* is a critical factor for weight gain and insulin resistance development (Liu, Hougen, Vollmer, & Hiebert, 2012; Yu et al., 2014). Menni et al. determined an association between DHA and 38 operational taxonomic units. The circulative level of DHA is related to the higher microbiome diversity and higher abundance of the *Lachnospiraceae* family (Menni et al., 2017). The influence of omega-3 fatty acids on the gut microbiota was revealed in details (Costantini, Molinari, Farinon, & Merendino, 2017).

## 2.4 Other nanoscale biomaterials

Besides the three main categories of the sENP in food, namely proteins, carbohydrates and lipids, other complex biopolymer nanoparticles are also widely used and investigated as food ingredient-based nanomaterials. In the context of this review, complex biopolymer nanoparticles refer to nanomaterials that are comprised of multiple soft materials as the matrix. Complex biopolymer matrices are able to provide increased protection against thermo- and photo-degradation (Patel, Hu, Tiwari, & Velikov, 2010). They can be prepared by coacervation *via* electrostatic interaction between proteins and polysaccharides that carry opposite charges. As a highly versatile cationic charged material, chitosan has been used frequently in preparation of complex biopolymer nanoparticles. It has been coupled with zein for encapsulation of α-tocopherol (Luo, Zhang, Whent, Yu, & Wang, 2011), and with soy protein for encapsulation of vitamin D3 (Teng, Luo, & Wang, 2013). It is also used to prepare chitosan-zinc oxide complex nanoparticles with antimicrobial applications (Perelshtein et al., 2013). The GI tract fate of complex biopolymer nanoparticles is very similar to proteins, carbohydrates and lipids nanoparticles, and is influenced heavily by the matrix composition. If the shell matrix materials are digestible and generally accepted as safe, it is very unlikely that they will cause any deleterious effects. However, similar to other sENP in food, the physiological fate of complex soft materials has yet to be fully investigated.

Engineered water nanostructures (EWNS) are a unique type of water-based nanomaterials with potentially broad applications. EWNS are prepared by spontaneously electrospraying and ionization of water with a high voltage source (6.7 kV) (Fig. 5). The current applications of EWNS are primarily on surface sanitation of food and food contact surfaces (Pyrgiotakis, McDevitt, Bordini, et al., 2014; Pyrgiotakis, McDevitt, Gao, et al., 2014; Pyrgiotakis, McDevitt, Yamauchi, & Demokritou, 2012). By enriching EWNS with modest amounts (0.1–1%) of commonly used sanitizers in the food industry, such as $H_2O_2$, citric acid, and lysozyme, the disinfection potential increases several orders of magnitude, providing disinfection within minutes. The amount of these active ingredients delivered to the disinfection surfaces is in the picogram range and practically negligible, making iEWNS a highly promising chemical-free sanitation technology for food

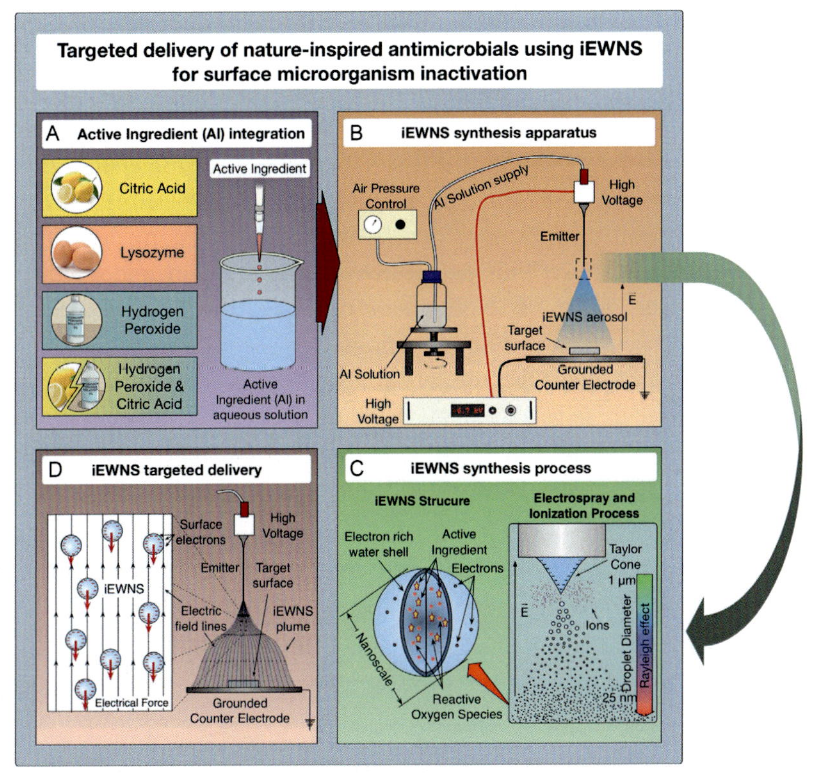


**Fig. 5** Synthesis and the targeted delivery of iEWNS. (A) The active ingredient is added to DI water to result in an aqueous solution. (B) The solution is transferred to the iEWNS emitter using an air compressor. The applied voltage between the capillary and grounded electrode results in the combined electrospray ionization process that produces the iEWNS nanoparticles. (C) The synthesized iEWNS have nanoscale size ($\sim$20 nm) and high electric charges. They also contain both the AI molecules and the ROS produced during the ionization process. (D) The targeted delivery of these particles to the surface of interest is performed utilizing the electric field and their inherent electric charges (Vaze et al., 2019).

safety applications. The synergistic effects of common sanitizers such as citric acid and ROS from EWNS are responsible for the enhanced desired antimicrobial effects of EWNS (Vaze et al., 2019).

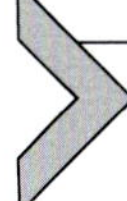

# 3. Methodologies for assessing pathophysiological effects of sENP

The availability of numerous well-established chemical and biological analytical methods today is an advantage for the assessment of pathophysiological effects of engineered nanomaterials. Yet, challenges are still no less than in other fields. Contrary to metal and metal oxide nanoparticle additives in foods, which can be traced *via* transmission and/or scanning electron microscopy or single particle inductively coupled plasma mass spectrometry (ICP-MS), sENP are relatively much harder to monitor and require different analytical tools, ones that are more suitable for organic molecules. Borel and Sabliov reviewed the toxicity and physiological effects of sENP from the ADME (absorption, distribution, metabolism, and excretion) perspective (Borel & Sabliov, 2014). This perspective is important in understanding the physiological effects of sENP incorporated in the food. Cockburn et al. propose a two-tiered system for nanomaterials safety assessment. Tier 1 comprises *in vitro*, and subacute *in vivo* studies and Tier 2 comprises more advanced acute *in vivo* studies (Cockburn et al., 2012). Carcinogenicity, teratogenicity, reproductive toxicity, developmental toxicity, and neurotoxicity also require considerations for certain nanomaterials.

## 3.1 Models and methods for toxic and pathophysiological effects assessment

sENP in foods, when ingested, can potentially lead to desirable or undesirable interactions with cellular components of the GI tract and thus, impacting uptake and biodistribution of other critical micro/nutrients in foods (Bettini et al., 2017). Most of these methods needed for the safety assessment of sENP in foods, could be adapted from inorganic nanoparticle exposure studies as well as from the field of cancer nanomedicine where several oral drug formulations containing lipid and/or polymeric nanoparticles have been investigated for their safety and efficacy. In addition, sENP interact with various components within complex food matrices and digestive fluids of the GI tract. Later, we discuss how such food matrix and GI tract interactions could play important roles in determining the extent and site of toxicity as well as other physiological implications of ingested exposure scenarios. These interactions have been frequently ignored in biological fate and transport studies of ingested sENP (Bellmann et al., 2015; DeLoid et al., 2017; McClements et al., 2016).

## 3.2 *In vitro* models of GI tract

To assess the toxicity of ingested nanoparticles, the most simplistic model is to treat *in vitro* cell lines representative of cell types associated with different anatomical compartments of the GI tract, with a range of concentrations of test nanomaterial. As most ingested food products eventually reach the small intestine where food resides for the longest time, it would be realistic to determine the effects of test nanomaterials on *in vitro* cell lines representative of the small intestine. The *in vitro* models can be simple, cost-effective, high throughput, and predictive of potential toxicities.

### 3.2.1 Monocultures

Caco-2 and a subclone of Caco-2, C2BBe1, originating from human epithelial colorectal adenocarcinoma cells, can differentiate into morphologically and functionally mature cells that resemble the enterocytes lining the small intestine. Indeed, during the past decade of inorganic nanotoxicity literature, the Caco-2 monoculture has been used by more than half of the studies (Sohal, O'Fallon, Gaines, Demokritou, & Bello, 2018). Although it has been suggested that C2BBe1 monolayers, the enterocytic differentiation subclone of Caco-2 cells, are more representative of the small intestinal epithelium than Caco-2, due to the more similar transepithelial electrical resistance, morphological homogeneity and BB myosin I expression levels similar to that of a human enterocyte, it has the potential to be a valuable tool in the study of physiological effects of sENP (Huang & Adams, 2003; Masuda, Kajikawa, & Igimi, 2011; Peterson & Mooseker, 1992). Moreover, both cell lines are cancerous (immortal), and their higher resilience compared to normal cells, might lead to underestimation of the true toxicity of nanomaterials. HIEC-6, a normal small intestinal epithelial cell line, or primary cells isolated from the small intestine, are more sensitive and preferred over cancerous cell lines. Additionally, epithelial cell lines used *in vitro* should be allowed to grow, form tight junctions and differentiate to enterocytes to form an intact barrier—representative of the GI tract epithelium, which should be verified by measuring TEER (Trans-epithelial electrical resistance) and expression of tight junction proteins, before commencing any toxicological assessment (Natoli et al., 2011; Natoli, Leoni, D'Agnano, Zucco, & Felsani, 2012; Sambuy et al., 2005). Other *in vitro* models include the cell lines representative of gastric epithelium (GES-1), mucus-secreting cells (HT29-MTX), colon epithelium (SW480, DLD-1), and mucus-secreting colon epithelium (NCM460).

### 3.2.2 Co-cultures

Monocultures are unable to represent the complexity of any segment of the GI tract. For instance, small intestinal epithelium not only includes enterocytes (epithelial cells) but also mucus-secreting goblet cells, as well as immune cells in certain regions (Peyer's patches). Caco-2/HT29-MTX and Caco-2/Raji-B cell lines in co-culture models have been used to represent mucus-secreting epithelium and lymphoid-associated epithelium, respectively (Brun et al., 2014). These more complex co-culture models offer better representation of the GI complexity and possible interactions between various cell types.

However, both monoculture and co-culture models have limitations. Other important factors associated with the GI tract include the gut microbiome, GI fluids of varying pH, underlying blood supply for nutrient absorption and several other direct or indirect interactions. Thus, any *in vitro* model, whether monoculture, co-culture or tri-culture, are unable to represent the true environment and complex interactions between cells and the environment, and findings from such toxicological studies should be interpreted with extreme caution. In addition, *in vitro* models cannot be used to determine the biokinetics of food nanoparticles *via* the ingested route. Furthermore, long-term adverse effects of chronic exposures are difficult to replicate in *in vitro* studies. Such test conditions would require culturing the cells for long periods of time, which itself can induce cellular stress. Additionally, compensatory homeostatic mechanisms that are typically involved in chronic *in vivo* exposures, are difficult to replicate for *in vitro* conditions.

## 3.3 *In vivo* models

The limitations of an *in vitro* model could be addressed using an *in vivo* model, which can not only inform about the physiological implications of sENP to a specific cell or tissue type, but can also demonstrate effects on the gut microbiome, effects of GI fluids on nanoparticle accumulation and biokinetics in the body, as well as overall health of an organism. In addition, *in vivo* rat and mice models are extensively used for testing chronic toxicity of food nanoparticles (especially inorganic NPs), as well as oral lipid and polymeric nanoparticle formulations for cancer treatment. In the toxicology literature of inorganic ingested nanoparticles, 13 out of 19 studies in the past decade used the Sprague Dawley rat model and CD-1 (ICR) mouse model (Sohal, O'Fallon, et al., 2018), while recent studies have also used adult male Wistar rats (Bettini et al., 2017). For assessing biokinetics of ingested nanoparticles during or after the treatment regiment, animals are sacrificed and tissues can be collected for elemental analysis using

ICP-MS (inductively-coupled plasma-mass spectrometry) (Ammendolia et al., 2017; Hong et al., 2017; Tassinari et al., 2014), identifying nanoparticles by visualizing tissue samples using isotopic TEM (transmission electron microscopy) (Susi et al., 2016), and nanoSIMS (nanoscale secondary ion mass spectrometry) (Bettini et al., 2017). For sENP, fluorescent or NIR (near infrared) spectroscopic techniques may be used in conjunction with confocal imaging of tissue sections or live animal imaging, respectively (Chen et al., 2016). For quantitative purposes, chromatography-mass spectrometry techniques should be considered. It should be noted that qualitative and quantitative characterization of sENP in food can be challenging. The building blocks of sENP are soft materials that typically exhibit less stability than inorganic nanomaterials, when exposed to different environmental factors, including food matrices, pH, temperature, water activity, etc. sENP also go through coagulation or degradation during extraction and sample preparation of different characterization techniques, which alters particle size, surface properties of target sENP.

The toxic effects caused by ingested nanoparticles *in vivo* can be assessed using several methods or panels of biomarkers of organ function. They can be categorized into broader effects—(1) changes in body weight, blood parameters, levels of alanine aminotransferase (ALT) and alkaline phosphatase (ALP) as an indicator of liver function; (2) granular effects, at the cellular or tissue level histopathological examination of various tissues to determine pathophysiology, permeability measurements assess for compromised intestinal epithelial barrier, cell isolation and flow cytometry analysis to determine changes in cell populations as part of a stress or immune response; and (3) cytokine assays to determine type of immune responses and cytotoxicity assays for assessing cytotoxic responses in specific cell types. Because sENPs are used to increase delivery efficacy of nutrients, antioxidants, cofactors, vitamins, etc., it is important to monitor circulating levels of these ingredients and other critical biochemical compounds in their respective metabolic pathways. For delivery of carbohydrate nanoparticles in foods, one should consider monitoring their impact on circulating carbohydrates and insulin release. Metabolomics studies using nanoparticles microarrays are efficient approachs to monitor gut microbiome changes (Hansen, Dueñas, Looft, & Lee, 2019).

Animal models provide valuable information about toxicity and pharmacokinetics of sENP, which otherwise is difficult to assess. However, results must be carefully extrapolated to human relevance as there are significant differences in physiology and nutrient uptake of the GI tract between humans and rats (DeSesso & Jacobson, 2001; Kararli, 1995). Additionally, it is challenging to investigate effects of the GI fluids of varying pH on sENP passing through the ingested route in a living animal, in which case, simulating the passage through the GI tract in an *in vitro* model could easily elucidate potential biotransformation ingested nanoparticles might undergo.

## 3.4 Assessment of nanoparticles induced oxidative stress using electrochemical sensors

Oxidative/nitrosative stress is among the major adverse pathophysiological effects from nanoparticles exposure (Nel, Xia, Mädler, & Li, 2006). Oxidative/nitrosative stress is a physiological redox disequilibrium state. Under this state, the generation of reactive oxygen and nitrogen species (ROS/RNS) overwhelms antioxidant defensive system comprised of antioxidants and enzymes including catalase and superoxide dismutase (Halliwell, 1994). Electrochemical biosensors are demonstrated to be effective and versatile for the assessment of nanoparticles induced oxidative stress (Özel, Liu, Alkasir, & Andreescu, 2014). The biorecognition system tailored for each specific analyte in sensor design ensures selectivity of these devices. Additionally, electrochemical biosensors are sensitive, rapid, portable, have real-time monitoring capability and temporal and spatial resolution (Liu, Dumitrescu, & Andreescu, 2015). More importantly, electrochemical biosensors provide a better mechanistic understanding of various phenomena. For example, a cytochrome *c* based biosensor has been used to study the mechanisms of antibiotics meditated oxidative stress at the cellular level (Liu, Marrakchi, Jahne, Rogers, & Andreescu, 2016; Takahashi et al., 2017). Fig. 6 showed a few examples of using electrochemical biosensors for nanotoxicity assessment.

Electrochemical biosensors are used to study pathophysiological effects of nanoparticles exposure in zebrafish embryo model *via* direct *in vivo* sensing (Özel, Alkasir, Ray, Wallace, & Andreescu, 2013; Özel, Wallace, & Andreescu, 2014). Despite advances in nanotoxicity applications of electrochemical biosensors, several major challenges need to be addressed. Electrochemical biosensors work on a rather limited and narrow selection of analytes. sENP and metabolites are mostly redox inactive and represent major challenges for electrochemical detection. In addition, background food components that are protein, carbohydrate, or lipid-based, can produce a huge systemic error or background noise when using the electrochemical biosensor to study sENP in food. Differentiation of sENP from the

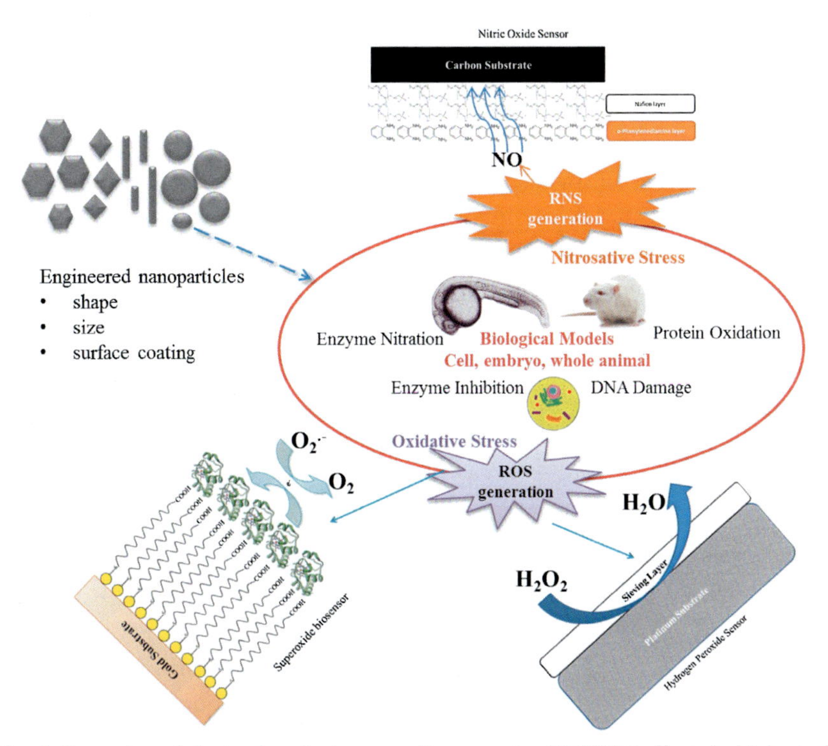


**Fig. 6** Examples of electrochemical sensors for assessing ROS/RNS effects in biological models exposed to NPs (Özel, Liu, et al., 2014).

background food component is required for distinguishing physiological responses exclusively generated by sENP.

Another issue is biofouling during *in vivo* and *in vitro* measurements. Biofouling is a common issue for biosensors, which may increase noise levels, jeopardizing their sensitivity. A recent *in vivo* study showed that the mice GI tract conditions and accumulation of blood from the sensor entrance caused higher noise and a ~40% decrease in signal response (Gubernatorova et al., 2017). Overall, electrochemical biosensors are versatile tools for nanotoxicity studies, although they present their own limitations. These devices can be used as complementary methods in assessment of toxicity and pathophysiological effects of sENP in food.

## 3.5 Simulating the GI tract passage

The interaction of food or digestive fluids with ingested nanoparticles may alter their physicochemical properties and therefore their biological fate and function (e.g., absorption and toxicity) (Utembe, Potgieter, Stefaniak, & Gulumian, 2015). Sohal et al. in a recent study determined the dissolution and bio-durability of major ingested inorganic nanoparticles in simulated GI fluids and a similar method can be used for organic nanoparticles (Sohal, Cho, et al., 2018). However, because organic nanoparticles' composition can be similar to that of the lipid, protein or carbohydrate components of the GI fluids, high-performance liquid chromatography-tandem mass spectrometry techniques and alike, are needed to accurately differentiate and quantify molecular components in sENP.

### 3.6 Gut microbiome and clinical studies

The small and large intestines are home to various microorganisms, which play an important role in human health (Hsiao et al., 2013; Lepage et al., 2011; Mondot et al., 2011). In fact, changes in diet alone cause rapid changes in activity and communal composition of the gut microbiota, and ingested nanomaterials are expected to have similar effects (David et al., 2013). It is, therefore, not only appropriate but critical to consider effects of ingested sENP on the gut microbiome in humans, as well as on *in vivo* or *in vitro* models' representative of small intestine and colon. Although studies on this front have just begun to emerge, metagenomics, also known as 16S rRNA sequencing, is extensively used to assess the identity of microbes and shifts in microbial communities upon exposure to ingested nanoparticles (Chen et al., 2017; Taylor, Marcus, Guysi, & Sharon, 2015). The emerging next generation sequencing (NGS) platform has also been explored, in a combination of bioinformatic tools, to study effects of ingested nanoparticles on the gut microbiome (Ogawa et al., 2018; van den Brule et al., 2016).

Most importantly, humans are the ultimate consumer of sENP in food. With minor exceptions, clinical studies that systematically investigate the links between sENP ingestion, diet, GI health, and health outcome and implications, are currently lacking. In this context, the greatest insight on the sENP and host-biome interactions can be obtained when deep sequencing is applied to characterize responses of both systems.

## 4. Summary

A nanometer is an intermediate length scale which is in between microscale and atomic scale. Recent advances in nanoscale science have revealed many superior properties and novel phenomena of nanomaterials that inspire numerous beneficial applications. Nanotechnology offers unprecedented solutions to meet challenges of ensuring more sustainable, healthier, safer and more joyous society by promoting food security along the supply chain, increasing the nutritional value of foods, and delivering specialty nutrients for individuals with unique dietary needs or medical conditions. It is important to note that nanoscale materials naturally exist in nature, such as the casein micelle. Many natural food components based on proteins, carbohydrates, lipids, and biopolymers are also in the nanometer scale. Traditional food preparations such as cooking and boiling and commonly used food processing technologies such as homogenization, thermal processing, and others result in soft engineered nanoparticles (sENP) or nanostructures with advantageous functional properties of better food products. The sENP found in our food supply has been consumed for a long time. There have been no reported safety concerns about them in literature. It is clear that the simplistic question of "Is nano safe to eat or not" is not a proper question in the broad discussion of risk assessment and management of engineered nanoparticles.

sENP (protein, carbohydrate, lipids and other biomaterials) have been widely used in food for their superior physio-chemical and functional properties. The current primary use of sENP is in nutrient delivery; meanwhile, other applications in food and agriculture are being explored. While the public perception, labeling, and safety issue are hurdles for wider applications of sENP, no toxic effects from ingested sENP have been reported to-date. Many beneficial effects have been achieved using thoughtfully designed functional sENP. To facilitate a faster acceptance of emerging applications of sENP in food and agriculture, systematic evaluation and risk assessment of such materials, particularly in foods, should be established. A series of traditional and novel analytical methodologies can be used to establish and support such evaluation systems.

Investigations of the nanoscale processes, phenomena, and properties of food materials are a new endeavor. Numerous lines of meaningful inquiries are apparent to promote scientific understanding of food science, nutrition, and safety. In addition, these inquiries can ultimately formulate and manufacture nutritious food with potential to improve human health and quality of life. We have articulated four major research priorities throughout the chapter and summarized below for the scientific community and public to consider.

**4.1.** Functionalities of sENP:

**a.** Advanced delivery systems, attributed to nanoscale dependent physiochemical properties, have been, and still are, a major research foci of new sENP development.

**b.** An emerging and very promising functionality for sENP is to remove unhealthy ingredients in food or prevent absorption of unhealthy ingredients (such as undesirable cholesterols and fatty acids) in the GI tract.

**4.2.** Safety assessment of sENP:

**a.** Improving payload and uptake of bioactive components in sENP delivery systems has been, for understandable reasons, a primary focus. Bioactive ingredients, such as vitamins and nutraceuticals, appear to have normal rage of optimal therapeutic/beneficial effects. Current efforts focus on increasing bioavailability. There is little information at present to establish upper therapeutic limits or margins of safety for these delivered bioactive components. This should be considered an important priority, best done early on in research and development phase before the product reaches the market, if one is to prevent overdose-induced toxicological effects.

**b.** Comprehensive assessment on the full spectrum of biological and physiological responses to sENP uptake has not been considered necessary in the development of new ingredients or additives. However, it may be necessary to develop a battery of effective assessment methods to respond to the needs of regulatory and public concerns for new sENP development and application in food and nutrition supplements.

**c.** Nutrition has emerged as an important arm in comprehensive therapies as well as disease prevention and intervention strategies. It is of vital importance to evaluate interactions among sENP, bioactive cargo, and other prescription or non-prescription medicines and nutraceuticals including dietary intervention strategies. This should also be a priority, especially if both substances are used to treat or offset similar diseases or conditions, such as anti-inflammatory nutraceuticals, corticosteroids, non-steroidal anti-inflammatory drug (NSAIDs), etc.

**d.** Unlike inorganic ENP, there is very little evidence suggesting sENP have any acute adverse physiological effects. However, it may be an interesting line of inquiry about health implications of long-term ingestion of sENP in food.

**4.3.** Advancement in state-of-the-art methodologies:

**a.** Recent advancement in omics techniques and bioinformatics empowers research in all aspects of health sciences. Genomics and microbiomics are powerful tools that can provide important mechanistic insights into understanding of sENP-host interactions, and how they impact the microbiome and subsequent human health, such as cognitive functions and aging.

**b.** Metabolomics, especially proteomics and lipidomics, also pose unparalleled power to investigate physiological pathways and implications of sENP and their bioactive cargo in human, especially if combined with microbiomics and genomics.

**c.** Human clinical trials on sENP in food or nutritional supplements have been very rare. Animal models may elucidate mechanistic and kinetic effects of sENP to potential health benefits and implications.

**4.4.** Health assessment for immunocompromised population:

**a.** The majority of sENP research and development has been tailored for health promotion and disease prevention in healthy adults. However, safety and physiological assessment of sENP is very important and often neglected for sensitive or immunocompromised populations.

**b.** The pediatric market is rapidly growing, and more regulation and safety assessment need to be enforced. There is overwhelming evidence suggesting infants and toddlers have dramatically different physiological pathways compared to adults.

**i.** There is a need in understanding the physiological and developmental effects of sENP and other nanoscale materials in infant or baby food, especially as it relates to neonate microbiome, interaction of sENP with intestinal stem cells, and activation or silencing of the expression of genes that are critical for maintaining epithelial barrier, intestinal motility and normal immune homeostasis later in life.

**ii.** In addition, sENP used in nutrition supplements and nutraceuticals for pregnant and lactating mothers should also be scrutinized for sensitivity concerns. Maternal intake may have cross-generational impacts on the fetus or the baby.

**c.** Geriatrics population also represents huge market shares in nutrition supplements and nutraceutical. Elderly population tends to have higher risks in developing cancer, diabetes, hypertension, cognitive decline, and other degenerative diseases, some of which have now been linked with the shifting microbiome. It is very important to understand the impact of sENP and their bioactive cargo on the gut microbiome, as well as subsequent health benefits and implications.

## References

Alger, H., Momcilovic, D., Carlander, D., & Duncan, T. V. (2014). Methods to evaluate uptake of engineered nanomaterials by the alimentary tract. *Comprehensive Reviews in Food Science and Food Safety*, *13*(4), 705–729.

Ammendolia, M. G., Iosi, F., Maranghi, F., Tassinari, R., Cubadda, F., Aureli, F., et al. (2017). Short-term oral exposure to low doses of nano-sized $TiO_2$ and potential modulatory effects on intestinal cells. *Food and Chemical Toxicology*, *102*, 63–75.

Anderson, J. W., Baird, P., Davis, J. R. H., Ferreri, S., Knudtson, M., Koraym, A., et al. (2009). Health benefits of dietary fiber. *Nutrition Reviews*, *67*(4), 188–205.

Astete, C. E., Dolliver, D., Whaley, M., Khachatryan, L., & Sabliov, C. M. (2011). Antioxidant poly(lactic-co-glycolic) acid nanoparticles made with α-tocopherol–ascorbic acid surfactant. *ACS Nano*, *5*(12), 9313–9325.

Bellmann, S., Carlander, D., Fasano, A., Momcilovic, D., Scimeca, J. A., Waldman, W. J., et al. (2015). Mammalian gastrointestinal tract parameters modulating the integrity, surface properties, and absorption of food-relevant nanomaterials. *Wiley Interdisciplinary Reviews: Nanomedicine and Nanobiotechnology*, 7(5), 609–622.

Beloglazova, N. V., Goryacheva, I. Y., Shmelin, P. S., Kurbangaleev, V., & De Saeger, S. (2015). Preparation and characterization of stable phospholipid–silica nanostructures loaded with quantum dots. *Journal of Materials Chemistry B*, *3*(2), 180–183. https://doi.org/10.1039/C4TB01662A.

Bettini, S., Boutet-Robinet, E., Cartier, C., Coméra, C., Gaultier, E., Dupuy, J., et al. (2017). Food-grade $TiO_2$ impairs intestinal and systemic immune homeostasis, initiates preneoplastic lesions and promotes aberrant crypt development in the rat colon. *Scientific Reports*, 7, 40373.

Borel, T., & Sabliov, C. M. (2014). Nanodelivery of bioactive components for food applications: Types of delivery systems, properties, and their effect on ADME profiles and toxicity of nanoparticles. *Annual Review of Food Science and Technology*, *5*(1), 197–213.

Brun, E., Barreau, F., Veronesi, G., Fayard, B., Sorieul, S., Chanéac, C., et al. (2014). Titanium dioxide nanoparticle impact and translocation through ex vivo, in vivo and in vitro gut epithelia. *Particle and Fibre Toxicology*, *11*(1), 13.

Bunjes, H. (2010). Lipid nanoparticles for the delivery of poorly water-soluble drugs. *Journal of Pharmacy and Pharmacology*, *62*(11), 1637–1645.

Chen, G., Wang, K., Zhou, Y., Ding, L., Ullah, A., Hu, Q., et al. (2016). Oral nanostructured lipid carriers loaded with near-infrared dye for image-guided photothermal therapy. *ACS Applied Materials & Interfaces*, *8*(38), 25087–25095.

Chen, H., Zhao, R., Wang, B., Cai, C., Zheng, L., Wang, H., et al. (2017). The effects of orally administered Ag, $TiO_2$ and $SiO_2$ nanoparticles on gut microbiota composition and colitis induction in mice. *NanoImpact*, *8*, 80–88.

Chen, J., Zheng, J., McClements, D. J., & Xiao, H. (2014). Tangeretin-loaded protein nanoparticles fabricated from zein/β-lactoglobulin: Preparation, characterization, and functional performance. *Food Chemistry*, *158*, 466–472.

Chua, B. Y., Al Kobaisi, M., Zeng, W., Mainwaring, D., & Jackson, D. C. (2012). Chitosan microparticles and nanoparticles as biocompatible delivery vehicles for peptide and protein-based immunocontraceptive vaccines. *Molecular Pharmaceutics*, *9*(1), 81–90.

Cockburn, A., Bradford, R., Buck, N., Constable, A., Edwards, G., Haber, B., et al. (2012). Approaches to the safety assessment of engineered nanomaterials (ENM) in food. *Food and Chemical Toxicology*, *50*(6), 2224–2242.

Cornacchia, L., & Roos, Y. H. (2011). Stability of β-carotene in protein-stabilized oil-in-water delivery systems. *Journal of Agricultural and Food Chemistry*, *59*(13), 7013–7020.

Costantini, L., Molinari, R., Farinon, B., & Merendino, N. (2017). Impact of omega-3 fatty acids on the gut microbiota. *International Journal of Molecular Sciences*, *18*(12), 2645.

da Silva Malheiros, P., Daroit, D. J., & Brandelli, A. (2010). Food applications of liposome-encapsulated antimicrobial peptides. *Trends in Food Science & Technology*, *21*(6), 284–292.

da Silva Malheiros, P., Daroit, D. J., da Silveira, N. P., & Brandelli, A. (2010). Effect of nanovesicle-encapsulated nisin on growth of listeria monocytogenes in milk. *Food Microbiology*, *27*(1), 175–178.

David, L. A., Maurice, C. F., Carmody, R. N., Gootenberg, D. B., Button, J. E., Wolfe, B. E., et al. (2013). Diet rapidly and reproducibly alters the human gut microbiome. *Nature*, *505*, 559.

DeLoid, G. M., Wang, Y., Kapronezai, K., Lorente, L. R., Zhang, R., Pyrgiotakis, G., et al. (2017). An integrated methodology for assessing the impact of food matrix and gastrointestinal effects on the biokinetics and cellular toxicity of ingested engineered nanomaterials. *Particle and Fibre Toxicology*, *14*(1), 40.

DeSesso, J. M., & Jacobson, C. F. (2001). Anatomical and physiological parameters affecting gastrointestinal absorption in humans and rats. *Food and Chemical Toxicology*, *39*(3), 209–228.

Dinesh Kumar, V., Verma, P. R. P., & Singh, S. K. (2016). Morphological and in vitro antibacterial efficacy of quercetin loaded nanoparticles against food-borne microorganisms. *LWT—Food Science and Technology*, *66*, 638–650.

D'Mello, S. R., Cruz, C. N., Chen, M.-L., Kapoor, M., Lee, S. L., & Tyner, K. M. (2017). The evolving landscape of drug products containing nanomaterials in the United States. *Nature Nanotechnology*, *12*(6), 523.

Dutta, R. K., & Sahu, S. (2012). Development of a novel probe sonication assisted enhanced loading of 5-FU in SPION encapsulated pectin nanocarriers for magnetic targeted drug delivery system. *European Journal of Pharmaceutics and Biopharmaceutics*, *82*(1), 58–65.

Farcas, M. T., Kisin, E. R., Menas, A. L., Gutkin, D. W., Star, A., Reiner, R. S., et al. (2016). Pulmonary exposure to cellulose nanocrystals caused deleterious effects to reproductive system in male mice. *Journal of Toxicology and Environmental Health, Part A*, *79*(21), 984–997.

Fathi, M., Martín, Á., & McClements, D. J. (2014). Nanoencapsulation of food ingredients using carbohydrate based delivery systems. *Trends in Food Science & Technology*, *39*(1), 18–39.

Fellows, P. J. (2016). *Food processing technology: Principles and practice* (4th ed.). Woodhead Publishing.

Fröhlich, E., & Fröhlich, E. (2016). Cytotoxicity of nanoparticles contained in food on intestinal cells and the gut microbiota. *International Journal of Molecular Sciences*, *17*(4), 509.

Ganea, G. M., Fakayode, S. O., Losso, J. N., Van Nostrum, C. F., Sabliov, C. M., & Warner, I. M. (2010). Delivery of phytochemical thymoquinone using molecular micelle modified poly (D, L lactide-co-glycolide) (PLGA) nanoparticles. *Nanotechnology*, *21*(28), 285104.

Giroux, H. J., & Britten, M. (2011). Encapsulation of hydrophobic aroma in whey protein nanoparticles. *Journal of Microencapsulation*, *28*(5), 337–343.

Gökmen, V., Mogol, B. A., Lumaga, R. B., Fogliano, V., Kaplun, Z., & Shimoni, E. (2011). Development of functional bread containing nanoencapsulated omega-3 fatty acids. *Journal of Food Engineering*, *105*(4), 585–591.

Gokulan, K., Bekele, A. Z., Drake, K. L., & Khare, S. (2018). Responses of intestinal virome to silver nanoparticles: Safety assessment by classical virology, whole-genome sequencing and bioinformatics approaches. *International Journal of Nanomedicine*, *13*, 2857.

Gubernatorova, E. O., Liu, X., Othman, A., Muraoka, W. T., Koroleva, E. P., Andreescu, S., et al. (2017). Europium-doped cerium oxide nanoparticles limit reactive oxygen species formation and ameliorate intestinal ischemia–reperfusion injury. *Advanced Healthcare Materials*, *6*(14), 1700176.

Gupta, A., Eral, H. B., Hatton, T. A., & Doyle, P. S. (2016). Nanoemulsions: Formation, properties and applications. *Soft Matter*, *12*(11), 2826–2841. https://doi.org/10.1039/C5SM02958A.

Halliwell, B. (1994). Free radicals, antioxidants, and human disease: Curiosity, cause, or consequence? *The Lancet*, *344*(8924), 721–724.

Hamaker, B. R., & Tuncil, Y. E. (2014). A perspective on the complexity of dietary fiber structures and their potential effect on the gut microbiota. *Journal of Molecular Biology*, *426*(23), 3838–3850.

Hans, M. L., & Lowman, A. M. (2002). Biodegradable nanoparticles for drug delivery and targeting. *Current Opinion in Solid State and Materials Science*, *6*(4), 319–327.

Hansen, R. L., Dueñas, M. E., Looft, T., & Lee, Y. J. (2019). Nanoparticle microarray for high-throughput microbiome metabolomics using matrix-assisted laser desorption ionization mass spectrometry. *Analytical and Bioanalytical Chemistry*, *411*(1), 147–156.

Hawkins, M. J., Soon-Shiong, P., & Desai, N. (2008). Protein nanoparticles as drug carriers in clinical medicine. *Advanced Drug Delivery Reviews*, *60*(8), 876–885.

He, X., & Hwang, H.-M. (2016). Nanotechnology in food science: Functionality, applicability, and safety assessment. *Journal of Food and Drug Analysis*, *24*(4), 671–681.

Hong, F., Wu, N., Zhou, Y., Ji, L., Chen, T., & Wang, L. (2017). Gastric toxicity involving alterations of gastritis-related protein expression in mice following long-term exposure to nano $TiO_2$. *Food Research International*, *95*, 38–45.

Hsiao, E. Y., McBride, S. W., Hsien, S., Sharon, G., Hyde, E. R., McCue, T., et al. (2013). Microbiota modulate behavioral and physiological abnormalities associated with neurodevelopmental disorders. *Cell*, *155*(7), 1451–1463.

Huang, Y., & Adams, M. C. (2003). An in vitro model for investigating intestinal adhesion of potential dairy propionibacteria probiotic strains using cell line C2BBe1. *Letters in Applied Microbiology*, *36*(4), 213–216.

Hulla, J., Sahu, S., & Hayes, A. (2015). Nanotechnology: History and future. *Human & Experimental Toxicology*, *34*(12), 1318–1321.

Itthisoponkul, T., Mitchell, J. R., Taylor, A. J., & Farhat, I. A. (2007). Inclusion complexes of tapioca starch with flavour compounds. *Carbohydrate Polymers*, *69*(1), 106–115.

Jain, A. K., Khar, R. K., Ahmed, F. J., & Diwan, P. V. (2008). Effective insulin delivery using starch nanoparticles as a potential trans-nasal mucoadhesive carrier. *European Journal of Pharmaceutics and Biopharmaceutics*, *69*(2), 426–435.

Jung, J., Deng, Z., Simonsen, J., Bastías, R. M., & Zhao, Y. (2016). Development and preliminary field validation of water-resistant cellulose nanofiber based coatings with high surface adhesion and elasticity for reducing cherry rain-cracking. *Scientia Horticulturae*, *200*, 161–169.

Kararli, T. T. (1995). Comparison of the gastrointestinal anatomy, physiology, and biochemistry of humans and commonly used laboratory animals. *Biopharmaceutics & Drug Disposition*, *16*(5), 351–380.

Khorshidi, Z., Sarvi Moghanlou, K., Imani, A., & Behrouzi, S. (2018). The interactive effect of dietary curcumin and silver nanoparticles on gut microbiota of common carp (Cyprinus carpio). *Iranian Journal of Science and Technology, Transactions A: Science*, *42*(2), 379–387.

Kovacs, T., Naish, V., O'Connor, B., Blaise, C., Gagné, F., Hall, L., et al. (2010). An ecotoxicological characterization of nanocrystalline cellulose (NCC). *Nanotoxicology*, *4*(3), 255–270.

Kraegeloh, A., Suarez-Merino, B., Sluijters, T., & Micheletti, C. (2018). Implementation of safe-by-design for nanomaterial development and safe innovation: Why we need a comprehensive approach. *Nanomaterials*, *8*(4), 239.

Lattimer, J. M., & Haub, M. D. (2010). Effects of dietary fiber and its components on metabolic health. *Nutrients*, *2*(12), 1266.

Lepage, P., Häsler, R., Spehlmann, M. E., Rehman, A., Zvirbliene, A., Begun, A., et al. (2011). Twin study indicates loss of interaction between microbiota and mucosa of patients with ulcerative colitis. *Gastroenterology*, *141*(1), 227–236.

Lesmes, U., Cohen, S. H., Shener, Y., & Shimoni, E. (2009). Effects of long chain fatty acid unsaturation on the structure and controlled release properties of amylose complexes. *Food Hydrocolloids*, *23*(3), 667–675.

Li, H., Zhao, X., Ma, Y., Zhai, G., Li, L., & Lou, H. (2009). Enhancement of gastrointestinal absorption of quercetin by solid lipid nanoparticles. *Journal of Controlled Release*, *133*(3), 238–244.

Lin, N., & Dufresne, A. (2014). Nanocellulose in biomedicine: Current status and future prospect. *European Polymer Journal*, *59*, 302–325.

Liu, X., Chen, X., Li, Y., Wang, X., Peng, X., & Zhu, W. (2012). Preparation of superparamagnetic Fe3O4@alginate/chitosan nanospheres for Candida rugosa lipase immobilization and utilization of layer-by-layer assembly to enhance the stability of immobilized lipase. *ACS Applied Materials & Interfaces*, *4*(10), 5169–5178.

Liu, X., Dumitrescu, E., & Andreescu, S. (2015). Electrochemical biosensors for real-time monitoring of reactive oxygen and nitrogen species. In M. Hepel & S. Andreescu (Eds.), *Vol. 1200. Oxidative stress: Diagnostics, prevention, and therapy volume 2.* (pp. 301–327). American Chemical Society.

Liu, T., Hougen, H., Vollmer, A. C., & Hiebert, S. M. (2012). Gut bacteria profiles of Mus musculus at the phylum and family levels are influenced by saturation of dietary fatty acids. *Anaerobe*, *18*(3), 331–337.

Liu, X., Marrakchi, M., Jahne, M., Rogers, S., & Andreescu, S. (2016). Real-time investigation of antibiotics-induced oxidative stress and superoxide release in bacteria using an electrochemical biosensor. *Free Radical Biology and Medicine*, *91*, 25–33.

Livney, Y. D. (2010). Milk proteins as vehicles for bioactives. *Current Opinion in Colloid & Interface Science*, *15*(1), 73–83.

Lohcharoenkal, W., Wang, L., Chen, Y. C., & Rojanasakul, Y. (2014). Protein nanoparticles as drug delivery carriers for cancer therapy. *BioMed Research International*, *2014*, 12.

Luo, Y., Zhang, B., Whent, M., Yu, L., & Wang, Q. (2011). Preparation and characterization of zein/chitosan complex for encapsulation of α-tocopherol, and its in vitro controlled release study. *Colloids and Surfaces B: Biointerfaces*, *85*(2), 145–152.

Mackie, A., & Macierzanka, A. (2010). Colloidal aspects of protein digestion. *Current Opinion in Colloid & Interface Science*, *15*(1), 102–108.

Maghsoudi, A., Shojaosadati, S. A., & Vasheghani Farahani, E. (2008). 5-Fluorouracil-loaded BSA nanoparticles: Formulation optimization and in vitro release study. *AAPS PharmSciTech*, *9*(4), 1092–1096.

Malheiros Pda, S., Sant'Anna, V., Barbosa, M. S., Brandelli, A., & Franco, B. D. (2012). Effect of liposome-encapsulated nisin and bacteriocin-like substance P34 on listeria monocytogenes growth in Minas frescal cheese. *International Journal of Food Microbiology*, *156*(3), 272–277.

Marciani, L., Faulks, R., Wickham, M. S. J., Bush, D., Pick, B., Wright, J., et al. (2008). Effect of intragastric acid stability of fat emulsions on gastric emptying, plasma lipid profile and postprandial satiety. *British Journal of Nutrition*, *101*(6), 919–928.

Masuda, K., Kajikawa, A., & Igimi, S. (2011). Establishment and evaluation of an in vitro M cell model using C2BBe1 cells and Raji cells. *Bioscience and Microflora*, *30*(2), 37–44.

Matalanis, A., Jones, O. G., & McClements, D. J. (2011). Structured biopolymer-based delivery systems for encapsulation, protection, and release of lipophilic compounds. *Food Hydrocolloids*, *25*(8), 1865–1880.

McClements, D. J. (2004). Protein-stabilized emulsions. *Current Opinion in Colloid & Interface Science*, *9*(5), 305–313.

McClements, D. J. (2013). Edible lipid nanoparticles: Digestion, absorption, and potential toxicity. *Progress in Lipid Research*, *52*(4), 409–423.
McClements, D. J., DeLoid, G., Pyrgiotakis, G., Shatkin, J. A., Xiao, H., & Demokritou, P. (2016). The role of the food matrix and gastrointestinal tract in the assessment of biological properties of ingested engineered nanomaterials (iENMs): State of the science and knowledge gaps. *NanoImpact*, *3–4*, 47–57.
McClements, D. J., & Rao, J. (2011). Food-grade nanoemulsions: Formulation, fabrication, properties, performance, biological fate, and potential toxicity. *Critical Reviews in Food Science and Nutrition*, *51*(4), 285–330.
McClements, D. J., & Xiao, H. (2012). Potential biological fate of ingested nanoemulsions: Influence of particle characteristics. *Food & Function*, *3*(3), 202–220. https://doi.org/10.1039/C1FO10193E.
McClements, D. J., & Xiao, H. (2017). Is nano safe in foods? Establishing the factors impacting the gastrointestinal fate and toxicity of organic and inorganic food-grade nanoparticles. *npj Science of Food*, *1*(1), 6.
Mehnert, W., & Mäder, K. (2001). Solid lipid nanoparticles: Production, characterization and applications. *Advanced Drug Delivery Reviews*, *47*(2), 165–196.
Menni, C., Zierer, J., Pallister, T., Jackson, M. A., Long, T., Mohney, R. P., et al. (2017). Omega-3 fatty acids correlate with gut microbiome diversity and production of N-carbamylglutamate in middle aged and elderly women. *Scientific Reports*, 7(1), 11079.
Mizrahy, S., & Peer, D. (2012). Polysaccharides as building blocks for nanotherapeutics. *Chemical Society Reviews*, *41*(7), 2623–2640. https://doi.org/10.1039/C1CS15239D.
Mondot, S., Kang, S., Furet, J. P., Aguirre de Carcer, D., McSweeney, C., Morrison, M., et al. (2011). Highlighting new phylogenetic specificities of Crohn's disease microbiota. *Inflammatory Bowel Diseases*, *17*(1), 185–192.
Mortensen, L. S., Hartvigsen, M. L., Brader, L. J., Astrup, A., Schrezenmeir, J., Holst, J. J., et al. (2009). Differential effects of protein quality on postprandial lipemia in response to a fat-rich meal in type 2 diabetes: Comparison of whey, casein, gluten, and cod protein. *The American Journal of Clinical Nutrition*, *90*(1), 41–48.
Murugeshu, A., Astete, C., Leonardi, C., Morgan, T., & Sabliov, C. M. (2011). Chitosan/PLGA particles for controlled release of α-tocopherol in the GI tract via oral administration. *Nanomedicine*, *6*(9), 1513–1528.
Myrick, J. M., Vendra, V. K., & Krishnan, S. (2014). Self-assembled polysaccharide nanostructures for controlled-release applications. *Nanotechnology Reviews*, *3*(4), 319–346.
National Science and Technology Council (NSTC), Committee on Technology, Subcommittee on Nanoscale Science, & Engineering and Technology. (2014). *Supplement to the president's budget for fiscal year 2015.*
National Science and Technology Council (NSTC), Committee on Technology, Subcommittee on Nanoscale Science, & Engineering and Technology. (2016). *National nanotechnology initiative strategic plan.*
Natoli, M., Leoni, B. D., D'Agnano, I., D'Onofrio, M., Brandi, R., Arisi, I., et al. (2011). Cell growing density affects the structural and functional properties of Caco-2 differentiated monolayer. *Journal of Cellular Physiology*, *226*(6), 1531–1543.
Natoli, M., Leoni, B. D., D'Agnano, I., Zucco, F., & Felsani, A. (2012). Good Caco-2 cell culture practices. *Toxicology In Vitro*, *26*(8), 1243–1246.
Nel, A., Xia, T., Mädler, L., & Li, N. (2006). Toxic potential of materials at the nanolevel. *Science*, *311*(5761), 622–627.
Noonan, G. O., Whelton, A. J., Carlander, D., & Duncan, T. V. (2014). Measurement methods to evaluate engineered nanomaterial release from food contact materials. *Comprehensive Reviews in Food Science and Food Safety*, *13*(4), 679–692.

Ogawa, A., Takakura, K., Sano, K., Kanematsu, H., Yamano, T., Saishin, T., et al. (2018). Microbiome analysis of biofilms of silver nanoparticle-dispersed silane-based coated carbon steel using a next-generation sequencing technique. *Antibiotics*, 7(4), 91.

Özel, R. E., Alkasir, R. S. J., Ray, K., Wallace, K. N., & Andreescu, S. (2013). Comparative evaluation of intestinal nitric oxide in embryonic zebrafish exposed to metal oxide nanoparticles. *Small*, *9*(24), 4250–4261.

Özel, R. E., Liu, X., Alkasir, R. S. J., & Andreescu, S. (2014). Electrochemical methods for nanotoxicity assessment. *TrAC Trends in Analytical Chemistry*, *59*, 112–120.

Özel, R. E., Wallace, K. N., & Andreescu, S. (2014). Alterations of intestinal serotonin following nanoparticle exposure in embryonic zebrafish. *Environmental Science: Nano*, *1*(1), 27–36.

Pan, X., Yao, P., & Jiang, M. (2007). Simultaneous nanoparticle formation and encapsulation driven by hydrophobic interaction of casein-graft-dextran and β-carotene. *Journal of Colloid and Interface Science*, *315*(2), 456–463.

Patel, A., Hu, Y., Tiwari, J. K., & Velikov, K. P. (2010). Synthesis and characterisation of zein–curcumin colloidal particles. *Soft Matter*, *6*(24), 6192–6199.

Perelshtein, I., Ruderman, E., Perkas, N., Tzanov, T., Beddow, J., Joyce, E., et al. (2013). Chitosan and chitosan–ZnO-based complex nanoparticles: Formation, characterization, and antibacterial activity. *Journal of Materials Chemistry B*, *1*(14), 1968–1976.

Peterson, M. D., & Mooseker, M. S. (1992). Characterization of the enterocyte-like brush border cytoskeleton of the C2BBe clones of the human intestinal cell line, Caco-2. *Journal of Cell Science*, *102*(3), 581–600.

Piercey, M. J., Mazzanti, G., Budge, S. M., Delaquis, P. J., Paulson, A. T., & Truelstrup Hansen, L. (2012). Antimicrobial activity of cyclodextrin entrapped allyl isothiocyanate in a model system and packaged fresh-cut onions. *Food Microbiology*, *30*(1), 213–218.

Pietroiusti, A., Magrini, A., & Campagnolo, L. (2016). New frontiers in nanotoxicology: Gut microbiota/microbiome-mediated effects of engineered nanomaterials. *Toxicology and Applied Pharmacology*, *299*, 90–95.

Piorkowski, D. T., & McClements, D. J. (2014). Beverage emulsions: Recent developments in formulation, production, and applications. *Food Hydrocolloids*, *42*, 5–41.

Portune, K. J., Beaumont, M., Davila, A.-M., Tomé, D., Blachier, F., & Sanz, Y. (2016). Gut microbiota role in dietary protein metabolism and health-related outcomes: The two sides of the coin. *Trends in Food Science & Technology*, *57*, 213–232.

Prabaharan, M. (2011). Prospective of guar gum and its derivatives as controlled drug delivery systems. *International Journal of Biological Macromolecules*, *49*(2), 117–124.

Pyrgiotakis, G., McDevitt, J., Bordini, A., Diaz, E., Molina, R., Watson, C., et al. (2014). A chemical free, nanotechnology-based method for airborne bacterial inactivation using engineered water nanostructures. *Environmental Science: Nano*, *1*(1), 15–26.

Pyrgiotakis, G., McDevitt, J., Gao, Y., Branco, A., Eleftheriadou, M., Lemos, B., et al. (2014). Mycobacteria inactivation using engineered water nanostructures (EWNS). *Nanomedicine: Nanotechnology, Biology and Medicine*, *10*(6), 1175–1183.

Pyrgiotakis, G., McDevitt, J., Yamauchi, T., & Demokritou, P. (2012). A novel method for bacterial inactivation using electrosprayed water nanostructures. *Journal of Nanoparticle Research*, *14*(8), 1027.

Qian, C., Decker, E. A., Xiao, H., & McClements, D. J. (2013). Impact of lipid nanoparticle physical state on particle aggregation and β-carotene degradation: Potential limitations of solid lipid nanoparticles. *Food Research International*, *52*(1), 342–349.

Rao, J., Decker, E. A., Xiao, H., & McClements, D. J. (2013). Nutraceutical nanoemulsions: Influence of carrier oil composition (digestible versus indigestible oil) on β-carotene bioavailability. *Journal of the Science of Food and Agriculture*, *93*(13), 3175–3183.

Sagiri, S. S., Anis, A., & Pal, K. (2016). Review on encapsulation of vegetable oils: Strategies, preparation methods, and applications. *Polymer-Plastics Technology and Engineering*, *55*(3), 291–311.

Sambuy, Y., De Angelis, I., Ranaldi, G., Scarino, M. L., Stammati, A., & Zucco, F. (2005). The Caco-2 cell line as a model of the intestinal barrier: Influence of cell and culture-related factors on Caco-2 cell functional characteristics. *Cell Biology and Toxicology*, *21*(1), 1–26.

Semete, B., Booysen, L., Lemmer, Y., Kalombo, L., Katata, L., Verschoor, J., et al. (2010). In vivo evaluation of the biodistribution and safety of PLGA nanoparticles as drug delivery systems. *Nanomedicine: Nanotechnology, Biology and Medicine*, *6*(5), 662–671.

Shatkin, J. A., & Kim, B. (2015). Cellulose nanomaterials: Life cycle risk assessment, and environmental health and safety roadmap. *Environmental Science: Nano*, *2*(5), 477–499.

Simon, L. C., & Sabliov, C. M. (2013). Time analysis of poly (lactic-co-glycolic) acid nanoparticle uptake by major organs following acute intravenous and oral administration in mice and rats. *Industrial Biotechnology*, *9*(1), 19–23.

Singh, G., Stephan, C., Westerhoff, P., Carlander, D., & Duncan, T. V. (2014). Measurement methods to detect, characterize, and quantify engineered nanomaterials in foods. *Comprehensive Reviews in Food Science and Food Safety*, *13*(4), 693–704.

Singh, V., Yeoh, B. S., Chassaing, B., Xiao, X., Saha, P., Aguilera Olvera, R., et al. (2018). Dysregulated microbial fermentation of soluble fiber induces cholestatic liver cancer. *Cell*, *175*(3), 679–694.

Sohal, I. S., Cho, Y. K., O'Fallon, K. S., Gaines, P., Demokritou, P., & Bello, D. (2018). Dissolution behavior and biodurability of ingested engineered nanomaterials in the gastrointestinal environment. *ACS Nano*, *12*(8), 8115–8128.

Sohal, I. S., O'Fallon, K. S., Gaines, P., Demokritou, P., & Bello, D. (2018). Ingested engineered nanomaterials: State of science in nanotoxicity testing and future research needs. *Particle and Fibre Toxicology*, *15*(1), 29.

Sozer, N., & Kokini, J. L. (2009). Nanotechnology and its applications in the food sector. *Trends in Biotechnology*, *27*(2), 82–89.

Strasdat, B., & Bunjes, H. (2013). Incorporation of lipid nanoparticles into calcium alginate beads and characterization of the encapsulated particles by differential scanning calorimetry. *Food Hydrocolloids*, *30*(2), 567–575.

Susi, T., Hofer, C., Argentero, G., Leuthner, G. T., Pennycook, T. J., Mangler, C., et al. (2016). Isotope analysis in the transmission electron microscope. *Nature Communications*, *7*, 13040.

Takahashi, N., Gruber, C. C., Yang, J. H., Liu, X., Braff, D., Yashaswini, C. N., et al. (2017). Lethality of MalE-LacZ hybrid protein shares mechanistic attributes with oxidative component of antibiotic lethality. *Proceedings of the National Academy of Sciences of the United States of America*, *114*(34), 9164–9169.

Tamjidi, F., Shahedi, M., Varshosaz, J., & Nasirpour, A. (2013). Nanostructured lipid carriers (NLC): A potential delivery system for bioactive food molecules. *Innovative Food Science & Emerging Technologies*, *19*, 29–43.

Tassinari, R., Cubadda, F., Moracci, G., Aureli, F., D'Amato, M., Valeri, M., et al. (2014). Oral, short-term exposure to titanium dioxide nanoparticles in Sprague-Dawley rat: Focus on reproductive and endocrine systems and spleen. *Nanotoxicology*, *8*(6), 654–662.

Taylor, A., Marcus, I., Guysi, R., & Sharon, L. (2015). Metal oxide nanoparticles induce minimal phenotypic changes in a model colon gut microbiota. *Environmental Engineering Science*, *32*(7), 602–612.

Teng, Z., Luo, Y., & Wang, Q. (2012). Nanoparticles synthesized from soy protein: Preparation, characterization, and application for nutraceutical encapsulation. *Journal of Agricultural and Food Chemistry*, *60*(10), 2712–2720.

Teng, Z., Luo, Y., & Wang, Q. (2013). Carboxymethyl chitosan–soy protein complex nanoparticles for the encapsulation and controlled release of vitamin D3. *Food Chemistry*, *141*(1), 524–532.

Trompette, A., Gollwitzer, E. S., Yadava, K., Sichelstiel, A. K., Sprenger, N., Ngom-Bru, C., et al. (2014). Gut microbiota metabolism of dietary fiber influences allergic airway disease and hematopoiesis. *Nature Medicine*, *20*, 159.

Utembe, W., Potgieter, K., Stefaniak, A. B., & Gulumian, M. (2015). Dissolution and biodurability: Important parameters needed for risk assessment of nanomaterials. *Particle and Fibre Toxicology*, *12*(1), 11.

van den Brule, S., Ambroise, J., Lecloux, H., Levard, C., Soulas, R., De Temmerman, P.-J., et al. (2016). Dietary silver nanoparticles can disturb the gut microbiota in mice. *Particle and Fibre Toxicology*, *13*(1), 38.

Vaze, N., Pyrgiotakis, G., Mena, L., Baumann, R., Demokritou, A., Ericsson, M., et al. (2019). A nano-carrier platform for the targeted delivery of nature-inspired antimicrobials using engineered water nanostructures for food safety applications. *Food Control*, *96*, 365–374.

Watson, H., Mitra, S., Croden, F. C., Taylor, M., Wood, H. M., Perry, S. L., et al. (2018). A randomised trial of the effect of omega-3 polyunsaturated fatty acid supplements on the human intestinal microbiota. *Gut*, *67*(11), 1974–1983.

Weber, C., Coester, C., Kreuter, J., & Langer, K. (2000). Desolvation process and surface characterisation of protein nanoparticles. *International Journal of Pharmaceutics*, *194*(1), 91–102.

Yada, R. Y., Buck, N., Canady, R., DeMerlis, C., Duncan, T., Janer, G., et al. (2014). Engineered nanoscale food ingredients: Evaluation of current knowledge on material characteristics relevant to uptake from the gastrointestinal tract. *Comprehensive Reviews in Food Science and Food Safety*, *13*(4), 730–744.

Yaméogo, J. B. G., Gèze, A., Choisnard, L., Putaux, J.-L., Gansané, A., Sirima, S. B., et al. (2012). Self-assembled biotransesterified cyclodextrins as artemisinin nanocarriers—I: Formulation, lyoavailability and in vitro antimalarial activity assessment. *European Journal of Pharmaceutics and Biopharmaceutics*, *80*(3), 508–517.

Yi, J., Lam, T. I., Yokoyama, W., Cheng, L. W., & Zhong, F. (2015). Beta-carotene encapsulated in food protein nanoparticles reduces peroxyl radical oxidation in Caco-2 cells. *Food Hydrocolloids*, *43*, 31–40.

Yu, H.-N., Zhu, J., Pan, W.-S., Shen, S.-R., Shan, W.-G., & Das, U. N. (2014). Effects of fish oil with a high content of n-3 polyunsaturated fatty acids on mouse gut microbiota. *Archives of Medical Research*, *45*(3), 195–202.

Zabar, S., Lesmes, U., Katz, I., Shimoni, E., & Bianco-Peled, H. (2009). Studying different dimensions of amylose–long chain fatty acid complexes: Molecular, nano and micro level characteristics. *Food Hydrocolloids*, *23*(7), 1918–1925.

Zhao, L., Ortiz, C., Adeleye, A. S., Hu, Q., Zhou, H., Huang, Y., et al. (2016). Metabolomics to detect response of lettuce (Lactuca sativa) to Cu(OH)2 nanopesticides: Oxidative stress response and detoxification mechanisms. *Environmental Science & Technology*, *50*(17), 9697–9707.

Zhao, Y., Simonsen, J., Cavender, G., Jung, J., & Fuchigami, L. H. (2018). *Nano-cellulose coatings to prevent damage in foodstuffs*. US Patent Application US20140272013A1.

Zimet, P., Rosenberg, D., & Livney, Y. D. (2011). Re-assembled casein micelles and casein nanoparticles as nano-vehicles for ω-3 polyunsaturated fatty acids. *Food Hydrocolloids*, *25*(5), 1270–1276.

Zohri, M., Shafiee Alavidjeh, M., Mirdamadi, S. S., Behmadi, H., Hossaini Nasr, S. M., Eshghi Gonbaki, S., et al. (2013). Nisin-loaded chitosan/alginate nanoparticles: A hopeful hybrid biopreservative. *Journal of Food Safety*, *33*(1), 40–49.